Safety and Human Resource Law for the Safety Professional

Occupational Safety and Health Guide Series

Series Editor: Thomas D. Schneid, Eastern Kentucky University, Richmond, Kentucky

The aim of this series is to develop books to address the myriad of unique topics which are encompassed within the areas of responsibility of a safety professional. The scope of the series is broad, addressing traditional occupational safety and health topics as well as emerging and correlating topic areas, such as ergonomics, security, labor, school safety, employment, workplace violence, and other topics. Safety professionals often assume a broad spectrum of responsibilities, depending on the organization, which go beyond the traditional safety professional's role. This series should become the "one stop" source for safety professionals to locate texts on traditional as well as emerging and correlating subjects related to the safety and health function.

Safety Law
Legal Aspects in Occupational Safety and Health
Thomas D. Schneid

Physical Hazards of the Workplace
Barry Spurlock

Loss Prevention and Safety Control
Terms and Definitions
Dennis P. Nolan

Discrimination Law Issues for the Safety Professional
Thomas D. Schneid

Labor and Employment Issues for the Safety Professional
Thomas D. Schneid

The Comprehensive Handbook of School Safety
Edited by E. Scott Dunlap

Creative Safety Solutions
Thomas D. Schneid

Physical Security and Safety
A Field Guide for the Practitioner
Edited by Truett A. Ricks, Bobby E. Ricks, and Jeffrey Dingle

Workplace Safety and Health
Assessing Current Practices and Promoting Change in the Profession
Thomas D. Schneid

For more information about this series, please visit: https://www.crcpress.com/Occupational-Safety--Health-Guide-Series/book-series/CRCOCCSHGS

Safety and Human Resource Law for the Safety Professional

Thomas D. Schneid

With Contributions by
Shelby L. Schneid

CRC Press
Taylor & Francis Group
Boca Raton London New York

CRC Press is an imprint of the
Taylor & Francis Group, an **informa** business

CRC Press
Taylor & Francis Group
6000 Broken Sound Parkway NW, Suite 300
Boca Raton, FL 33487-2742

© 2019 by Taylor & Francis Group, LLC
CRC Press is an imprint of Taylor & Francis Group, an Informa business

No claim to original U.S. Government works

Printed on acid-free paper

International Standard Book Number-13: 978-1-138-19621-6 (Hardback)

Library of Congress Cataloging-in-Publication Data

Names: Schneid, Thomas D., author.
Title: Safety and human resource law for the safety professional / authored
by Thomas D. Schneid.
Description: Boca Raton : Taylor & Francis, 2019. | Series: Occupational
safety and health guide series
Identifiers: LCCN 2018055113| ISBN 9781138196216 (hardback : alk. paper) |
ISBN 9781315302713 (e-book)
Subjects: LCSH: Industrial safety--Law and legislation--United States.
Classification: LCC KF3570 .S363 2019 | DDC 344.7304/65--dc23
LC record available at https://lccn.loc.gov/2018055113

Visit the Taylor & Francis Website at
http://www.taylorandfrancis.com

and the CRC Press Website at
http://www.crcpress.com

Contents

Foreword

In today's highly specialized society where we educate our safety professionals on all of the fundamentals of safety compliance, auditing, safety metrics, machine guarding, accident investigations, and chemical process safety (to name a few), we now have a very timely book from Dr. Thomas D. Schneid that addresses many of those issues that may, at first blush, appear merely ancillary to the function of the average safety professional. However, these are truly key areas of knowledge that all safety professionals need to be aware of, be able to recognize, and deal with to become a more well- rounded safety professional.

To give readers just one example of how this may apply, consider the following scenario I had a few years ago, when I was brought in as the outside legal labor and employment counsel to a manufacturing company that was starting to hear rumors of a union trying to organize its non-union workforce. We scheduled a meeting with upper management to begin training them on how to deal with union organizing, the dos and don'ts of when a union comes calling on your employees. In fact, what happened that first day I showed up was that the company received a letter from the union, asking them if they wanted to voluntarily recognize the union because they claimed to already have a majority of employees who had signed union authorization cards. A few days later, one of the lead employees of the organizing campaign stopped the company president in the plant and told him that the company's new safety manager, who had been hired only a few weeks before, was going around the plant asking employees as to which employees were key players in the union election and why they wanted a union. Of course, he thought he was being proactive for the company. In fact, he was violating the National Labor Relations Act (NLRA), which prohibits employers from coercively questioning employees about their activities, support, or sympathies for a union. It also prohibits polling employees about their support for a union. This is a violation of the employees' Section 7 rights and is an unfair labor practice under Section 8(a)(1) of the Act.

This book also deals with those other key issues such as the FMLA, ADA, and Workers' Compensation that has a huge impact on dealing with workers and injured workers. I highly recommend this book to any safety professional who wants to broaden their knowledge and become a true well-rounded safety professional.

Michael S. Schumann, M.S., J.D., PhD, CSHM
Professor
Occupational Safety
Department of Safety, Security & Emergency Management
Eastern Kentucky University, Richmond, Kentucky

Preface

ANALYZING AND BRIEFING A COURT DECISION

Safety professionals work with and for the law on a daily basis. In the United States, laws can be made by the executive branch (i.e., executive orders) or the legislative branch (i.e., new laws) and the judicial branch reviews and assesses the validity and applicability of these laws. This assessment by the judicial branch of our government is often called "case law" and is the evaluation, assessment, and decisions of the courts at all levels up to, and including, the U.S. Supreme Court.

Safety professionals should be aware that there are state courts as well as federal courts, and even specialty courts, such as Family Law Courts and Traffic Courts. Although the court name may vary, both the state and federal courts possess a hierarchy wherein each court decision can be appealed to a higher court by any of the parties involved in the actions. The highest court in most state judiciary systems as well as the federal judiciary system is the Supreme Court. For safety professionals who may be unfamiliar with their individual state judiciary system or the federal judiciary system, it is important to acquire a basic knowledge of the levels of the courts, namely the specialty courts, such as the tax court, the district, or trial courts, where most trials take place, the appellate courts, and the state top court or the U.S. Supreme Court.

When reading a court decision, safety professionals may wish utilize the following outline:

1. Identify the court and date of the case at the top of the decision.
2. Look up any legal terms you are not familiar with.
3. Read the case in total.
4. Determine the type of case you are reading.
5. Review the case summary or headnotes.
6. Read the case again and identify the court's decision.
7. If an appellate case, identify if the decision was unanimous or a split decision.
8. Did the minority provide a dissenting opinion?
9. Re-read the case. Identify the parties, issues, and facts of the case.
10. Brief your case in writing so you will remember the issues, facts, and decision at a later date.

Safety professionals should be aware that although the courts are not supposed to make law, their decisions are in fact shaping and making new law. The decisions of the courts are often called "case law," which is the accumulation of court decisions that provide guidance and direction on current and future cases and decisions. As identified above, cases usually start at the lowest level in either the state or federal judiciary system and are appealed upward within the system to the top court, but can stop at any level. The decision of the higher court usually supersedes the decision of the lower court in whole or in part.

It is important that safety professionals acquire the skill and ability to carefully analyze these court decisions in order to know the status of the law on any given day. Safety professionals can find the court's decisions at most courthouses or law libraries; however, databases, such as Westlaw and Lexis, provide all cases to the safety professional at near as his/her computer. Court decisions are identified by name, *Jones v. Smith*, as well as the volume and page number within the identified location of the case and the year of the decision. As an example: Jones v. Smith, 22 U.S. 25 (2011). The parties are Jones bringing the action against Smith. Jones would be the plaintiff and Smith would be the defendant. The case can be located in volume 22 within the U.S. Supreme Court cases at page 25. The decision was rendered by the U.S. Supreme Court in 2011.

As identified above, it is important that the safety professional first read the case in full to identify the type of case, e.g., criminal or civil, as well as acquiring a flavor of the case. Safety and health professionals should identify the parties, the type of case, and the broad issues and defenses, as well as the court's decision. After the initial reading of the case, safety professionals should re-read the case with an eye to the detail provided within the case. On the third reading, the safety professional should takes notes and begin to assemble the structure of the case brief which the safety professional can use to refresh his/her memory of the case or for use in whatever activity is at hand.

The primary reason safety professionals should brief a case is to provide an understanding of the particular issues and decisions in the cases, as well as to provide a method of remembering the cases when a large number of cases are involved in the situation. Although there are various methods of briefing a case, the following method is provided as one example. Safety professionals should find a method in which they are comfortable and utilize this method consistently in their work.

Briefing a Case – Methodology:

1. CASE – List the case name, the court, and the date of the case at the top of the brief.
2. ISSUE – In a clear and concise manner of no more than 1–3 sentences, completely explain the issue(s) in the case.
3. FACTS – In a clear and concise manner of no more than 1–2 paragraphs, identify all of the pertinent facts of the case.
4. HOLDING OR DECISION – Clearly and concisely identify the decision of the court.
5. DISSENT OR DISSENTING OPINION – If the minority provided a dissenting opinion, provide a clear and concise explanation of the dissenting judge's position.
6. YOUR OPINION – It is important for safety professionals to identify why they agree or disagree with the decision of the court.

Most case briefs should be one page in length, but no more than two pages. Safety professionals who exceed the one-page limit may want to re-assess their analysis and reduce the verbiage to address only the issues, facts, and decision provided by the court. In essence, the brief is a short, concise and "to the point" document which safety professionals can use to remember the case and have a quick review of the important aspects of the case. If additional details are needed, the safety professional should go back and review the entire case.

Below please find an example of a very basic case brief for your review and use:

Case Name:	*Smith v. Jones, Inc,* 10 Anystate Court, 21 (2011)
Issues:	Smith alleged she was discriminated against by her employer, Jones, Inc., in violation of the Pregnancy Discrimination Act and Title VII of the Civil Rights Act.
Facts:	Smith, employed by Jones, Inc. for five years, was terminated from her employment. Jones alleges Smith was terminated for theft of company property. Smith alleges she was terminated when she informed her boss that she was pregnant.
Holding:	For the Defendant, Jones, Inc. The court found that the termination of Smith for theft was appropriate and found no discrimination based on her pregnancy.
My Opinion:	I disagree with the decision of the court. Jones possessed a history of terminating employees who files for any type of leave of absence. Smith was constructively terminated under the precursor of theft which was simply a company pen she forgot in her purse.

Acknowledgments

I would like to thank my mother and father for their support of education throughout our lives. Without their constant support and motivation, their children, including myself, would not have reached the level we have achieved today.

I would like to thank my wife, Jani, for her patience in the writing of this book. I would also like to thank Shelby who works in the human resource field for her contribution to this book, as well as Madison, who works in the safety and health field, and Kasi who is pursuing her undergraduate degree.

I would also like to than Alexander (Alex) Dougherty, a graduate assistant in the MSSSEM program, for his assistance in researching the issues and cases in this book.

Author

Thomas D. Schneid is a tenured professor in the Department of Safety, Security, and Emergency Management at Eastern Kentucky University and serves as the Chair of the Department. Schneid earned a BS in education, MS and CAS in safety, and M.S. in international business, and PhD in environmental engineering as well as his Juris Doctor (JD in Law) from West Virginia University and LL.M (Graduate Law) from the University of San Diego. He is a member of the Bar for the U.S. Supreme Court, Sixth Circuit Court of Appeals, and a member of the Bars of a number of federal districts, as well as the Kentucky and West Virginia State Bar associates. Schneid has authored and coauthored 20 texts and more than 100 articles on various safety, fire, and medical service (EMS), disaster management, and legal topics.

1 The Safety Function and the Law

******

Chapter Objectives:

1. Understand the safety and health functions.
2. Understand the correlation and relationship between the human resource and safety functions.
3. Identify the impacts on the safety and health functions.

******

The safety and health function in most organizations is all-inclusive. Safety and health professionals do not work in a bubble focusing solely on safety and health regulations and compliance therewith by employees, but function under and are impacted by many different laws and regulations beyond their central focus. Safety and health professionals perform an expansive and important role to safeguard employees as well as equipment, structures, and other assets of the organization. Identification of risks and eliminating or minimizing their impact involves knowledge of the potential risks created through peripheral but potentially impactful laws and regulations beyond the safety and health scope.

The safety and health function, although it may be only one professional, does not work in a silo. In most organizations, the safety and health professional is part of the management team and serves in a leadership role to guide the safety and health efforts of the organization. However, in performing the functions required to ensure the safety and health of all employees, the safety and health function must interact with the laws, regulations, and policies of other functions within the organization, as well as outside of the organization. Given the fact that the safety and health function interacts with employees at all levels within the organizational structure, the safety and health function often works closely with the human resource or personnel function on a daily basis. This frequent interaction between the safety and health function and the human resource function is essential in order to provide the appropriate leadership and guidance to employees with regard not only to workplace safety and health issues, but also operational, quality, and related expectations for each employee by the organization.

The location of the safety and health function within the organizational structure of the company is often unique to the individual company and often has developed throughout the history of the company. For more established companies injecting formalized safety and health after 1970, the safety and health function often started as a compliance-ensuring entity and was aligned with production or quality. As the safety and health function expanded beyond compliance to become a vital component within the organizational structure, companies often designed an internal hierarchy for the safety and health function within the overall organizational structure, ranging from the operational level through the corporate level safety and health functions. Given the expanded role of the safety and health professional and the daily interaction with the human resource function, many companies aligned their reporting structure, especially at operational levels, to permit the safety and health function to report directly to the human resource function, with a secondary reporting line to a corporate safety and health function.

To ensure the efficiency and effectiveness of the safety and health function, many companies found that it made operational "sense" to align the safety and health function with the human resource function, due to the overlapping responsibilities and expertise requirements. In order to achieve a proactive environment which prevents not only accidents but also violations of specific laws relevant to the human resource function, alignment of the reporting structure merged not only the functions, but also the experience and expertise of the functions, permitting overlapping compliance requirements to be addressed in an efficient and effective manner. However, education and training are essential for the safety and health function to become knowledgeable about the laws, regulations, and policies within the human resource function, and for the human resource function to become knowledgeable about the laws, regulations, and policies within the safety and health function.

Often as the result of difficult economic or organizational circumstances resulting in downsizing within operations, safety and health professionals are often tasked with peripheral functions in addition to their safety and health responsibilities, such as security or environmental functions. These peripheral functions often were individual functions under the human resource umbrella prior to the economic downturn; however, they require consolidation for cost-saving reasons. These "slash" positions (e.g., Safety/Security Coordinator) often require new functions to integrate within the human resource function for operational purposes. Although these commingled functions often require a minimalist approach due simply to the number of responsibilities and the time limitations, the survival of the company is often at risk, and the safety and health professional can integrate and apply his/her managerial skills and abilities to the correlated function.

The safety and health professional is often the unofficial "eyes and ears" of the human resource function within the operational setting. Safety and health professionals frequently work within operations, conducting assessments, observations, and related activities within the safety and health function. However, while working within operations, the safety and health professional often observes employee activities which may impact laws, regulations, or policies within the human resource function. It is imperative that the safety and health professional possesses the knowledge and ability to be able to recognize the potential violation and the ability to communicate this information to the human resource function in order that preventative action can be initiated. For example, while the safety and health professional is conducting his/her compliance-related inspection, he/she learns that an employee is upset within the operation. In talking with the employee, the employee explains that she has a very ill family member and asked her supervisor for time off work. The supervisor denied this request. The safety and health professional, with knowledge of the Family and Medical Leave Act (FMLA) and identifying that this could be a FMLA issue, informs the human resource function who can further investigate and talk through the appropriate actions to avoid potential liability for the company.

It is imperative that safety and health professionals possess a working knowledge of the laws, regulations, and policies within the human resource function. Additionally, if working in a unionized operation, it is essential that the safety and health professional have a working knowledge of the collective bargaining agreement and, in a non-union operation, knowledge of the employee handbook. In general, safety and health professionals should possess a working knowledge of discrimination law and be able to recognize situations involving potential discrimination involving pregnancy, age, religion, gender, disability, race, and national origin. Additionally, safety and health professionals should be able to recognize issues involving leave from the workplace including FMLA, short-term and long-term disability (by policy), Americans with Disabilities Act (ADA) and workers' compensation overlap, and related return to work issues. Safety and health professionals should have a firm grasp of the issues related to employee discipline, employee involuntary termination due to direct impact on the safety and health function, and controlled substance testing.

Safety and health professionals are often tasked with the management of the individual state's workers' compensation function. Although the workers' compensation function is reactive (i.e., the injury or illness has already occurred), whereas the safety and health function is proactive, many organizations tend to commingle these functions within their corporate structure. Safety and health professionals should be aware that each state possesses individual state laws, regulations, and systems, and these laws can overlap with other state and federal laws, as well as internal human resource or company protocols. Given the nature of most state workers' compensation systems, the state legislatures create the laws and an administrative structure is established within the state government which establishes rules and regulations for virtually every aspect of the workers' compensation function from rates to duration to care. Safety and health professionals tasked with this responsibility must adhere to these laws and regulations while assisting the injured or ill employee and balancing this with compliance with other laws and regulations. As noted above, it is essential that safety and health professionals have a working knowledge of these laws and regulations so as not to create situations involving non-compliance. For example, an employee is injured on-the-job on a weekend. He/she goes to the emergency room after work and is instructed by the emergency room physician to remain off work for a week. The employee remains at home without notifying his/her supervisor or the company for three days and is subsequently terminated in accordance with the company absenteeism policy (No Call-No Show). Should the doctor's directive or the company policy take precedence? If the employee is terminated, how will this impact the company's workers' compensation costs? Following this case further, the employee has reached maximum medical recovery and is rated at permanency rating of 40% of a whole person. The employee is released to return to work and requests an accommodation to be able to perform the job. Is the employee qualified under the

Americans with Disabilities Act? Is the company required to provide accommodation to the employee?

As noted above, the safety and health function is definitely not "siloed" within the organization, and every situation and decision possesses the potential of being impacted other laws and regulations depending on the circumstances and facts of the situation. Given the fact that safety and health professionals work with employees and the human resource function is the primary managing function for employees in many companies, the potential for conflict among and between functions exists in virtually every situation involving any employee. It is essential that safety and health professionals become knowledgeable regarding the policies, protocol, and personnel within the human resource function and investigate each situation thoroughly before finalizing a decision which may have human resource implications. With virtually every decision made by the safety and health professional impacting employees directly or indirectly, careful analysis of potential conflicts with human resource laws, regulations, and policies is essential.

In general, safety and health professionals should possess a working knowledge of the major laws and regulations impacting the safety and health function, as well as an expert knowledge of the specific laws and regulations which are within his/her sphere of direct responsibilities. In general, safety and health professionals should possess an expert level of knowledge with regards to the OSH Act and the numerous standards. This expert level must be maintained on a daily basis with all changes, modifications, or decisions with application within the safety and health structure being noted. Safety and health professionals with concurrent environmental or workers' compensation responsibilities should possess expert knowledge in the various laws, regulations, and court decisions. Further, in addition to the expert level within these specific spheres of direct responsibilities, safety and health professionals should have a working knowledge of the potential impacting laws and regulations which can be embedded within the specific situation and decision-making process. In general, these laws and regulations include, but are not limited to, the areas of discrimination in the workplace, labor laws, the ADA and laws governing handicap or disability, leave laws (such as FMLA), hiring and recruitment laws, confidentiality laws, and employee records laws.

In the area of hiring, most safety and health professionals do not possess direct responsibilities for the hiring and placement of employees; however, this function, or the impacts from this function, can have a direct correlation to the safety and health function. Hiring, recruitment, and placement errors can result in increased risk probability of workplace accidents and injuries: for example, if an individual is hired for a very physical job function when the individual does not possess the physical capabilities of performing the current job function. Although the new employee may be able to perform the job function successfully over a short period of time, the risk probabilities are substantially increased that the new employee may incur a work-related injury if appropriate job modifications or adjustments are not made to the job function.

The laws and regulations surrounding the hiring and recruitment function have substantially changed as a result of new and impacting laws and regulations as well as technology. Areas such as pre-employment background checks, especially in the areas of criminal background check and financial background checks, have changed as a result of not only federal laws, but also individual state laws. Questions which were historically asked in within the interview process may now be prohibited and can serve as the foundation for discrimination actions. The laws and regulations regarding medical screening, controlled substance testing, and confidentiality of records have substantially changed. Technology, especially in the area of social media, as well as recent legislation and court decisions, have impacted the hiring and recruitment process. At a minimum, safety and health professionals should be aware of the best practices in this important area and ensure that compliance is assured.

As noted above, the incorrect placement of new employees within the operational structure can detrimentally impact safety and health performance. Safety and health professionals should take an active role in the development of written job descriptions which are key within the hiring and placement process. The written job description should address the specific activities required within the job, appropriate and specific physical demands, specific job duties and responsibilities, and, above all, be completely accurate and up-to-date. As safety and health professionals are well aware, as soon as the candidate becomes an employee, he/she is your responsibility and should be appropriately placed and prepared for long-term success within the job position.

Given the fact that progressive disciplinary action is one of the primary methods through which behavior modification is achieved in the workplace, safety and health professionals should be knowledgeable about the process and impacts of each level within this important human resource policy. Under most progressive disciplinary policies or protocols, disciplinary action for failure to comply with a safety and health requirement is usually part-and-parcel of the policy. Whether proscribed within a policy and employee handbook within non-unionized operations, or within the collective bargaining agreement in unionized operations, the safety and health professional often is actively involved in disciplinary investigations, as well as throughout the progressive disciplinary process.

An area where a confluence of many different laws and regulations impacting a situation can amass is the area of restricted duty return to work programs. Safety and health professionals should be aware of the many possible protections afforded to the injured employee beyond simply workers' compensation coverage. Although many companies have initiated restricted duty programs to reduce costs within the category of workers' compensation time loss payments to the employee and maintaining the employee within the routine of the job function, mismanagement of this function can create impacts not only within the workers' compensation area, but also in the areas of ADA and state protections, internal short- and long-term disability policies, and related laws and regulations.

Safety and health professionals seldom are disciplined or terminated as a result of their direct efforts within the safety and health realm. However, when a safety and health professional "steps on his/her tail" in areas encompassed by other laws, companies are often placed in the awkward position of being required to act against a management team member. Safety and health professionals should be aware that, although they are a valued member of the management team, they are still employees of the company and are required to adhere to all laws, regulations, and company policies. Areas of specific focus for safety and health professionals often include, but are not limited to: violations of discrimination laws and policies; relationships in violation of company policies; improper initiation of actions in violation of a law; or inappropriate or unprofessional conduct within the scope of the position.

In summation, the safety and health professional often works hand-in-hand with the human resource function on a daily basis within many operations. Safety and health professionals should recognize this relationship and be aware that the safety and health function does not "work in a vacuum" and is impacted, directly or indirectly, by the laws, regulations, and policies with which the human resource function is tasked with ensuring compliance. Safety and health professionals should not only be experts in their field, but also knowledgeable about and able to recognize the various laws, regulations, and policies which could impact their function. It is the laws and regulations of which the safety and health professional is not aware or has not recognized which will arise to create the greatest difficulties. Preparation, knowledge, and recognition is key for safety and health professionals to avoid missteps or conflicts in the effective and efficient management of your safety and health program.

Chapter Questions:

1. Safety and HR often work together. Identify one area where these functions perform together to achieve an objective.
2. Identify one (1) law which impacts the hiring process.
3. Identify one (1) personnel-related issue which impacts the safety and health function.

Case Study—Please be aware that this case may have been modified for the purposes of this text.

98 S.Ct. 1816

Supreme Court of the United States

Ray MARSHALL, Secretary of Labor, et al., Appellants,

v.

BARLOW'S, INC., Appellee

No. 76-1143.

|

Argued Jan. 9, 1978.

|

Decided May 23, 1978.

SYNOPSIS

Action was brought by an employer to enjoin enforcement of inspection provisions of the Occupational Safety and Health Act of 1970. The Three-Judge District Court, 424 F. Supp. 437, held that the Fourth Amendment required a warrant for the type of OSHA search involved and that statutory authorization for warrantless inspections was unconstitutional, and the Secretary of Labor appealed. The Supreme Court,

Mr. Justice White, held that the Occupational Safety and Health Act, which empowers the Secretary's agents to search the work area of any employment facility within OSHA's jurisdiction for safety hazards and violations, is unconstitutional insofar as it purports to authorize inspections without warrant or its equivalent.

Affirmed.

Mr. Justice Stevens filed a dissenting opinion in which Mr. Justice Blackmun and Mr. Justice Rehnquist joined.

1818 *307 *Syllabus

Appellee brought this action to obtain injunctive relief against a warrantless inspection of its business premises pursuant to § 8(a) of the Occupational Safety and Health Act of 1970 (OSHA), which empowers agents of the Secretary of Labor to search the work area of any employment facility within OSHA's jurisdiction for safety hazards and violations of OSHA regulations. A three-judge District Court ruled in the appellee's favor, concluding, in reliance on *Camara v. Municipal Court*, 387 U.S. 523, 528–529, 87 S.Ct. 1727, 1730, 1731, 18 L.Ed.2d 930, and *See v. City of Seattle*, 387 U.S. 541, 543, 87 S.Ct. 1737, 1739, 18 L.Ed.2d 943, that the Fourth Amendment required a warrant for the type of search involved and that the statutory authorization for warrantless inspections was unconstitutional.

Held: The inspection without a warrant or its equivalent pursuant to § 8(a) of OSHA violated the Fourth Amendment Pp. 1819–1827.

(a) The rule that warrantless searches are generally unreasonable applies to commercial premises as well as homes. *Camara v. Municipal Court, supra,* and *See v. City of Seattle, supra.* Pp. 1819–1821.

(b) Though an exception to the search warrant requirement has been recognized for "closely regulated" industries "long subject to close supervision and inspection," *Colonnade Catering Corp. v. United States*, 397 U.S. 72, 74, 77, 90 S.Ct. 774, 775, 777, 25 L.Ed.2d 60, that exception does not apply simply because the business is in interstate commerce. Pp. 1820–1821.

(c) Nor does the fact that an employer by the necessary utilization of employees in his operation mean that he has opened areas where the employees alone are permitted to the warrantless scrutiny of Government agents. Pp. 1821–1822.

(d) Insofar as experience to date indicates, requiring warrants to make OSHA inspections will impose no serious burdens on the inspections system or the courts. The advantages of surprise through the opportunity of inspecting without prior notice will not be lost if, after entry to an inspector is refused, an *ex parte* warrant can be obtained, facilitating an inspector's reappearance at the premises without further notice; and the appellant Secretary's entitlement to a warrant will not depend on his demonstrating probable cause to believe that conditions on the premises ***308** violate

---

* The syllabus constitutes no part of the opinion of the Court but has been prepared by the Reporter of Decisions for the convenience of the reader. See *United States v. Detroit Timber & Lumber Co.*, 200 U.S. 321, 337, 26 S.Ct. 282, 287, 50 L.Ed.2d 499.

OSHA, but merely that reasonable legislative or administrative standards for conducting an inspection are satisfied with respect to a particular establishment. Pp. 1822–1825.

(e) Requiring a warrant for OSHA inspections does not mean that, as a practical matter, warrantless search provisions in other regulatory statutes are unconstitutional, as the reasonableness of those provisions depends upon the specific enforcement needs and privacy guarantees of each statute. P. 1825.

424 F. Supp. 437, affirmed.

OPINION

*309 Mr. Justice WHITE delivered the opinion of the Court.

Section 8(a) of the Occupational Safety and Health Act of 1970 (OSHA or Act)* **1819 empowers agents of the Secretary of Labor (Secretary) to search the work area of any employment facility within the Act's jurisdiction. The purpose of the search is to inspect for safety hazards and violations of OSHA regulations. No search warrant or other process is expressly required under the Act.

On the morning of September 11, 1975, an OSHA inspector entered the customer service area of Barlow's, Inc., an electrical and plumbing installation business located in Pocatello, Idaho. The president and general manager, Ferrol G. "Bill" Barlow, was on hand, and the OSHA inspector, after showing his credentials,[†] informed Mr. Barlow that he wished to conduct a *310 search of the working areas of the business. Mr. Barlow inquired whether any complaint had been received about his company. The inspector answered no, but that Barlow's Inc., had simply turned up in the agency's selection process. The inspector again asked to enter the nonpublic area of the business; Mr. Barlow's response was to inquire whether the inspector had a search warrant. The inspector had none. Thereupon, Mr. Barlow refused the inspector admission to the employee area of his business. He said he was relying on his rights as guaranteed by the Fourth Amendment of the United States Constitution.

Three months later, the Secretary petitioned the United States District Court for the District of Idaho to issue an order compelling Mr. Barlow to admit the inspector.[‡]

---

* In order to carry out the purposes of this chapter, the Secretary, upon presenting appropriate credentials to the owner, operator, or agent in charge, is authorized–
  "(1) to enter without delay and at reasonable times any factory, plant, establishment, construction site, or other area, workplace or environment where work is performed by an employee of an employer; and
  "(2) to inspect and investigate during regular working hours and at other reasonable times, and within reasonable limits and in a reasonable manner, any such place of employment and all pertinent conditions, structures, machines, apparatus, devices, equipment, and materials therein, and to question privately any such employer, owner, operator, agent, or employee." 84 Stat. 1598, 29 U.S.C. § 657(a).
† This is required by the Act. See n. 2, *supra*.
‡ A regulation of the Secretary, 29 CFR § 1903.4 (1977), requires an inspector to seek compulsory

The requested order was issued on December 30, 1975, and was presented to Mr. Barlow on January 5, 1976. Mr. Barlow again refused admission, and he sought his own injunctive relief against the warrantless searches assertedly permitted by OSHA. A three-judge court was convened. On December 30, 1976, it ruled in Mr. Barlow's favor (424 F. Supp. 437). Concluding that *Camara v. Municipal Court*, 387 U.S. 523, 528-529, 87 S.Ct. 1727, 1730, 1731, 18 L.Ed.2d 930 (1967), and *See v. City of Seattle*, 387 U.S. 541, 543, 87 S.Ct. 1737, 1739, 18 L.Ed.2d 943 (1967), controlled this case, the court held that the Fourth Amendment required a warrant for the type of search involved here* and that the statutory authorization for warrantless inspections was unconstitutional. An injunction against searches or inspections pursuant to § 8(a) was entered. The Secretary appealed, challenging the judgment, and we noted probable jurisdiction (430 U.S. 964, 98 S.Ct. 474, 54 L.Ed.2d 309).

*311 I

The Secretary urges that warrantless inspections to enforce OSHA are reasonable within the meaning of the Fourth Amendment. Among other things, he relies on § 8(a) of the Act, 29 U.S.C. § 657(a), which authorizes inspection of business premises without a warrant and which the Secretary urges represents a congressional construction of the Fourth Amendment that the courts should not reject. Regrettably, we are unable to agree.

[1] The Warrant Clause of the Fourth Amendment protects commercial buildings as well as private homes. To hold otherwise would belie the origin of that Amendment, **1820 and the American colonial experience. An important forerunner of the first ten Amendments to the United States Constitution, the Virginia Bill of Rights, specifically opposed "general warrants, whereby an officer or messenger may be commanded to search suspected places without evidence of a fact committed."† The general warrant was a recurring point of contention in the Colonies immediately preceding the Revolution.‡ The particular offensiveness it engendered was acutely felt by the merchants and businessmen whose premises and products were inspected for compliance with the several parliamentary revenue measures that most irritated the colonists.§ "[T]he Fourth Amendment's commands grew in large measure out of the colonists' experience with the writs of assistance ... [that] granted sweeping power to customs officials and other agents of the King to search at large for smuggled goods." *United States v. Chadwick*, 433 U.S. 1, 7–8, 97 S.Ct. 2476, 2481, 53 L.Ed.2d 538 (1977). *312 See also *G. M. Leasing Corp. v. United States*, 429 U.S.

---

process if an employer refuses a requested search. See *infra*, at 1823, and n. 13.
* No *res judicata* bar arose against Mr. Barlow from the December 30, 1975 order authorizing a search, because the earlier decision reserved the constitutional issue. See 424 F. Supp. 437.
† H. Commager, *Documents of American History* 104 (8th ed. 1968).
‡ See, e.g., Dickerson, *Writs of Assistance as a Cause of the Revolution in The Era of the American Revolution* 40 (R. Morris ed. 1939).
§ The Stamp Act of 1765, the Townshend Revenue Act of 1767, and the tea tax of 1773 are notable examples. See Commager, *supra*, n. 6, at 53, 63. For commentary, see 1 S. Morison, H. Commager, and W. Leuchtenburg, *The Growth of the American Republic* 143, 149, 159 (1969).

338, 355, 97 S.Ct. 619, 630, 50 L.Ed.2d 530 (1977). Against this background, it is untenable that the ban on warrantless searches was not intended to shield places of business as well as of residence.

This Court has already held that warrantless searches are generally unreasonable, and that this rule applies to commercial premises as well as homes. In *Camara v. Municipal Court, supra*, 387 U.S., at 528–529, 87 S.Ct., at 1731, we held:

"[E]xcept in certain carefully defined classes of cases, a search of private property without proper consent is 'unreasonable' unless it has been authorized by a valid search warrant."

On the same day, we also ruled:

"As we explained in *Camara*, a search of private houses is presumptively unreasonable if conducted without a warrant. The businessman, like the occupant of a residence, has a constitutional right to go about his business free from unreasonable official entries upon his private commercial property. The businessman, too, has that right placed in jeopardy if the decision to enter and inspect for violation of regulatory laws can be made and enforced by the inspector in the field without official authority evidenced by a warrant." *See v. City of Seattle, supra*, 387 U.S., at 543, 87 S.Ct., at 1739.

These same cases also held that the Fourth Amendment prohibition against unreasonable searches protects against warrantless intrusions during civil as well as criminal investigations. *Ibid.* The reason is found in the "basic purpose of this Amendment ... [which] is to safeguard the privacy and security of individuals against arbitrary invasions by governmental officials." *Camara, supra*, 387 U.S., at 528, 87 S.Ct. at 1730. If the government intrudes on a person's property, the privacy interest suffers whether the government's motivation is to investigate violations of criminal laws or breaches of other statutory or *313 regulatory standards. It therefore appears that unless some recognized exception to the warrant requirement applies, *See v. City of Seattle*, would require a warrant to conduct the inspection sought in this case.

[2] The Secretary urges that an exception from the search warrant requirement has been recognized for "pervasively regulated business[es]," *United States v. Biswell*, 406 U.S. 311, 316, 92 S.Ct. 1593, 1596, 32 L.Ed.2d 87 (1972), and for "closely regulated" industries "long subject to close supervision and inspection." *Colonnade Catering **1821 Corp. v. United States*, 397 U.S. 72, 74, 77, 90 S.Ct. 774, 777, 25 L.Ed.2d 60 (1970). These cases are indeed exceptions, but they represent responses to relatively unique circumstances. Certain industries have such a history of government oversight that no reasonable expectation of privacy, see *Katz v. United States*, 389 U.S. 347, 351–352, 88 S.Ct. 507, 511, 19 L.Ed.2d 576 (1967), could exist for a proprietor over the stock of such an enterprise. Liquor (*Colonnade*) and firearms (*Biswell*) are industries of this type; when an entrepreneur embarks upon such a

business, he has voluntarily chosen to subject himself to a full arsenal of governmental regulation.

Industries such as these fall within the "certain carefully defined classes of cases," referenced in *Camara*, 387 U.S., at 528, 87 S.Ct., at 1731. The element that distinguishes these enterprises from ordinary businesses is a long tradition of close government supervision, of which any person who chooses to enter such a business must already be aware. "A central difference between those cases [*Colonnade* and *Biswell*] and this one is that businessmen engaged in such federally licensed and regulated enterprises accept the burdens as well as the benefits of their trade, whereas the petitioner here was not engaged in any regulated or licensed business. The businessman in a regulated industry in effect consents to the restrictions placed upon him." *Almeida-Sanchez v. United States*, 413 U.S. 266, 271, 93 S.Ct. 2535, 2538, 37 L.Ed.2d 596 (1973).

The clear import of our cases is that the closely regulated industry of the type involved in *Colonnade* and *Biswell* is the exception. The Secretary would make it the rule. Invoking ***314** the Walsh–Healey Act of 1936, 41 U.S.C. § 35 *et seq.*, the Secretary attempts to support a conclusion that all businesses involved in interstate commerce have long been subjected to close supervision of employee safety and health conditions. But the degree of federal involvement in employee working circumstances has never been of the order of specificity and pervasiveness that OSHA mandates. It is quite unconvincing to argue that the imposition of minimum wages and maximum hours on employers who contracted with the Government under the Walsh–Healey Act prepared the entirety of American interstate commerce for regulation of working conditions to the minutest detail. Nor can any but the most fictional sense of voluntary consent to later searches be found in the single fact that one conducts a business affecting interstate commerce; under current practice and law, few businesses can be conducted without having some effect on interstate commerce.

The Secretary also attempts to derive support for a *Colonnade–Biswell*-type exception by drawing analogies from the field of labor law. In *Republic Aviation Corp. v. NLRB*, 324 U.S. 793, 65 S.Ct. 982, 89 L.Ed. 1372 (1945), this Court upheld the rights of employees to solicit for a union during nonworking time where efficiency was not compromised. By opening up his property to employees, the employer had yielded so much of his private property rights as to allow those employees to exercise § 7 rights under the National Labor Relations Act. But this Court also held that the private property rights of an owner prevailed over the intrusion of nonemployee organizers, even in nonworking areas of the plant and during nonworking hours. *NLRB v. Babcock & Wilcox Co.*, 351 U.S. 105, 76 S.Ct. 679, 100 L.Ed. 975 (1956).

[3] The critical fact in this case is that entry over Mr. Barlow's objection is being sought by a Government agent.* Employees ***315** are not being prohibited from reporting

---

* The Government has asked that Mr. Barlow be ordered to show cause why he should not be held in contempt for refusing to honor the inspection order, and its position is that the OSHA inspector is now entitled to enter at once, over Mr. Barlow's objection.

OSHA violations. What they observe in their daily functions is undoubtedly beyond the employer's reasonable expectation of privacy. The Government inspector, however, is not an employee. Without a warrant, **1822 he stands in no better position than a member of the public. What is observable by the public is observable, without a warrant, by the Government inspector as well.* The owner of a business has not, by the necessary utilization of employees in his operation, thrown open the areas where employees alone are permitted to the warrantless scrutiny of Government agents. That an employee is free to report, and the Government is free to use, any evidence of noncompliance with OSHA that the employee observes furnishes no justification for federal agents to enter a place of business from which the public is restricted and to conduct their own warrantless search.†

II

[4] The Secretary nevertheless stoutly argues that the enforcement scheme of the Act requires warrantless searches, and that the restrictions on search discretion contained in the Act and its regulations already protect as much privacy as a warrant would. The Secretary thereby asserts the actual reasonableness of OSHA searches, whatever the general rule against warrantless searches might be. Because "reasonableness is still the ultimate standard," *316 Camara v. Municipal Court, 387 U.S., at 539, 87 S.Ct., at 1736, the Secretary suggests that the Court decide whether a warrant is needed by arriving at a sensible balance between the administrative necessities of OSHA inspections and the incremental protection of privacy of business owners a warrant would afford. He suggests that only a decision exempting OSHA inspections from the Warrant Clause would give "full recognition to the competing public and private interests here at stake." Ibid.

The Secretary submits that warrantless inspections are essential to the proper enforcement of OSHA because they afford the opportunity to inspect without prior notice and hence to preserve the advantages of surprise. While the dangerous conditions outlawed by the Act include structural defects that cannot be quickly hidden or remedied, the Act also regulates a myriad of safety details that may be amenable to speedy alteration or disguise. The risk is that during the interval between an inspector's initial request to search a plant and his procuring a warrant following the owner's refusal of permission, violations of this latter type could be corrected and thus escape the inspector's notice. To the suggestion that warrants may be issued *ex parte*

---

* Cf. *Air Pollution Variance Bd. v. Western Alfalfa Corp.*, 416 U.S. 861, 94 S.Ct. 2114, 40 L.Ed.2d 607 (1974).
† The automobile-search cases cited by the Secretary are even less helpful to his position than the labor cases. The fact that automobiles occupy a special category in Fourth Amendment case law is by now beyond doubt due, among other factors, to the quick mobility of a car, the registration requirements of both the car and the driver, and the more available opportunity for plain-view observations of a car's contents. *Cady v. Dombrowski*, 413 U.S. 433, 441–442, 93 S.Ct. 2523, 2528, 37 L.Ed.2d 706 (1973); see also *Chambers v. Maroney*, 399 U.S. 42, 48–51, 90 S.Ct. 1975, 1979–1981, 26 L.Ed.2d 419 (1970). Even so, probable cause has not been abandoned as a requirement for stopping and searching an automobile.

and executed without delay and without prior notice, thereby preserving the element of surprise, the Secretary expresses concern for the administrative strain that would be experienced by the inspection system, and by the courts, should *ex parte* warrants issued in advance become standard practice.

We are unconvinced, however, that requiring warrants to inspect will impose serious burdens on the inspection system or the courts, will prevent inspections necessary to enforce the statute, or will make them less effective. In the first place, the great majority of businessmen can be expected in normal course to consent to inspection without warrant; the Secretary has not brought to this Court's attention any widespread pattern of refusal.* In those cases where **1823 an owner does insist *317 on a warrant, the Secretary argues that inspection efficiency will be impeded by the advance notice and delay. The Act's penalty provisions for giving advance notice of a search, 29 U.S.C. § 666(f), and the Secretary's own regulations, 29 CFR § 1903.6 (1977), indicate that surprise searches are indeed contemplated. However, the Secretary has also promulgated a regulation providing that upon refusal to permit an inspector to enter the property or to complete his inspection, the inspector shall attempt to ascertain the reasons for the refusal and report to his superior, who shall "promptly take appropriate action, including compulsory process, if necessary." 29 CFR § 1903.4 (1977).† The regulation represents a choice to proceed by *318 process where entry is refused; and on the basis of evidence available from present practice, the Act's effectiveness has not been crippled by providing those owners who wish to refuse an initial requested entry with a time lapse while the inspector obtains the necessary

---

* We recognize that today's holding itself might have an impact on whether owners choose to resist requested searches; we can only await the development of evidence not present on this record to determine how serious an impediment to effective enforcement this might be.

† It is true, as the Secretary asserts, that § 8(a) of the Act, 29 U.S.C. § 657(a), purports to authorize inspections without warrant; but it is also true that it does not forbid the Secretary from proceeding to inspect only by warrant or other process. The Secretary has broad authority to prescribe such rules and regulations as he may deem necessary to carry out his responsibilities under this chapter, "including rules and regulations dealing with the inspection of an employer's establishment." § 8(g)(2), 29 U.S.C. § 657(g)(2). The regulations with respect to inspections are contained in 29 CFR Part 1903 (1977). Section 1903.4, referred to in the text, provides as follows:

"Upon a refusal to permit a Compliance Safety and Health Officer, in the exercise of his official duties, to enter without delay and at reasonable times any place of employment or any place therein, to inspect, to review records, or to question any employer, owner, operator, agent, or employee, in accordance with § 1903.3, or to permit a representative of employees to accompany the Compliance Safety and Health Officer during the physical inspection of any workplace in accordance with § 1903.8, the Compliance Safety and Health Officer shall terminate the inspection or confine the inspection to other areas, conditions, structures, machines, apparatus, devices, equipment, materials, records, or interviews concerning which no objection is raised. The Compliance Safety and Health Officer shall endeavor to ascertain the reason for such refusal, and he shall immediately report the refusal and the reason therefor to the Area Director. The Area Director shall immediately consult with the Assistant Regional Director and the Regional Solicitor, who shall promptly take appropriate action, including compulsory process, if necessary."

When his representative was refused admission by Mr. Barlow, the Secretary proceeded in federal court to enforce his right to enter and inspect, as conferred by 29 U.S.C. § 657.

process.* Indeed, the kind of process sought in this case and apparently anticipated by the regulation provides notice to the business operator.†

---

* A change in the language of the Compliance Operations Manual for OSHA inspectors supports the inference that, whatever the Act's administrators might have thought at the start, it was eventually concluded that enforcement efficiency would not be jeopardized by permitting employers to refuse entry, at least until the inspector obtained compulsory process. The 1972 Manual included a section specifically directed to obtaining "warrants," and one provision of that section dealt with *ex parte* warrants: "In cases where a refusal of entry is to be expected from the past performance of the employer, or where the employer has given some indication prior to the commencement of the investigation of his intention to bar entry or limit or interfere with the investigation, a warrant should be obtained before the inspection is attempted. Cases of this nature should also be referred through the Area Director to the appropriate Regional Solicitor and the Regional Administrator alerted." Dept. of Labor, *OSHA Compliance Operations Manual* V-7 (Jan. 1972).

The latest available manual, incorporating changes as of November 1977, deletes this provision, leaving only the details for obtaining "compulsory process" *after* an employer has refused entry. Dept. of Labor, *OSHA Field Operations Manual*, Vol. V, pp. V-4–V-5. In its present form, the Secretary's regulation appears to permit establishment owners to insist on "process"; and hence their refusal to permit entry would fall short of criminal conduct within the meaning of 18 U.S.C. §§ 111 and 1114 (1976 ed.), which make it a crime forcibly to impede, intimidate, or interfere with federal officials, including OSHA inspectors, while engaged in or on account of the performance of their official duties.

† The proceeding was instituted by filing an "Application for Affirmative Order to Grant Entry and for an Order to show cause why such affirmative order should not issue." The District Court issued the order to show cause, the matter was argued, and an order then issued authorizing the inspection and enjoining interference by Barlow's. The following is the order issued by the District Court: "IT IS HEREBY ORDERED, ADJUDGED AND DECREED that the United States of America, United States Department of Labor, Occupational Safety and Health Administration, through its duly designated representative or representatives, are entitled to entry upon the premises known as Barlow's Inc., 225 West Pine, Pocatello, Idaho, and may go upon said business premises to conduct an inspection and investigation as provided for in Section 8 of the Occupational Safety and Health Act of 1970 (29 U.S.C. 651, *et seq.*), as part of an inspection program designed to assure compliance with that Act; that the inspection and investigation shall be conducted during regular working hours or at other reasonable times, within reasonable limits and in a reasonable manner, all as set forth in the regulations pertaining to such inspections promulgated by the Secretary of Labor, at 29 C.F.R., Part 1903; that appropriate credentials as representatives of the Occupational Safety and Health Administration, United States Department of Labor, shall be presented to the Barlow's Inc. representative upon said premises and the inspection and investigation shall be commenced as soon as practicable after the issuance of this Order and shall be completed within reasonable promptness; that the inspection and investigation shall extend to the establishment or other area, workplace, or environment where work is performed by employees of the employer, Barlow's Inc., and to all pertinent conditions, structures, machines, apparatus, devices, equipment, materials, and all other things therein (including but not limited to records, files, papers, processes, controls, and facilities) bearing upon whether Barlow's Inc. is furnishing to its employees employment and a place of employment that are free from recognized hazards that are causing or are likely to cause death or serious physical harm to its employees, and whether Barlow's Inc. is complying with the Occupational Safety and Health Standards promulgated under the Occupational Safety and Health Act and the rules, regulations, and orders issued pursuant to that Act; that representatives of the Occupational Safety and Health Administration may, at the option of Barlow's Inc., be accompanied by one or more employees of Barlow's Inc., pursuant to Section 8(e) of that Act; that Barlow's Inc., its agents, representatives, officers, and employees are hereby enjoined and restrained from in any way whatsoever interfering with the inspection and investigation authorized by this Order and, further, Barlow's Inc. is hereby ordered and directed to, within five working days from the date of this Order, furnish a copy of this Order to its officers and managers, and, in addition, to post a copy of this Order at its employee's bulletin board located upon the business premises; and Barlow's Inc. is hereby ordered and directed to comply in all respects with this order and allow the inspection and investigation to take place without delay and forthwith."

*319 If this safeguard ****1824** endangers the efficient administration of OSHA, the Secretary should never have adopted it, particularly when the Act does not require it. Nor is it immediately ***320** apparent why the advantages of surprise would be lost if, after being refused entry, procedures were available for the Secretary to seek an *ex parte* warrant and to reappear at the premises without further notice to the establishment being inspected.*

[5] Whether the Secretary proceeds to secure a warrant or other process, with or without prior notice, his entitlement to inspect will not depend on his demonstrating probable cause to believe that conditions in violation of OSHA exist on the premises. Probable cause in the criminal law sense is not required. For purposes of an administrative search such as this, probable cause justifying the issuance of a warrant may be based not only on specific evidence of an existing violation,[†] but also on a showing that "reasonable legislative or administrative standards for conducting an ... inspection are satisfied with respect to a particular [establishment]." ***321** *Camara v. Municipal Court*, 387 U.S., at 538, 87 S.Ct., at 1736. A warrant showing that a specific ****1825** business has been chosen for an OSHA search on the basis of a general administrative plan for the enforcement of the Act derived from neutral sources such as, for example, dispersion of employees in various types of industries across a given area, and the desired frequency of searches in any of the lesser divisions of the area, would protect an employer's Fourth Amendment rights.[‡] We doubt that the consumption of enforcement energies in the obtaining of such warrants will exceed manageable proportions.

[6] Finally, the Secretary urges that requiring a warrant for OSHA inspectors will mean that, as a practical matter, warrantless search provisions in other regulatory statutes are also constitutionally infirm. The reasonableness of a warrantless search, however, will depend upon the specific enforcement needs and privacy guarantees of each statute. Some of the statutes cited apply only to a single industry, where regulations might already be so pervasive that a *Colonnade–Biswell* exception to the warrant requirement could apply. Some statutes already envision resort to federal court

---

* Insofar as the Secretary's statutory authority is concerned, a regulation expressly providing that the Secretary could proceed *ex parte* to seek a warrant or its equivalent would appear to be as much within the Secretary's power as the regulation currently in force and calling for "compulsory process."

† Section 8(f)(1), 29 U.S.C. § 657(f)(1), provides that employees or their representatives may give written notice to the Secretary of what they believe to be violations of safety or health standards and may request an inspection. If the Secretary then determines that "there are reasonable grounds to believe that such violation or danger exists, he shall make a special inspection in accordance with the provisions of this section as soon as practicable." The statute thus purports to authorize a warrantless inspection in these circumstances.

‡ The Secretary, Brief for Petitioner 9 n. 7, states that the Barlow inspection was not based on an employee complaint but was a "general schedule" investigation. "Such general inspections," he explains, "now called Regional Programmed Inspections, are carried out in accordance with criteria based upon accident experience and the number of employees exposed in particular industries. U. S. Department of Labor, Occupational Safety and Health Administration, *Field Operations Manual, supra*, 1 CCH Employment Safety and Health Guide ¶ 4327.2 (1976)."

enforcement when entry is refused, employing specific language in some cases* and general language in others.† In short, we base ***322** today's opinion on the facts and law concerned with OSHA and do not retreat from a holding appropriate to that statute because of its real or imagined effect on other, different administrative schemes.

[7] [8] Nor do we agree that the incremental protections afforded the employer's privacy by a warrant are so marginal that they fail to justify the administrative burdens that may be entailed. ***323** The authority ****1826** to make warrantless searches devolves almost unbridled discretion upon executive and administrative officers, particularly those in the field, as to when to search and whom to search. A warrant, by contrast, would provide assurances from a neutral officer that the inspection is reasonable under the Constitution, is authorized by statute, and is pursuant to an administrative plan containing specific neutral criteria.‡ Also, a warrant would then and there advise the owner of the scope and objects of the search, beyond which limits the inspector is not expected to proceed.§ These are important functions for

---

* The Federal Metal and Nonmetallic Mine Safety Act provides: "Whenever an operator ... refuses to permit the inspection or investigation of any mine which is subject to this chapter ... a civil action for preventive relief, including an application for a permanent or temporary injunction, restraining order, or other order, may be instituted by the Secretary in the district court of the United States for the district ... ." 30 U.S.C. § 733(a). "The Secretary may institute a civil action for relief, including a permanent or temporary injunction, restraining order, or any other appropriate order in the district court ... whenever such operator or his agent ... refuses to permit the inspection of the mine .... Each court shall have jurisdiction to provide such relief as may be appropriate." 30 U.S.C. § 818. Another example is the Clean Air Act, which grants federal district courts jurisdiction "to require compliance" with the Administrator of the Environmental Protection Agency's attempt to inspect under 42 U.S.C. § 7414 (1976 ed., Supp. I), when the Administrator has commenced "a civil action" for injunctive relief or to recover a penalty. 42 U.S.C. § 7413(b)(4) (1976 ed., Supp. I).

† Exemplary language is contained in the Animal Welfare Act of 1970 which provides for inspections by the Secretary of Agriculture; federal district courts are vested with jurisdiction "specifically to enforce, and to prevent and restrain violations of this chapter, and shall have jurisdiction in all other kinds of cases arising under this chapter." 7 U.S.C. § 2146(c) (1976 ed.). Similar provisions are included in other agricultural inspection Acts; see, e.g., 21 U.S.C. § 674 (meat product inspection); 21 U.S.C. § 1050 (egg product inspection). The Internal Revenue Code, whose excise tax provisions requiring inspections of businesses are cited by the Secretary, provides: "The district courts ... shall have such jurisdiction to make and issue in civil actions, writs and orders of injunction ... and such other orders and processes, and to render such ... decrees as may be necessary or appropriate for the enforcement of the internal revenue laws." 26 U.S.C. § 7402(a). For gasoline inspections, federal district courts are granted jurisdiction to restrain violations and enforce standards (one of which, 49 U.S.C. § 1677, requires gas transporters to permit entry or inspection). The owner is to be afforded the opportunity for notice and response in most cases, but "failure to give such notice and afford such opportunity shall not preclude the granting of appropriate relief [by the district court]." 49 U.S.C. § 1679(a).

‡ The application for the inspection order filed by the Secretary in this case represented that "the desired inspection and investigation are contemplated as a part of an inspection program designed to assure compliance with the Act and are authorized by Section 8(a) of the Act." The program was not described, however, nor any facts presented that would indicate why an inspection of Barlow's establishment was within the program. The order that issued concluded generally that the inspection authorized was "part of an inspection program designed to assure compliance with the Act."

§ Section 8(a) of the Act, as set forth in 29 U.S.C. § 657(a), provides that "in order to carry out the purposes of this chapter" the Secretary may enter any establishment, area, workplace or environment "where work is performed by an employee of an employer" and "inspect and investigate" any such place of employment and all "pertinent conditions, structures, machines, apparatus, devices, equipment, and materials therein, and ... question privately any such employer, owner, operator, agent, or

a warrant to perform, functions which underlie the Court's prior decisions that the Warrant Clause applies to ***324** inspections for compliance with regulatory statutes.* *Camara v. Municipal Court*, 387 U.S. 523, 87 S.Ct. 1727, 18 L.Ed.2d 930 (1967); *See v. City of Seattle*, 387 U.S. 541, 87 S.Ct. 1737, 18 L.Ed.2d 943 (1967). We conclude that the concerns expressed by the Secretary do not suffice to ****1827** justify warrantless inspections under OSHA or vitiate the general constitutional requirement that for a search to be reasonable a warrant must be obtained.

***325** III

[9] We hold that Barlow's was entitled to a declaratory judgment that the Act is unconstitutional insofar as it purports to authorize inspections without warrant or its equivalent and to an injunction enjoining the Act's enforcement to that extent.† The judgment of the District Court is therefore affirmed.

---

employee." Inspections are to be carried out "during regular working hours and at other reasonable times, and within reasonable limits and in a reasonable manner." The Secretary's regulations echo the statutory language in these respects. 29 CFR § 1903.3 (1977). They also provide that inspectors are to explain the nature and purpose of the inspection and to "indicate generally the scope of the inspection." 29 CFR § 1903.7(a) (1977). Environmental samples and photographs are authorized, 29 CFR § 1903.7(b) (1977), and inspections are to be performed so as "to preclude unreasonable disruption of the operations of the employer's establishment." 29 CFR § 1903.7(d) (1977). The order that issued in this case reflected much of the foregoing statutory and regulatory language.

* Delineating the scope of a search with some care is particularly important where documents are involved. Section 8(c) of the Act, 29 U.S.C. § 657(c), provides that an employer must "make, keep and preserve, and make available to the Secretary [of Labor] or to the Secretary of Health, Education and Welfare" such records regarding his activities relating to OSHA as the Secretary of Labor may prescribe by regulation as necessary or appropriate for enforcement of the statute or for developing information regarding the causes and prevention of occupational accidents and illnesses. Regulations requiring employers to maintain records of and to make periodic reports on "work-related deaths, injuries and illnesses" are also contemplated, as are rules requiring accurate records of employee exposure to potential toxic materials and harmful physical agents.

In describing the scope of the warrantless inspection authorized by the statute, § 8(a) does not expressly include any *records* among those items or things that may be examined, and § 8(c) merely provides that the employer is to "make available" his pertinent records and to make periodic reports.

The Secretary's regulation, 29 CFR § 1903.3 (1977), however, expressly includes among the inspector's powers the authority "to review records required by the Act and regulations published in this chapter, and other records which are directly related to the purpose of the inspection." Further, § 1903.7 requires inspectors to indicate generally "the records specified in § 1903.3 which they wish to review," but "such designations of records shall not preclude access to additional records specified in § 1903.3." It is the Secretary's position, which we reject, that an inspection of documents of this scope may be effected without a warrant. The order that issued in this case included among the objects and things to be inspected "all other things therein (including but not limited to records, files, papers, processes, controls and facilities) bearing upon whether Barlow's, Inc. is furnishing to its employees employment and a place of employment that are free from recognized hazards that are causing or are likely to cause death or serious physical harm to its employees, and whether Barlow's, Inc. is complying with ..." the OSHA regulations.

† The injunction entered by the District Court, however, should not be understood to forbid the Secretary from exercising the inspection authority conferred by § 8 pursuant to regulations and judicial process that satisfy the Fourth Amendment. The District Court did not address the issue whether the order for inspection that was issued in this case was the functional equivalent of a warrant, and the Secretary has limited his submission in this case to the constitutionality of a warrantless search of the Barlow establishment authorized by § 8(a). He has expressly declined to rely on 29 CFR § 1903.4 (1977) and upon the order obtained in this case: Tr. of Oral Arg. 19. Of course, if the process obtained here, or obtained in other cases under revised regulations, would satisfy the Fourth Amendment, there would

So ordered.

Mr. Justice BRENNAN took no part in the consideration or decision of this case.

Mr. Justice STEVENS, with whom Mr. Justice BLACKMUN and Mr. Justice REHNQUIST join, dissenting.

Congress enacted the Occupational Safety and Health Act to safeguard employees against hazards in the work areas of businesses subject to the Act. To ensure compliance, Congress authorized the Secretary of Labor to conduct routine, nonconsensual inspections. Today the Court holds that the Fourth Amendment prohibits such inspections without a warrant. The Court also holds that the constitutionally required warrant may be issued without any showing of probable cause. I disagree with both of these holdings.

The Fourth Amendment contains two separate Clauses, each ***326** flatly prohibiting a category of governmental conduct. The first Clause states that the right to be free from unreasonable searches "shall not be violated";* the second unequivocally prohibits the issuance of warrants except "upon probable cause."† In this case the ultimate question is whether the category of warrantless searches authorized by the statute is "unreasonable" within the meaning of the first Clause.

In cases involving the investigation of criminal activity, the Court has held that the reasonableness of a search generally depends upon whether it was conducted pursuant to a valid warrant. See, e.g., *Coolidge v. New Hampshire*, 403 U.S. 443, 91 S.Ct. 2022, 29 L.Ed.2d 564. There is, however, also a category of searches which are reasonable within the meaning of the first Clause, even though the probable cause requirement of the Warrant Clause cannot be satisfied. See *United States v. Martinez-Fuerte*, 428 U.S. 543, 96 S.Ct. 3074, 49 L.Ed.2d 1116; *Terry v. Ohio*, 392 U.S. 1, 88 S.Ct. 1868, 20 L.Ed.2d 889; *South Dakota v. Opperman*, 428 U.S. 364, 96 S.Ct. 3092, 49 L.Ed.2d 1000; *United States v. Biswell*, 406 U.S. 311, 92 S.Ct. 1593, 32 L.Ed.2d 87. The regulatory inspection program challenged in this case, in my judgment, falls within this category.

I

The warrant requirement is linked "textually ... to the probable-cause concept" in the Warrant Clause. ****1828** *South Dakota v. Opperman, supra*, 428 U.S., at 370 n.5, 96 S.Ct. at 3097. The routine OSHA inspections are, by definition, not based

---

be no occasion for enjoining the inspections authorized by § 8(a).

* "The right of the people to be secure in their persons, houses, papers, and effects, against unreasonable searches and seizures, shall not be violated ... ."

† "[A]nd no Warrants shall issue, but upon probable cause, supported by Oath or affirmation, and particularly describing the place to be searched, and the persons or things to be seized."

on cause to believe there is a violation on the premises to be inspected. Hence, if the inspections were measured against the requirements of the Warrant Clause, they would be automatically and unequivocally unreasonable.

*327 Because of the acknowledged importance and reasonableness of routine inspections in the enforcement of federal regulatory statutes such as OSHA, the Court recognizes that requiring full compliance with the Warrant Clause would invalidate all such inspection programs. Yet, rather than simply analyzing such programs under the "reasonableness" Clause of the Fourth Amendment, the Court holds the OSHA program invalid under the Warrant Clause and then avoids a blanket prohibition on all routine, regulatory inspections by relying on the notion that the "probable cause" requirement in the Warrant Clause may be relaxed whenever the Court believes that the governmental need to conduct a category of "searches" outweighs the intrusion on interests protected by the Fourth Amendment.

The Court's approach disregards the plain language of the Warrant Clause and is unfaithful to the balance struck by the Framers of the Fourth Amendment—"the one procedural safeguard in the Constitution that grew directly out of the events which immediately preceded the revolutionary struggle with England."* This preconstitutional history includes the controversy in England over the issuance of general warrants to aid enforcement of the seditious libel laws and the colonial experience with writs of assistance issued to facilitate collection of the various import duties imposed by Parliament. The Framers' familiarity with the abuses attending the issuance of such general warrants provided the principal stimulus for the restraints on arbitrary governmental intrusions embodied in the Fourth Amendment.

"[O]ur constitutional fathers were not concerned about warrantless searches, but about overreaching warrants. It is perhaps too much to say that they feared the warrant more than the search, but it is plain enough that the warrant was the prime object of their concern. Far from *328 looking at the warrant as a protection against unreasonable searches, they saw it as an authority for unreasonable and oppressive searches … ."†

Since the general warrant, not the warrantless search, was the immediate evil at which the Fourth Amendment was directed, it is not surprising that the Framers placed precise limits on its issuance. The requirement that a warrant only issue on a showing of particularized probable cause was the means adopted to circumscribe the warrant power. While the subsequent course of Fourth Amendment jurisprudence in this Court emphasizes the dangers posed by warrantless searches conducted without probable cause, it is the general reasonableness standard in the first Clause, not the Warrant Clause, that the Framers adopted to limit this category of searches. It is, of course, true that the existence of a valid warrant normally satisfies the reasonableness requirement under the Fourth Amendment. But we should not dilute the

---

* J. Landynski, *Search and Seizure and the Supreme Court* 19 (1966).
† T. Taylor, *Two Studies in Constitutional Interpretation* 41 (1969).

requirements of the Warrant Clause in an effort to force every kind of governmental intrusion which satisfies the Fourth Amendment definition of a "search" into a judicially developed, warrant-preference scheme.

Fidelity to the original understanding of the Fourth Amendment, therefore, leads to the conclusion that the Warrant Clause has no application to routine, regulatory inspections of commercial premises. If such inspections are valid, it is because they comport with the ultimate reasonableness standard of the Fourth Amendment. If the Court were correct in its view that such inspections, if undertaken without a warrant, **1829 are unreasonable in the constitutional sense, the issuance of a "new-fangled warrant"—to use Mr. Justice Clark's characteristically expressive term—without any true showing of particularized probable cause would not be sufficient to validate them.*

*329 II

Even if a warrant issued without probable cause were faithful to the Warrant Clause, I could not accept the Court's holding that the Government's inspection program is constitutionally unreasonable because it fails to require such a warrant procedure. In determining whether a warrant is a necessary safeguard in a given class of cases, "the Court has weighed the public interest against the Fourth Amendment interest of the individual ... ." *United States v. Martinez-Fuerte*, 428 U.S., at 555, 96 S.Ct., at 3081. Several considerations persuade me that this balance should be struck in favor of the routine inspections authorized by Congress.

Congress has determined that regulation and supervision of safety in the workplace furthers an important public interest and that the power to conduct warrantless searches is necessary to accomplish the safety goals of the legislation. In assessing the public interest side of the Fourth Amendment balance, however, the Court today substitutes its judgment for that of Congress on the question of what inspection authority is needed to effectuate the purposes of the Act. The Court states that if surprise is truly an important ingredient of an effective, representative inspection program, it can be retained by obtaining *ex parte* warrants in advance. The Court assures the Secretary that this will not unduly burden enforcement resources because most employers will consent to inspection.

The Court's analysis does not persuade me that Congress' determination that the warrantless inspection power as a necessary adjunct of the exercise of the regulatory power is unreasonable. It was surely not unreasonable to conclude that the rate at which employers deny entry to inspectors would increase if covered businesses, which may have safety violations on their premises, have a right to deny warrantless entry to a compliance inspector. The Court is correct that this problem could be avoided by requiring inspectors to obtain a warrant prior to every inspection visit. But the adoption of *330 such a practice undercuts the Court's explanation of why

---

* *See v. City of Seattle*, 387 U.S. 541, 547, 87 S.Ct. 1737, 1741, 18 L.Ed.2d 943 (Clark, J., dissenting).

a warrant requirement would not create undue enforcement problems. For, even if it were true that many employers would not exercise their right to demand a warrant, it would provide little solace to those charged with administration of OSHA; faced with an increase in the rate of refusals and the added costs generated by futile trips to inspection sites where entry is denied, officials may be compelled to adopt a general practice of obtaining warrants in advance. While the Court's prediction of the effect a warrant requirement would have on the behavior of covered employers may turn out to be accurate, its judgment is essentially empirical. On such an issue, I would defer to Congress' judgment regarding the importance of a warrantless search power to the OSHA enforcement scheme.

The Court also appears uncomfortable with the notion of second-guessing Congress and the Secretary on the question of how the substantive goals of OSHA can best be achieved. Thus, the Court offers an alternative explanation for its refusal to accept the legislative judgment. We are told that, in any event, the Secretary, who is charged with enforcement of the Act, has indicated that inspections without delay are not essential to the enforcement scheme. The Court bases this conclusion on a regulation prescribing the administrative response when a compliance inspector is denied entry. It provides: "The Area Director shall immediately consult with the Assistant Regional **1830 Director and the Regional Solicitor, who shall promptly take appropriate action, including compulsory process, if necessary." 29 CFR § 1903.4 (1977). The Court views this regulation as an admission by the Secretary that no enforcement problem is generated by permitting employers to deny entry and delaying the inspection until a warrant has been obtained. I disagree. The regulation was promulgated against the background of a statutory right to immediate entry, of which covered employers are presumably *331 aware and which Congress and the Secretary obviously thought would keep denials of entry to a minimum. In these circumstances, it was surely not unreasonable for the Secretary to adopt an orderly procedure for dealing with what he believed would be the occasional denial of entry. The regulation does not imply a judgment by the Secretary that delay caused by numerous denials of entry would be administratively acceptable.

Even if a warrant requirement does not "frustrate" the legislative purpose, the Court has no authority to impose an additional burden on the Secretary unless that burden is required to protect the employer's Fourth Amendment interests.* The essential function of the traditional warrant requirement is the interposition of a neutral magistrate between the citizen and the presumably zealous law enforcement officer so that there might be an objective determination of probable cause. But this purpose is not served by the newfangled inspection warrant. As the Court acknowledges, the inspector's "entitlement to inspect will not depend on his demonstrating probable cause to believe that conditions in violation of OSHA exist on the premises. ...

---

* When it passed OSHA, Congress was cognizant of the fact that in light of the enormity of the enforcement task "the number of inspections which it would be desirable to have made will undoubtedly for an unforeseeable period, exceed the capacity of the inspection force ... ." Senate Committee on Labor and Public Welfare, *Legislative History of the Occupational Safety and Health Act of 1970*, 92d Cong., 1st Sess., 152 (Comm. Print 1971).

For purposes of an administrative search such as this, probable cause justifying the issuance of a warrant may be based … on a showing that 'reasonable legislative or administrative standards for conducting an … inspection are satisfied with respect to a particular [establishment].'" *Ante,* at 1824. To obtain a warrant, the inspector need only show that "a specific business has been chosen for an OSHA search on the basis of a general administrative plan for the enforcement of the Act derived *332 from neutral sources … ." *Ante,* at 1825. Thus, the only question for the magistrate's consideration is whether the contemplated inspection deviates from an inspection schedule drawn up by higher level agency officials.

Unlike the traditional warrant, the inspection warrant provides no protection against the search itself for employers who the Government has no reason to suspect are violating OSHA regulations. The Court plainly accepts the proposition that random health and safety inspections are reasonable. It does not question Congress' determination that the public interest in workplaces free from health and safety hazards outweighs the employer's desire to conduct his business only in the presence of permittees, except in those rare instances when the Government has probable cause to suspect that the premises harbor a violation of the law.

What purposes, then, are served by the administrative warrant procedure? The inspection warrant purports to serve three functions: to inform the employer that the inspection is authorized by the statute, to advise him of the lawful limits of the inspection, and to assure him that the person demanding entry is an authorized inspector: *Camara v. Municipal Court,* 387 U.S. 523, 532, 87 S.Ct. 1727, 1732, 18 L.Ed.2d 930. An examination of these functions in the OSHA context reveals that the inspection warrant adds little to the protections already afforded by the statute and pertinent regulations, and the slight additional benefit it might provide is insufficient to identify a constitutional violation or to justify **1831 overriding Congress' judgment that the power to conduct warrantless inspections is essential.

The inspection warrant is supposed to assure the employer that the inspection is in fact routine, and that the inspector has not improperly departed from the program of representative inspections established by responsible officials. But to the extent that harassment inspections would be reduced by the necessity of obtaining a warrant, the Secretary's present enforcement scheme would have precisely the same effect. *333 The representative inspections are conducted "in accordance with criteria based upon accident experience and the number of employees exposed in particular industries." *Ante,* at 1825 n.17. If, under the present scheme, entry to covered premises is denied, the inspector can gain entry only by informing his administrative superiors of the refusal and seeking a court order requiring the employer to submit to the inspection. The inspector who would like to conduct a nonroutine search is just as likely to be deterred by the prospect of informing his superiors of his intention and of making false representations to the court when he seeks compulsory process as by the prospect of having to make bad faith representations in an *ex parte* warrant proceeding.

The other two asserted purposes of the administrative warrant are also adequately achieved under the existing scheme. If the employer has doubts about the official status of the inspector, he is given adequate opportunity to reassure himself in this regard before permitting entry. The OSHA inspector's statutory right to enter the premises is conditioned upon the presentation of appropriate credentials. 29 U.S.C. § 657(a)(1). These credentials state the inspector's name, identify him as an OSHA compliance officer, and contain his photograph and signature. If the employer still has doubts, he may make a toll-free call to verify the inspector's authority (*Usery v. Godfrey Brake & Supply Service, Inc.*, 545 F. 2d 52, 54 (CA 8 1976)), or simply deny entry and await the presentation of a court order.

The warrant is not needed to inform the employer of the lawful limits of an OSHA inspection. The statute expressly provides that the inspector may enter all areas in a covered business "where work is performed by an employee of an employer," 29 U.S.C. § 657(a)(1), "to inspect and investigate during regular working hours and at other reasonable times, and within reasonable limits and in a reasonable manner ... all pertinent conditions, structures, machines, apparatus, ***334** devices, equipment, and materials therein ... ." 29 U.S.C. § 657(a)(2). See also 29 CFR § 1903 (1977). While it is true that the inspection power granted by Congress is broad, the warrant procedure required by the Court does not purport to restrict this power but simply to ensure that the employer is apprised of its scope. Since both the statute and the pertinent regulations perform this informational function, a warrant is superfluous.

Requiring the inspection warrant, therefore, adds little in the way of protection to that already provided under the existing enforcement scheme. In these circumstances, the warrant is essentially a formality. In view of the obviously enormous cost of enforcing a health and safety scheme of the dimensions of OSHA, this Court should not, in the guise of construing the Fourth Amendment, require formalities which merely place an additional strain on already overtaxed federal resources.

Congress, like this Court, has an obligation to obey the mandate of the Fourth Amendment. In the past the Court "has been particularly sensitive to the Amendment's broad standard of 'reasonableness' where ... authorizing statutes permitted the challenged searches." *Almeida-Sanchez v. United States*, 413 U.S. 266, 290, 93 S.Ct. 2535, 2548, 37 L.Ed.2d 596 (WHITE, J., dissenting). In *United States v. Martinez-Fuerte*, 428 U.S. 543, 96 S.Ct. 3074, 49 L.Ed.2d 1116, for example, respondents challenged the routine stopping of vehicles to check for aliens at permanent checkpoints located away from the border. ****1832** The checkpoints were established pursuant to statutory authority and their location and operation were governed by administrative criteria. The Court rejected respondents' argument that the constitutional reasonableness of the location and operation of the fixed checkpoints should be reviewed in a *Camara* warrant proceeding. The Court observed that the reassuring purposes of the inspection warrant were adequately served by the visible manifestations of authority exhibited at the fixed checkpoints.

*335 Moreover, although the location and method of operation of the fixed check-points were deemed critical to the constitutional reasonableness of the challenged stops, the Court did not require Border Patrol officials to obtain a warrant based on a showing that the checkpoints were located and operated in accordance with administrative standards. Indeed, the Court observed that "[t]he choice of check-point locations must be left largely to the discretion of Border Patrol officials, to be exercised in accordance with statutes and regulations that may be applicable ... [and] [m]any incidents of checkpoint operation also must be committed to the discretion of such officials." 428 U.S., at 559–560, n.13, 96 S.Ct., at 3083. The Court had no difficulty assuming that those officials responsible for allocating limited enforcement resources would be "unlikely to locate a checkpoint where it bears arbitrarily or oppressively on motorists as a class." *Id.* at 559, 96 S.Ct. at 3083.

The Court's recognition of Congress' role in balancing the public interest advanced by various regulatory statutes and the private interest in being free from arbitrary governmental intrusion has not been limited to situations in which, for example, Congress is exercising its special power to exclude aliens. Until today, we have not rejected a congressional judgment concerning the reasonableness of a category of regulatory inspections of commercial premises.* While businesses are unquestionably entitled to Fourth Amendment protection, we have "recognized that a business by its special nature and voluntary existence, may open itself to intrusions that would not be permissible in a purely private context.": ***336** *G. M. Leasing Corp. v. United States*, 429 U.S. 338, 353, 97 S.Ct. 619, 629, 50 L.Ed.2d 530. Thus, in *Colonnade Catering Corp. v. United States*, 397 U.S. 72, 90 S.Ct. 774, 25 L.Ed.2d 60, the Court recognized the reasonableness of a statutory authorization to inspect the premises of a caterer dealing in alcoholic beverages, noting that "Congress has broad power to design such powers of inspection under the liquor laws as it deems necessary to meet the evils at hand." *Id.*, at 76, 90 S.Ct. at 777. And in *United States v. Biswell*, 406 U.S. 311, 92 S.Ct. 1593, 32 L.Ed.2d 87, the Court sustained the authority to conduct warrantless searches of firearm dealers under the Gun Control Act of 1968 primarily on the basis of the reasonableness of the congressional evaluation of the interests at stake.[†]

The Court, however, concludes that the deference accorded Congress in *Biswell* and *Colonnade* should be limited to situations ****1833** where the evils addressed by the

---

* The Court's rejection of a legislative judgment regarding the reasonableness of the OSHA inspection program is especially puzzling in light of recent decisions finding law enforcement practices constitutionally reasonable, even though those practices involved significantly more individual discretion than the OSHA program. See, e.g., *Terry v. Ohio*, 392 U.S. 1, 88 S.Ct. 1868, 20 L.Ed.2d 889; *Adams v. Williams*, 407 U.S. 143, 92 S.Ct. 1921, 32 L.Ed.2d 612; *Cady v. Dombrowski*, 413 U.S. 433, 93 S.Ct. 2523, 37 L.Ed.2d 706; *South Dakota v. Opperman*, 428 U.S. 364, 96 S.Ct. 3092, 49 L.Ed.2d 1000.

† The Court held:
"In the context of a regulatory inspection system of business premises that is carefully limited in time, place, and scope, the legality of the search depends ... on the authority of a valid statute.
"We have little difficulty in concluding that where, as here, regulatory inspections further urgent federal interest, and the possibilities of abuse and the threat to privacy are not of impressive dimensions, the inspection may proceed without a warrant where specifically authorized by statute." 406 U.S., at 315, 317, 92 S.Ct., at 1596.

regulatory statute are peculiar to a specific industry and that industry is one which has long been subject to Government regulation. The Court reasons that only in those situations can it be said that a person who engages in business will be aware of and consent to routine, regulatory inspections. I cannot agree that the respect due the congressional judgment should be so narrowly confined.

In the first place, the longevity of a regulatory program does not, in my judgment, have any bearing on the reasonableness of routine inspections necessary to achieve adequate enforcement of that program. Congress' conception of what constitute *337 urgent federal interests need not remain static. The recent vintage of public and congressional awareness of the dangers posed by health and safety hazards in the workplace is not a basis for according less respect to the considered judgment of Congress. Indeed, in *Biswell*, the Court upheld an inspection program authorized by a regulatory statute enacted in 1968. The Court there noted that "[f]ederal regulation of the interstate traffic in firearms is not as deeply rooted in history as is governmental control of the liquor industry, but close scrutiny of this traffic is undeniably" an urgent federal interest: 406 U.S., at 315, 92 S.Ct. at 1596. Thus, the critical fact is the congressional determination that federal regulation would further significant public interests, not the date that determination was made.

In the second place, I see no basis for the Court's conclusion that a congressional determination that a category of regulatory inspections is reasonable need only be respected when Congress is legislating on an industry-by-industry basis. The pertinent inquiry is not whether the inspection program is authorized by a regulatory statute directed at a single industry, but whether Congress has limited the exercise of the inspection power to those commercial premises where the evils at which the statute is directed are to be found. Thus, in *Biswell*, if Congress had authorized inspections of all commercial premises as a means of restricting the illegal traffic in firearms, the Court would have found the inspection program unreasonable; the power to inspect was upheld because it was tailored to the subject matter of Congress' proper exercise of regulatory power. Similarly, OSHA is directed at health and safety hazards in the workplace, and the inspection power granted the Secretary extends only to those areas where such hazards are likely to be found.

Finally, the Court would distinguish the respect accorded Congress' judgment in *Colonnade* and *Biswell* on the ground that businesses engaged in the liquor and firearms industry "'accept the burdens as well as the benefits of their trade ... .'" *338 *Ante*, at 1821. In the Court's view, such businesses consent to the restrictions placed upon them, while it would be fiction to conclude that a businessman subject to OSHA consented to routine safety inspections. In fact, however, consent is fictional in both contexts. Here, as well as in *Biswell*, businesses are required to be aware of and comply with regulations governing their business activities. In both situations, the validity of the regulations depends not upon the consent of those regulated, but on the existence of a federal statute embodying a congressional determination that the public interest in the health of the Nation's workforce or the limitation of illegal firearms traffic outweighs the businessman's interest in preventing a Government inspector from viewing those areas of his premises which relate to the subject matter of the regulation.

The case before us involves an attempt to conduct a warrantless search of the working area of an electrical and plumbing contractor. The statute authorizes such an inspection during reasonable hours. The inspection is limited to those areas over which Congress has exercised its proper legislative authority.* The area is also one to **1834 which employees *339 have regular access without any suggestion that the work performed or the equipment used has any special claim to confidentiality.† Congress has determined that industrial safety is an urgent federal interest requiring regulation and supervision, and further, that warrantless inspections are necessary to accomplish the safety goals of the legislation. While one may question the wisdom of pervasive governmental oversight of industrial life, I decline to question Congress' judgment that the inspection power is a necessary enforcement device in achieving the goals of a valid exercise of regulatory power.‡

I respectfully dissent.

ALL CITATIONS

436 U.S. 307, 98 S. Ct. 1816, 56 L.Ed.2d 305, 8 Envtl. L. Rep. 20,434, 6 O.S.H. Cas. (BNA) 1571, 1978 O.S.H.D. (CCH) P 22,735

---

* What the Court actually decided in *Camara v. Municipal Court*, 387 U.S. 523, 87 S.Ct. 1727, 18 L.Ed.2d 930, and *See v. City of Seattle*, 387 U.S. 541, 87 S.Ct. 1737, 18 L.Ed.2d 943, does not require the result it reaches today. *Camara* involved a residence, rather than a business establishment; although the Fourth Amendment extends its protection to commercial buildings, the central importance of protecting residential privacy is manifest. The building involved in *See* was, of course, a commercial establishment, but a holding that a locked warehouse may not be entered pursuant to a general authorization to "enter all buildings and premises, except the interior of dwellings, as often as may be necessary," 387 U.S., at 541, 87 S.Ct. at 1738, need not be extended to cover more carefully delineated grants of authority. My view that the *See* holding should be narrowly confined is influenced by my favorable opinion of the dissent written by Mr. Justice Clark and joined by Justices Harlan and Stewart. As *Colonnade* and *Biswell* demonstrate, however, the doctrine of *stare decisis* does not compel the Court to extend those cases to govern today's holding.

† The Act and pertinent regulation provide protection for any trade secrets of the employer: 29 U.S.C. §§ 664–665; 29 CFR § 1903.9 (1977).

‡ The decision today renders presumptively invalid numerous inspection provisions in federal regulatory statutes, e.g., 30 U.S.C. § 813 (Federal Coal Mine Health and Safety Act of 1969); 30 U.S.C. §§ 723, 724 (Federal Metal and Nonmetallic Mine Safety Act); 21 U.S.C. § 603 (inspection of meat and food products). That some of these provisions apply only to a single industry, as noted above, does not alter this fact. And the fact that some "envision resort to federal-court enforcement when entry is refused" is also irrelevant, since the OSHA inspection program invalidated here requires compulsory process when a compliance inspector has been denied entry. *Ante*, at 1825.

2 Talent Acquisition and Education

Interviewing, Hiring, and Training

******

Chapter Objectives:

1. Achieve a general knowledge of the talent acquisition and education process.
2. Acquire an understanding of the safety and health function's role in the talent acquisition and education process.
3. Identify the "pitfalls" and potential legal issues involved in the talent acquisition and education process.

******

Turnover costs money! When an employee who has been trained and educated and is working successfully within your organization "walks out the door," whether voluntarily or involuntarily, the organization incurs all the costs of replacing this employee. Costs can include, but are not limited to, the cost of hiring, the cost of alcohol and controlled substance testing, the cost of other testing, the cost of training, the cost of the loss in quality and productivity, and other costs. When the cost of replacement is multiplied by the number of employees leaving an organization, this cost can have a significant impact on the organization's bottom line.

Safety and health professionals should be aware that there can be many "pitfalls" to watch for within the hiring process. Although the safety and health function may possess input into the hiring process, in most organizations the human resource function is responsible for the actual hiring process and procedures. Given the number of laws and regulations governing the hiring process, it is imperative that the hiring process has been carefully vetted and skillfully managed in order to avoid possible legal or other issues arising.

In today's workplace, safety and health professionals may not be readily able to identify the differences between an employee of the company, a temporary employee who is working at the site but is an employee of a staffing company, and a subcontractor working on site. From a safety and health viewpoint, these individuals are all encompassed under the safety and health umbrella while working within the operation. However, from a human resource and legal viewpoint, these positions are substantially different. Additionally, safety and health professionals should be aware that there are different categories among and between employees, including probational employees, senior employees, leadpersons, and other categories. Management and supervisory personnel are usually exempt employees, while hourly employees are usually non-exempt for compensation and hours purposes. An injured employee is usually covered by the company's workers' compensation coverage. For a temporary employee, this is often a contractual arrangement between the company and a third-party staffing organization. The company pay the staffing company and they provide employees. These temporary employees are employees of the staffing company and are paid by the staffing company. Injured temporary employees are usually covered by the staffing company's workers' compensation coverage. A contractor is again employed under a contractual relationship between the company (or is a general contractor with a contract with the company) and the subcontractor to perform specific work. Injured subcontractors are usually required to provide their own workers' compensation coverage. Safety and health professionals should be aware that this chapter is focused on employees of the company; however, these lines between employment statuses can often cause confusion within the safety and health function.

The hiring process may vary from employer to employer. Some employers market and hire on site, while others may hire through a local unemployment office. One recent trend for employers is to hire through a temporary employment agency wherein the individual works through the temporary agency as an employee for the agency for a period of time and the primary employer hires from those individuals working as temporary employees. It is important for safety and health professionals to acquire a working knowledge of the hiring and employment process for their organization in order to provide input into the process and selections.

For safety and health professionals, the screening procedures for potential candidates are key to ensuring a successful hiring and placement of the individual within their organizations. Not all candidates for employment are a good "fit" with the specific job function in which they potentially could be hired. An improper "fit" of the individual with the job function can not only create dissatisfaction for the employee, but also can create production and quality deficiencies, training difficulties, and a high safety risk within the overall safety and health program. Safety and health professionals should be actively involved in the hiring and placement process to ensure appropriate ergonomic adjustment are made, Americans with Disabilities Act (ADA) accommodations are addressed, appropriate training is scheduled, and other adjustments are made to ensure the safety and health of the selected candidate. One of the highest safety and health risk periods for employees is during the initial stages of a new job function.

Can an employer make a "bad hire"? In general, if the candidate meets all of the requirements and successfully completes all of the required screenings, the candidate should possess the requisite skills and abilities to be successful in the job function. However, if the placement of the candidate improperly aligns the specific skill set of the job function, and/or the employer does not prepare the candidate for success due to lack of training, improper job structure, and other factors, the probabilities of success are substantially lowered, and there is a high probability that the candidate will leave the employment within a short time period. Safety and health professionals play a key role in this component of the employment process in ensuring appropriate job training, job alignment, and guidance within the job function.

At the time of writing, the unemployment rate in the United States is 3.9% which is one of the lowest rates in decades. Additionally, with the opioid crisis and active use of other controlled substances which are disqualifiers to most employment, the pool of potential candidates is substantially small. To this end, prospective candidates are assessing potential employers as much as employers are assessing potential candidates. In addition to wages, hours, and benefits, safety and health professionals should be aware that safe and healthful working conditions are also high on most candidates' list of requirements when considering employment with an organization.

Safety and health professionals should acquire a general understanding of the hiring process at their organizations, as well as any specific components of the hiring process which are within their direct or indirect responsibilities. For many organizations, the first step with any position is the development of a specific job description identifying the physical, educational, and skill requirements. Although there are several legal reasons (ADA, Title VII of the Civil Rights Act 1964, etc.) for the development of a written job description, this careful and thorough analysis of the job function can also assist the safety and health professional in placement of a candidate and in identification of potential risks within the job function. As identified in a later chapter, this written job description often includes a job task analysis, specific lifting or other physical requirements, education or training specifications, and other aspects which the safety and health professional, as well as the candidate, can utilize to ensure a good job function "fit" and prepare the candidate for success.

After the development of the written job description, the next step in the process is usually the advertisement of the job position. Although safety and health professionals usually have limited, if any, input into this phase of the process (except for safety and health specific hires), it is important to recognize the structure and legal requirements when publicizing an open job position. Of particular importance is the avoidance of any statement which could be construed as discriminatory, statements addressing preferred sex, statements addressing disability, and related statements within the advertisement.

General Steps in Talent Acquisition and Preparation

1. Job Analysis and Job Description.
2. Marketing to Appropriate Audience.
3. Application Selection Process.
4. Candidate Screening and Testing.
5. Hiring the Candidate and Initial Documentation Completion.
6. Preparing the Worksite.
7. Training and Preparation of the Newly Hired Employee.
8. Follow-up and Feedback.

The next phase of the hiring and screening process usually includes submission of a cover letter, resumé, or application, depending on the type of position. Supervisory and managerial personnel often require a cover letter, resumé, and references while "blue-collar" positions often require an application and references. Of specific importance for safety and health professionals are the prohibited questions on a job application. Under the ADA and Equal Opportunity Employment Commission (EEOC) interpretations, specific questions, such as "Have you ever filed a workers' compensation claim?" can be a violation and can be considered discriminatory. Careful review of a job application and specifically the questions provided on a job application should be assessed prior to utilization of this document with prospective candidates.

Correlating with the application is the location for candidates to submit the application and where the interviewer is conducting of interviews. Under the ADA, a violation can occur when a qualified individual with a disability (QID) is unable to submit the application or participate in the interview due to barriers, such as steps. Safety and health professionals should be aware of the requirements of the ADA as well as individual state handicap or disabilities laws and regulations. Careful analysis of the hiring process as well as barriers and obstacles for individuals with disabilities can avoid potential liability for the employer.

Although the safety and health professional is usually not involved in the actual interviewing process, it is important to understand that there are many questions which cannot and should not be asked in the interview. Skilled interviewers should be utilized for this important process.

An area of the hiring process which often impacts the safety and health function is the screening process. In many organizations, alcohol and controlled substance testing is required. The acquisition of the sample for testing, whether hair, blood, or urine, can take place at the in-house medical facility or be contracted to a local medical facility. Safety and health professionals may be responsible for the management of this testing or oversight of the outside medical facility. A key component of this screening is the chain of custody of the sample, as well as the type of screening being performed on the sample. Safety and health professionals should be aware that applicants who test positive can challenge not only the individual test, but also every aspect of the testing process.

Either directly or indirectly, the safety and health professional is usually involved in the physical assessment process, i.e., the physical. This medical process can include a number of different screenings ranging from an X-ray to strength-testing. Although this evaluation is performed by a medical professional in a confidential manner, safety and health professionals are often involved with the review of the medical findings, as well as maintenance of the physical examination documents.

Safety and health professionals may also play a role in the physical testing of applicants. From the design of the physical testing to the actual testing itself, safety and health professionals should be aware that this type of testing has encountered numerous challenges primarily focused on potential discrimination based upon race, sex, color, and disability. Physical testing should be job-focused, correlated with the job description, and carefully analyzed prior to initiation to ensure the elimination of all potential bias within all phases of the program design through actual testing.

Many employers require different forms of psychological testing. Although this is often performed by medical professionals, safety and health professionals should be aware of the confidentiality and requirements for medically related documents. Safety and health professionals are often responsible for the first aid, medical dispensaries, and other medically related functions and strict adherence with the OSHA recordkeeping requirements, as well as medical records requirements, is imperative.

In many organizations, a criminal background check is required and this function is usually performed by the human resource department. The criminal background check is often a paid service through a third-party or through a local law enforcement agency. Additionally, safety and health professionals should be aware that some companies also conduct social media searches and, in specific job functions, financial searches on the candidates. Safety and health professionals are usually not involved in these activities. Safety and health professionals should be aware of the numerous laws and regulations, as well as the potential liability, involved in these types of searches.

Although safety and health professionals are seldom involved in compensation issues (unless relating to a direct report), it is important to recognize issues or circumstances involving compensation inequities. In companies where labor organizations represent employees, the collective bargaining process usually yields a contract wherein compensation levels are established by seniority or other criteria. In companies where there are no labor organizations, the human resource function is usually the entity establishing pay rates and compensation levels. Compensation inequity issues based upon race, sex, or other protected classes are currently a hotbed of litigation in our courts.

Safety and health professionals are often the primary company representative in developing and presenting new employee training due to the focus on safety and health issues. This training often occurs immediately or shortly after the employee accepts the position and completes the necessary documentation (e.g., tax forms, confidentiality agreements, insurance documents, etc.). Safety and health professionals are often provided a specific time period to address a myriad of generalized safety related issues ranging from required personal protective equipment (PPE) to workplace safety and health policies. Safety and health professionals should utilize this time efficiently and effectively to prepare the new employees for all aspects of safety and health within the new employee's job function as well as overall safety and health rules and policies.

General Elements of New Employee Safety and Health Orientation/Training

1. Company's Emphasis on Safety and Health.
2. Company Safety and Health Policies and Procedures.
3. Disciplinary Actions for Failure to Follow Safety and Health Policies and Procedures.
4. General Safety and Health Rules and Regulations.
5. Specific Safety and Health Issues within the Job Function.
6. Specific PPE Requirements and PPE Training.
7. Injury Reporting, First Aid Locations, and Related Information.
8. Emergency Evacuation Training.
9. Other Job Function or Location Specific Safety Training.

As part of this new employee training, the safety and health professional should prepare the new employee's workspace with consideration being provided to ergonomic factors, machine-guarding, or other adjustments or modifications which create a safer or more healthful work environment. Additionally, the safety and health professional may assist the supervisor or trainer who is performing the hands-on training to the employee performing the specific job functions to ensure the proper equipment, functions, and activities maintain safety as the highest priority.

Once the training is completed, the new employee is often permitted to perform the job function. Safety and health professionals should be aware that one of the highest rates of work-related injuries are incurred during this initial learning phase for new employees. Safety and health professionals may want to observe this new employee often and provide feedback and encouragement during this initial probationary period.

In summation, talent acquisition is attempting to acquire the best possible candidate to perform the job function. Talent education is preparing and training the selected candidate to perform the job function safely and in a healthful manner. In an ideal world, all employees would work safely, efficiently, and productively at all times. In reality, this is why companies need safety and health professionals. The talent acquisition function has a number of laws and regulations governing this important process and is usually performed by the human resource function. However, it is important that safety and health professionals are provided the ability to provide input to ensure that new employees are properly prepared to perform the job function safely at all times. As identified, turnover costs money. Acquisition of the right employee, providing a safe and healthful work environment and preparing the employee to work in a safe and healthful manner goes a long way toward maintaining the candidate for the long run!

Case Study: Please be advised that this case may have been modified for the purposes of this text.

271 F.Supp.3d 119

United States District Court, District of Columbia.

Sheila J. LAWSON, Plaintiff,

v.

Jefferson B. SESSIONS, U.S. Attorney General, et al., Defendants.

No. 15–cv–1723 (KBJ)

|

Signed 09/22/2017

SYNOPSIS

Background: Former employee of the Federal Bureau of Investigation (FBI) brought pro se action against the FBI, the Department of Justice, the Attorney General, and the director of the FBI, under Title VII of the Civil Rights Act of 1964 (Title VII) and the Age Discrimination in Employment Act (ADEA), alleging that the FBI's refusal to reinstate her after she resigned constituted discrimination on the basis of her age, sex, and race, and was also retaliation for an Equal Employment Opportunity (EEO) complaint that she had previously filed, and that the FBI further retaliated against her by improperly processing another EEO complaint. Defendants moved to dismiss.

Holdings: The District Court, Ketanji Brown Jackson, J., held that:

[1] Title VII's 45-day reporting deadline applicable to federal employees for contacting the EEO counselor under Title VII would not be tolled on employee's claim for discrimination and retaliation, with regard to first three times she was denied reinstatement;

[2] dismissal of employee's claims of age discrimination and retaliation under the ADEA, for failure to exhaust administrative remedies, was not warranted;

[3] employee plausibly alleged that her age was a factor in the FBI's decision not to reinstate her, in violation of the ADEA;

[4] employee plausibly alleged that the FBI's decision not to reinstate her on four separate occasions was unlawful retaliation under the ADEA; and

[5] employee plausibly alleged that she was retaliated against in violation of Title VII and the ADEA by her former supervisor's alleged interference with, and improper processing of, a prior EEO complaint.

Motion granted in part and denied in part.

MEMORANDUM OPINION

KETANJI BROWN JACKSON, United States District Judge

During the summer of 2006, pro se plaintiff Sheila Lawson resigned from the Federal Bureau of Investigation ("FBI") following a nearly 11–year tenure as a Special Agent. (First Am. Compl. ("Compl."), ECF No. 5, 10, 13.) Shortly after her resignation, Lawson had a change of heart, and between 2007 and 2010, she repeatedly asked to be reinstated to her former position. (*See id.* 18, 24, ***124** 27, 30.) The FBI denied each of Lawson's four requests for reinstatement. (*See id.* 20, 25, 28, 32.) In the instant lawsuit, Lawson alleges that the FBI's refusal to reinstate her as a Special Agent constitutes discrimination on the basis of her age, sex, and race, and was also retaliation for an Equal Employment Opportunity ("EEO") complaint that Lawson had filed in 2006. (*See id.* 1.) The instant complaint separately alleges that the FBI retaliated against Lawson by improperly processing another one of her EEO complaints; specifically, Lawson contends that an FBI employee interfered with the processing of an EEO complaint she filed in 2010 in order to retaliate against her for filing the 2006 EEO complaint. (*See id.* 106–10, 147–51.)

Notably, this legal action consists of seven separate discrimination or retaliation counts, and each of these counts has been brought under either Title VII of the Civil Rights Act of 1964 ("Title VII"), 42 U.S.C. §§ 2000e to 2000e–17 (*see* Counts V–VII), or the Age Discrimination in Employment Act ("ADEA"), 29 U.S.C. §§ 621–34 (*see* Counts I–IV). Furthermore, each count relates either to the FBI's refusal to reinstate Lawson as an SA (Counts I, II, III, V, and VI (referred to herein, collectively, as the "failure-to-hire claims")), or the alleged improper processing of Lawson's 2010 administrative complaint (Counts IV and VII (collectively, the "retaliatory interference claims")).

[1] [2] [3] Before this Court at present is the motion to dismiss Lawson's complaint that the FBI, the Department of Justice ("DOJ"), Attorney General Jefferson Sessions, and FBI Director Christopher Wray (collectively, "Defendants") have filed. (*See generally* Defs.' Mot. to Dismiss ("Defs.' Mot"), ECF No. 9.)* Defendants argue that several of Lawson's failure-to-hire claims are unexhausted (*see id.* at 13–15), that any exhausted claims were not timely presented to this Court (*see id.* at 12–13), and that all of the claims in the complaint fail to state valid grounds for relief (*see id.* at 15–21).† Defendants' arguments for dismissal generally treat the discrimination and retaliation claims that Lawson brings under Title VII as largely interchangeable with those that she brings under the ADEA; however, as explained below, there are critical differences between the procedures that a plaintiff must follow with respect to exhaustion and timeliness under those two statutes. Consequently, although the Court largely agrees with Defendants' exhaustion and timeliness arguments as they apply to Lawson's Title VII failure-to-hire *125 claims (with an exception discussed below), the Court concludes that Defendants have not demonstrated that Lawson's ADEA failure-to-hire claims are unexhausted or untimely. The Court also concludes that the ADEA failure-to-hire counts state valid claims for discrimination and retaliation, because the complaint plausibly alleges both (1) that age was a factor in the FBI's refusal to reinstate Lawson, and (2) that the FBI's refusal was causally related to an EEO complaint that Lawson previously filed in 2006. Finally, the Court concludes that Lawson's retaliatory interference claims state valid grounds for relief, because Lawson has plausibly alleged that interference in the processing of her EEO complaint was a materially adverse action of the sort that can substantiate retaliation claims under both Title VII and the ADEA.

Accordingly, Defendants' motion to dismiss will be **GRANTED IN PART AND DENIED IN PART**. Lawson's Title VII failure-to-hire claims (Counts V and VI) will be largely dismissed for failure to exhaust, while the corresponding ADEA failure-to-hire claims (Counts II and III), as well as her Title VII and ADEA retaliatory interference claims (Counts IV and VII), may proceed. With respect to the failure-to-hire allegations that Lawson makes in Count I, the Court will permit Lawson to

---

* Lawson's complaint actually names former Attorney General Loretta Lynch and former FBI Director James Comey as the officer defendants (*see* Compl. 1), but pursuant to Federal Rule of Civil Procedure 25(d), their respective successors in office—Attorney General Jefferson Sessions and FBI Director Christopher Wray—have since been automatically substituted as defendants. Furthermore, because "the only proper defendant in suits brought under [Title VII and the ADEA] is the head of the department or agency being sued[,]" *Wilson v. Dep't of Transp.*, 759 F.Supp.2d 55, 67 (D.D.C. 2011); *see also* 42 U.S.C. § 2000e-16(c), it is hereby **ORDERED** that all of the defendants in this action other than Attorney General Sessions are **DISMISSED**. The proper defendant in a Title VII action filed against the FBI is the head of DOJ. *See Mulhall v. Ashcroft*, 287 F.3d 543, 550 (6th Cir. 2002) ("[I]n the present case [the plaintiff] alleges Title VII retaliation by the FBI; the FBI is a subunit of the Justice Department. Therefore, the proper defendant is the Attorney General, the head of the Justice Department."). However, for the sake of convenience, this Court will persist in using the plural term "Defendants" when referring to the movant in this Memorandum Opinion.

† Page number citations to documents the parties have filed refer to the page numbers that the Court's electronic filing system automatically assigns.

amend her complaint to clarify the claim, and Lawson can also amend Counts V and VI to address deficiencies in the surviving portions of those claims, as outlined below. A separate Order consistent with this Memorandum Opinion will follow.

I. BACKGROUND

A. Facts Pertaining To Lawson's Failure-To-Hire Claims*

Sheila Lawson is an African–American woman who began her employment as a Special Agent ("SA") with the FBI on October 15, 1995. (*See* Compl. 9–10.) At some unspecified point in 2006, Lawson "initiated the EEOC discrimination complaint process" (*id.* 12), and filed a formal complaint of discrimination (*see id.* 90). The exact substance of Lawson's 2006 grievance is not apparent from her complaint in the instant case, although Lawson does allege that the EEO claims were brought "against [Robert Enriquez, her former supervisor] and other FBI employees[.]" (*Id.* 107.) On July 7, 2006, after serving nearly 11 years as an SA, Lawson resigned from her position (*see id.* 13), and the following year, she withdrew the 2006 EEO complaint (*see id.* 14).

Following Lawson's resignation, the FBI Human Resources office sent Lawson an electronic communication that outlined the agency's reinstatement policy for former SAs. (*See id.* 15.) This message "stated that if an individual took a refund of the retirement contributions made to the FERS pension account, that individual is prohibited by federal law from repaying that amount to get credit for their prior service and would, therefore, be ineligible for reinstatement if they are already older than age 37." (*Id.* 16 (internal quotation marks omitted).)[†] Lawson received this ***126** message on March 20, 2007. (*See id.* 15–16.) Ten days later—on March 30, 2007—Lawson requested reinstatement as an FBI SA. (*See* July 7, 2015 EEOC Decision ("Final EEOC Decision"), Ex. A to Compl., ECF No. 5–1, at 3.) And five days after the reinstatement request—on April 5, 2007—Lawson "took a refund of the retirement contributions in her FERS account." (Compl. 17.)

According to Lawson, on at least four different occasions between May 31, 2007, and March 26, 2010, the FBI denied her formal requests for reinstatement, and Lawson alleges that the FBI refused to rehire her because of her age, sex, and race,

---

* The following facts are drawn from Lawson's first amended complaint. Although the complaint is at times difficult to follow, the Court believes that the following recitation accurately represents the substance of Lawson's allegations and claims.

† The reinstatement policy provides that all reinstatement candidates must be able to complete 20 years of service by the mandatory retirement age, which at that time was 57. (*See* FBI Special Agent Reinstatement Policy, Ex. A to Defs.' Mot, ECF No. 9–1, at 2; Letter of May 31, 2007, Ex. C to Defs.' Mot., ECF No. 9–3, at 1.) When making this years-of-service calculation, the agency typically credits a reinstatement applicant with her years of prior service, *unless* that "individual took a refund of the retirement contributions [she] made to FERS," in which case the reinstatement applicant is prohibited "from repaying that amount to get credit for [her] prior service[.]" (FBI Special Agent Reinstatement Policy at 2.) Accordingly, "if an individual took a refund of [her] retirement contributions[,]" she could not receive credit for prior years of service and would thus "be ineligible for reinstatement if [she is] already older than age 37." (*Id.*) The Court includes this information, which is contained in exhibits to Defendants' motion, to provide additional context for its explanation of the facts at issue, and has not otherwise relied upon Defendants' exhibits in resolving the instant motion to dismiss.

and also in retaliation for her filing of the 2006 EEO complaint. The first denial occurred on May 31, 2007, when the Chief of Human Resources allegedly "denied [Lawson] the FBI SA position because she was 41 years old" (*id.* 20), and therefore could not accumulate 20 years of service before the FBI's mandatory-retirement age of 57 (*see id.* 22; *see also supra* note 4). Undaunted, Lawson again requested reinstatement, and enclosed with her reinstatement request was a letter that she addressed to the Director of the FBI and that asked for an age waiver. (*See id.* 24.)* In correspondence dated September 2, 2008, the FBI again denied Lawson's request, explaining that "the FBI Director could give 'no further consideration' because the FBI Director could only grant age waivers up to age 60" (*id.* 25), and as a 41–year-old requester, Lawson could not accumulate 20 years of service before that cutoff.

Lawson subsequently submitted two more reconsideration requests, both of which the agency swiftly denied in a letter dated January 7, 2009. (*See id.* 27–28.) In this denial letter—the agency's third in less than two years—the agency purportedly advised Lawson "that she had 'reached the age' where she could no longer be reinstated in the FBI SA position" (*id.* 28), and further instructed her to direct age waiver requests to the Attorney General (*see id.* 29). Lawson followed this instruction approximately four months later by sending "a letter to Attorney General Eric H. Holder, Jr. requesting a decision regarding her application for reinstatement in the FBI SA position." (*Id.* 30.) This request was subsequently forwarded to the FBI's Human Resources office, and in a letter dated March 26, *127 2010, the agency, for the fourth time, declined to reinstate Lawson. (*See id.* 31–32.)

B. Facts Pertaining To Lawson's Retaliatory Interference Claims

At some point in 2010, Lawson "contacted an EEO counselor" and "initiated the informal discrimination complaint counseling phase[.]" (*Id.* 106.) On July 10, 2010, Lawson filed a formal complaint with the EEOC in which she claimed that the FBI had discriminated against her on the basis of sex and age, and had retaliated against her for prior protected activity, when it refused to grant her reinstatement requests between May 23, 2007 and March 26, 2010. (*See* Final EEOC Decision at 1.) Lawson alleges that while she was "participat[ing] in the EEOC formal discrimination complaint process[,]" Robert Enriquez—Lawson's former FBI Unit Chief, "who knew [Lawson had] filed a prior discrimination complaint against him in 2006" (Compl. 48)—got involved with Lawson's EEO case and purportedly "interfered" with her administrative complaint "through improper complaint processing, an incomplete investigation of Plaintiff's claims of discrimination, and the omission of any investigation of Plaintiff's claims of retaliation." (*Id.* 49; *see also id.* 48.) Enriquez's actions allegedly prompted Lawson to file "a spin-off EEOC complaint" regarding

---

* Pursuant to "Human Resources Order–DOJ 1200.1[,]" age waivers are available to individuals who otherwise exceed the maximum permissible age for reinstatement in cases involving "especially qualified individuals; shortage of highly qualified individuals for specific law enforcement positions[;] ... [and] situations where tentative selectees for law enforcement positions have passed the maximum entry age due to unavoidable or unexpectedly lengthy clearance or processing requirements[.]" (Compl. 37 (internal quotation marks omitted) (quoting HR Order–DOJ 100.1, Chap. 1–6, Maximum Entry Age And Mandatory Retirement of Law Enforcement Officers, found at www.justice.gov/jmd/hr-order-doj120 01-part-1-employment-1).)

Enriquez's conduct during the administrative proceedings for Lawson's July 2010 complaint. (*Id.* 50.) The instant complaint provides no additional details regarding the timing, content, or disposition of Lawson's "spin-off" administrative complaint.

On July 7, 2015, the EEOC issued its final decision dismissing Lawson's July 2010 complaint. (*See generally* Final EEOC Decision.) At the end of its decision letter, the Commission informed Lawson that she had the right to file a civil action in federal court "within ninety (90) calendar days from the date that" she received its decision, and further explained that, "[f]or timeliness purposes, the Commission will presume that this decision was received within five (5) calendar days after it was mailed." (*Id.* at 6, 8.)

C. Procedural History

Lawson initiated the instant lawsuit on October 19, 2015—104 days after the EEOC issued its decision of July 7, 2015. Lawson subsequently filed an amended complaint, which is the current operative complaint in this matter, asserting seven separate causes of action that, as explained above, arise from two distinct categories of acts. (*See generally* Compl.)

The claims in the first category, which this Court calls the "failure-to-hire claims," challenge the FBI's repeated refusal to reinstate Lawson as an SA. Lawson alleges that the agency's four denial letters constitute disparate treatment due to age, race, and sex under the ADEA (Count II) and Title VII (Count V), respectively (*see id.* 71–85, 116–27), and Lawson also contends that the agency refused to reinstate her on these occasions in retaliation for her prior EEO activity, in violation of the ADEA (Count III) and Title VII (Count VI) (*see id.* 86–99, 128–40). Lawson's first failure-to-hire claim (Count I) is more difficult to characterize. This cause of action—which is brought under the ADEA and is captioned, "Unlawful Discrimination Because of Age in FBI Reinstatement Policy"—at times appears to challenge the FBI's reinstatement policy as facially discriminatory (*see id.* 67 ("The hiring policy ... unlawfully excluded Plaintiff because of age.")), and at other times appears to raise a disparate treatment claim (*see id.* 65 (alleging that the "discriminatory age-based policy was not ***128** applied to every over age 37 reinstatement applicant who depleted the FERS pension account but was applied to disadvantage Plaintiff because of her age")).

The second category of claims in Lawson's complaint, which the Court refers to as the "retaliatory interference claims," challenges Enriquez's purported interference with, and improper processing of, Lawson's EEO complaint. (*See id.* 100–15, 141–55.) The complaint contends that Enriquez's conduct amounted to retaliation in violation of the ADEA (Count IV) and Title VII (Count VII).

On June 15, 2016, Defendants filed a motion to dismiss Lawson's complaint. (*See generally* Defs.' Mot.) Largely without differentiating between Lawson's various claims and the asserted legal bases for them, Defendants argue that Lawson's "case" should be dismissed as untimely because Lawson filed the complaint more than 90 days after receiving her EEOC right-to-sue letter (*see id.* at 12–13), and because Lawson failed to exhaust any claims that are based on acts that occurred prior to November 17, 2009 (*see id.* at 15). Defendants also contend that none of Lawson's disparate treatment or retaliation allegations state a valid claim for discrimination or retaliation in violation of Title VII or the ADEA. (*See id.* at 15–21.) For her part,

Lawson responds that she timely filed her complaint within 90 days of receiving the right-to-sue letter (*see* Pl.'s Suppl. Opp'n to Defs.' Mot. to Dismiss ("Pl.'s Opp'n"), ECF No. 15–1, at 10–12), and Lawson also insists that she has exhausted all available administrative remedies (*see id.* at 13–14).* Lawson further maintains that the complaint adequately alleges discriminatory treatment and retaliation in violation of the law. (*See id.* at 14–23.)

Defendants' motion to dismiss is now ripe for this Court's review. (*See* Defs.' Mot; Pl.'s Opp'n; Defs.' Reply in Supp. of Defs.' Mot. to Dismiss ("Defs.' Reply"), ECF No. 16.)

II. LEGAL STANDARD

A motion to dismiss a complaint pursuant to Federal Rule of Civil Procedure 12(b)(6) challenges the adequacy of the complaint on its face, testing whether the pleading "state[s] a claim upon which relief can be granted[.]" Fed. R. Civ. P. 12(b)(6). Although a complaint does not require detailed factual allegations, it must contain sufficient factual matter, accepted as true, to "state a claim to relief that is plausible on its face." *Bell Atl. Corp. v. Twombly*, 550 U.S. 544, 570, 127 S.Ct. 1955, 167 L.Ed.2d 929 (2007). "[M]ere conclusory statements" are not enough to make out a cause of action against a defendant, *Ashcroft v. Iqbal*, 556 U.S. 662, 678, 129 S.Ct. 1937, 173 L.Ed.2d 868 (2009); instead, the facts alleged "must be enough to raise a right to relief above the speculative level," *Twombly*, 550 U.S. at 555, 127 S.Ct. 1955. "In determining whether a complaint states a claim, the court may consider the facts alleged in the complaint, documents attached thereto or incorporated therein, and matters of which it may take judicial notice." *Abhe & Svoboda, Inc. v. Chao*, 508 F.3d 1052, 1059 (D.C. Cir. 2007).

[4] Of course, this Court is mindful that Lawson is proceeding in this matter ***129** pro se, and that the pleadings of pro se parties are to be "liberally construed" and "held to less stringent standards than formal pleadings drafted by lawyers[.]" *Erickson v. Pardus*, 551 U.S. 89, 94, 127 S.Ct. 2197, 167 L.Ed.2d 1081 (2007). "This benefit is not, however, a license to ignore the Federal Rules of Civil Procedure[,]" *Sturdza v. United Arab Emirates*, 658 F.Supp.2d 135, 137 (D.D.C. 2009), and "even a pro se plaintiff must meet his burden of stating a claim for relief[,]" *Horsey v. Dep't of State*, 170 F.Supp.3d 256, 263–64 (D.D.C. 2016).

III. ANALYSIS

As explained above, Lawson's complaint raises seven causes of action that arise from two distinct categories of acts. First, Lawson's failure-to-hire claims

---

* Lawson filed an initial brief in opposition to Defendants' motion on July 15, 2016 (*see* Pl.'s Mem. in Opp'n to Defs.' Mot. to Dismiss, ECF No. 13), and thereafter sought, and received, leave to file a supplemental opposition brief (*see* Pl.'s Mot. for Leave to File Suppl. Mem. in Opp'n to Defs.' Mot. to Dismiss, ECF No. 15; Min. Order of Aug. 1, 2016). Because the supplemental memorandum that Lawson has filed effectively supplants, rather than supplements, her initial opposition brief, this Court will refer exclusively to the 'supplemental' memorandum when recounting Lawson's arguments.

challenge the FBI's repeated refusals to reinstate her as an SA, on the grounds that these refusals constitute discrimination and retaliation in violation of Title VII (Counts V, VI) and the ADEA (Counts I, II, III). By contrast, Lawson's second group of claims challenges the agency's interference with, and improper processing of, her EEO complaint, which, Lawson contends, amounts to retaliation in violation of Title VII (Count VII) and the ADEA (Count IV). This parsing of the claims is important because, for the reasons set forth below, the Court concludes that Lawson's failure-to-hire claims brought pursuant to Title VII must be dismissed in part for failure to exhaust, while those brought under the ADEA either suffice to state a claim or may be amended to clarify Lawson's theory of liability. The Court further finds that Lawson's retaliatory interference claims survive Defendants' motion to dismiss.

A. Lawson's Title VII Failure-To-Hire Claims Must Be Largely Dismissed

[5] Although Defendants conflate the failure-to-hire claims that Lawson has brought under Title VII with the similar claims that she has brought under the ADEA (*see* Defs.' Mot. at 12–15), the exhaustion and timeliness "rules relating to Title VII and ADEA claims … are not identical[,]" *Achagzai v. Broad. Bd. of Govs.*, 170 F.Supp.3d 164, 171 (D.D.C. 2016), and as a result, the two causes of action must be analyzed separately. So analyzed, it is clear from the face of Lawson's complaint that she failed to exhaust three of the four discrete acts that form the basis of her Title VII failure-to-hire claims (with respect to the fourth act, this Court will give Lawson permission to amend her complaint to address the deficiencies that Defendants have identified). With respect to Lawson's corresponding ADEA claims, the Court concludes that there is no exhaustion problem and that Lawson has alleged sufficient facts to support the age discrimination and retaliation claims stemming from the FBI's refusal to reinstate her. The Court will also allow Lawson to amend her complaint to clarify the ADEA claim she intends to raise in Count I.

1. The Failure-To-Hire Claims That Arise Under Title VII Must Be Dismissed For Failure To Exhaust Administrative Remedies With Respect To Three Out Of The Four Denial Letters Upon Which Those Claims Are Based, But Lawson May Amend Her Complaint With Respect To The Fourth Letter

[6] Lawson's Title VII failure-to-hire claims (Counts V and VI) arise out of the FBI's refusal on four separate occasions (May 31, 2007; September 2, 2008; January 7, 2009; and March 26, 2010) to reinstate Lawson as an FBI SA. As noted, Lawson contends that these denials constitute disparate treatment based on her sex *130 and race, and in addition, that the FBI issued these denials in retaliation for her 2006 EEOC complaint. It is well-established that, if Lawson is correct that the FBI's denials were discriminatory, each refusal is treated as a *separate* discriminatory act for purposes of Title VII's exhaustion requirements. *See Nat'l R.R. Passenger Corp. v. Morgan*, 536 U.S. 101, 113, 122 S.Ct. 2061, 153 L.Ed.2d 106 (2002) ("Each discrete discriminatory act starts a new clock for filing charges alleging that act."). For the reasons that follow, this Court finds that, with respect to the first three denial letters from the FBI, it is clear from the face of Lawson's complaint that she did not contact

an EEO counselor within 45 days of these purported violations, and as a result, failed to exhaust her administrative remedies in regard to those discrimination claims.

"Before a federal employee can file suit against a federal agency for violation of Title VII, the employee must run a gauntlet of agency procedures and deadlines to administratively exhaust his or her claims." *Crawford v. Duke*, 867 F.3d 103, 105 (D.C. Cir. 2017). First, an employee must contact the agency's Equal Employment Opportunity ("EEO") counselor to initiate informal counseling "within 45 days of the date of the matter alleged to be discriminatory or, in the case of personnel action, within 45 days of the effective date of the action." 29 C.F.R. § 1614.105(a)(1). If the matter is not resolved informally within 30 days, the employee then has 15 days to file a formal complaint with the agency. *See id.* §§ 1614.105(d), 1614.106(a). Once an employee has filed a formal complaint, the agency must "conduct an impartial and appropriate investigation of the complaint within 180 days" of that filing, *id.* § 1614.106(e)(2), and the employee may subsequently file suit in federal district court, but must do so within 90 days of receipt of the agency's final determination, or if the agency does not take final action, after 180 days have elapsed since the filing of the complaint with the agency, *see* 42 U.S.C. § 2000e–16(c); 29 C.F.R. § 1614.407(c).

[7] [8] "[T]he administrative time limits … erect no jurisdictional bars to bringing suit[,]" *Bowden v. United States*, 106 F.3d 433, 437 (D.C. Cir. 1997); instead, these requirements function like statutes of limitations, and as such, are subject to waiver, estoppel, and equitable tolling, *see Horsey*, 170 F.Supp.3d at 264–65; *see also Rann v. Chao*, 346 F.3d 192, 195 (D.C. Cir. 2003) (reiterating that "the timeliness and exhaustion requirements of § 633a(d) are subject to equitable defenses and are in that sense non-jurisdictional"). However, in order to receive the benefit of equitable tolling, a tardy plaintiff must show "(1) that [s]he has been pursuing h[er] rights diligently, and (2) that some extraordinary circumstance stood in h[er] way and prevented timely filing[.]" *Horsey*, 170 F. Supp.3d at 267 (internal quotation marks and citation omitted). The Supreme Court has suggested that equitable tolling might be available where a claimant "received inadequate notice," where "a motion for appointment of counsel is pending[,]" or "where the court has led the plaintiff to believe that she had done everything required of her[.]" *Baldwin Cty. Welcome Ctr. v. Brown*, 466 U.S. 147, 151, 104 S.Ct. 1723, 80 L.Ed.2d 196 (1984) (per curiam).

[9] In the instant case, it is evident on the face of Lawson's complaint that she did not exhaust administrative remedies with respect to the first three denial letters and that no grounds for equitable tolling exist. Lawson's complaint alleges that the agency denied her requests for reinstatement by issuing letters dated May 31, 2007; September 2, 2008; January 7, 2009; and March 26, 2010 (*see* Compl. *131 20, 25, 28, 32), and that she initiated contact with an EEO counselor at some unspecified time in 2010 (*see id.* 106). While it is possible that Lawson's 2010 EEO contact occurred within 45 days of the March 26, 2010 letter, the same cannot be said with respect to the first three denial letters. That is, even assuming that Lawson initiated contact with the EEO counselor at the earliest possible time in 2010 with respect to the 2007, 2008, and 2009 denial letters (i.e., on January 1, 2010), this contact

occurred 946 days after the May 31, 2007 letter, 486 days after the September 2, 2008 letter, and 359 days after the January 7, 2009 letter, respectively—far beyond the applicable 45-day reporting period. Therefore, it is clear on the face of Lawson's complaint that she did not timely contact an EEO counselor with respect to the first three denial letters.

Nor has Lawson established that this is one of the "extraordinary and carefully circumscribed instances" in which equitable tolling is warranted. *Washington v. WMATA*, 160 F.3d 750, 753 (D.C. Cir. 1998) (internal quotation marks and citation omitted). Lawson suggests that tolling is justified simply and solely because she "could not have known about the discrimination or retaliation upon receipt of a letter at issue[.]" (Pl.'s Opp'n at 13–14.) Critically, however, Lawson fails to articulate precisely *why* she could not possibly have known that the denial letters were discriminatory and retaliatory when she received them, and her contention in this regard is particularly odd given that her current discrimination and retaliation claims appear to rest solely upon the FBI's issuance of these same letters. What is more, because the denial letters at issue are dated between one and three years before Lawson initiated contact with an EEO counselor (again, construing Lawson's "2010" contact with an EEO counselor as having occurred on January 1, 2010 (Compl. 106)), Lawson has failed to demonstrate diligence in pursuing her administrative remedies by any stretch of the imagination. *See Washington*, 160 F.3d at 753 (finding that a complainant's lack of diligence precluded equitable tolling where the complainant filed an EEOC complaint "over a year" (13 months) after the alleged act of discrimination, in violation of an 180-day filing deadline); *Dyson v. District of Columbia*, 710 F.3d 415, 421–22 (D.C. Cir. 2013) (finding that a complainant's lack of diligence precluded equitable tolling where the complainant missed a filing deadline by 38 days due to circumstances that were within her control). In the absence of any evidence that Lawson exercised the requisite due diligence in pursuing her administrative remedies, or that some extraordinary circumstances impeded Lawson's ability to pursue those rights, this Court declines to toll the 45-day reporting deadline.

[10] In a final effort to avoid the dismissal of her Title VII failure-to-hire claims, Lawson argues that all four denial letters at issue constitute one continuous discriminatory and retaliatory action. (*See* Pl.'s Opp'n at 13 ("These letters mailed to Plaintiff from the FBI Human Resources Division continually provided reasons that denied Plaintiff's reinstatement in the FBI SA position.").) As this Court has already explained, however, "[d]iscrete acts such as termination, failure to promote, denial of transfer, or refusal to hire are" acts that occur at a fixed time, and thus an employee must adhere to the established administrative process for *each* discrete action for which she seeks to bring a claim. *Morgan*, 536 U.S. at 114, 122 S.Ct. 2061; *see also id.* at 113, 122 S.Ct. 2061; *Nguyen v. Mabus*, 895 F.Supp.2d 158, 172 (D.D.C. 2012) ("[S]ince the Supreme Court's decision in *Morgan*, the continuing violation ***132** theory is restricted to claims akin to hostile work environment claims because those violations—unlike a discrete act such as firing or failing to promote an employee— cannot be said to occur on any particular day." (internal quotation marks and citation omitted)).

Therefore, the Court rejects Lawson's argument that her claims are exhausted with respect to all four denial letters because they constitute a continuous discriminatory or retaliatory event, and the Court also finds it apparent from the face of Lawson's complaint that she failed to contact an EEO counselor within 45 days of the May 31, 2007, September 2, 2008, and January 7, 2009 denial letters. Thus, the Title VII discrimination and retaliation claims that arise from these denial letters must be dismissed. *See Fortune v. Holder*, 767 F.Supp.2d 116, 122–23 (D.D.C. 2011) (collecting cases in which courts dismissed Title VII claims on exhaustion grounds where it was clear that the complainant had not contacted an EEO counselor within 45 days of the alleged act of discrimination). Lawson's Title VII failure-to-hire claims that stem from the denial letter of March 26, 2010, will not be dismissed on exhaustion grounds, because at this juncture, Defendants cannot demonstrate that Lawson failed to contact an EEO counselor within 45 days of that letter, as Defendants themselves concede. (*See* Defs.' Mot. at 15.)

[11] Defendants do *not* concede that any exhausted claims were timely presented to this Court. In this regard, Lawson has requested the opportunity to amend the complaint "to reflect the date of receipt of the [right-to-sue letter] should the Court hold in abeyance a decision in regard to this issue" (Pl.'s Opp'n at 12), which this Court has hereby decided to allow (*see* the accompanying Order). With this opportunity to amend her complaint, Lawson will also have a chance to add any additional allegations of fact that pertain to the March 26, 2010, denial letter—which is the sole remaining basis for her Title VII failure-to-hire claims (Counts V, VI)—before the Court considers Defendants' arguments regarding the sufficiency of the factual allegations in support of these claims. Once Lawson files the anticipated amended complaint, the Court will permit Defendants to file a partial motion to dismiss that addresses the Title VII failure-to-hire claims arising out of the denial letter of March 26, 2010.

2. Lawson's Failure–To–Hire Claims Survive The Motion To Dismiss To The Extent That They Arise Under The ADEA

In their motion to dismiss, Defendants insist that the failure-to-hire claims that Lawson has brought under the ADEA (Counts I, II, and III) suffer from the same fundamental flaws as her Title VII claims—namely, that these claims are both unexhausted and untimely. (*See* Defs.' Mot. at 12–15.) In the alternative, Defendants argue that Lawson's complaint fails to state any claim for discrimination or retaliation under the ADEA. (*See id.* at 15–20.) For the reasons set forth below, this Court disagrees on both fronts.

a. *Defendants Have Not Demonstrated That Lawson's Age Discrimination And Retaliation Claims Must Be Dismissed On Exhaustion Or Timeliness Grounds*

[12] [13] [14] Although Lawson's ADEA failure-to-hire claims arise from the same set of underlying facts as the corresponding claims that Lawson has brought under Title VII, the pre-filing procedures that apply under these two statutes are analytically

distinct. Stated simply, Title VII requires a plaintiff to navigate a "maze of administrative processes[,]" **133** *Niskey v. Kelly*, 859 F.3d 1, 5 (D.C. Cir. 2017), while an ADEA plaintiff has a much easier row to hoe when filing a discrimination action, *see Chennareddy v. Bowsher*, 935 F.2d 315, 318 (D.C. Cir. 1991). This is because "[a]n ADEA plaintiff has two means of pursuing his age discrimination claim." *Id.* First, a federal employee has the option of bypassing the administrative process altogether and suing directly in federal court, subject to certain notice requirements. *See* 29 U.S.C. § 633a(d). However, an employee who selects this option must give notice of the lawsuit to the EEOC "within one hundred and eighty days after the alleged unlawful practice occurred" and "not less than thirty days[ ]" prior to the commencement of the action. *Id.*; *see also Stevens v. Dep't of the Treasury*, 500 U.S. 1, 6–7, 111 S.Ct. 1562, 114 L.Ed.2d 1 (1991) (clarifying that a plaintiff is required to file her *notice*—not the civil action itself—within 180 days of the conduct at issue and at least 30 days prior to the commencement of suit, and indicating that a plaintiff may wait considerably more than 30 days after filing her notice of intent to sue before filing her lawsuit). Alternatively, an ADEA plaintiff "may [opt to] invoke the EEOC's administrative process, and then sue if dissatisfied with the results." *Rann*, 346 F.3d at 195 (citing 29 U.S.C. § 633a(b), (c)).

[15] In the instant case, Defendants do not address whether Lawson actually provided the EEOC with the requisite notice before seeking to obtain direct judicial review of her ADEA claims in this Court, as would be required under the first of these two avenues. Instead, Defendants frame their exhaustion and timeliness arguments solely in terms of the deadlines applicable within the context of the EEOC administrative process, and they note in passing that Lawson's complaint "does not allege that Lawson relied on any avenue other than the FBI's administrative avenue for exhaustion of her alleged ADEA claims." (Defs.' Mot. at 14.) But the burden of establishing exhaustion does not lie with the plaintiff. That is, while it is true that Lawson's complaint does not allege that she provided an intent-to-sue notice to the EEOC within 180 days of the alleged discriminatory acts, or that she waited at least 30 days after doing so to commence suit, "untimely exhaustion of administrative remedies is an affirmative defense," and thus "*the defendant* bears the burden of pleading and proving it[.]" *Bowden*, 106 F.3d at 437 (emphasis added). Thus, at this early stage of the instant litigation, dismissal of the action is only appropriate if the plaintiff's failure to comply with established procedural prerequisites is evident on the face of the complaint. *See Horsey*, 170 F.Supp.3d at 265. In any event, Defendants here have not meaningfully addressed the separate ADEA framework or the facts in the complaint that implicate it (if any); therefore, they have not carried their burden of proving the lack of exhaustion or the untimeliness of Lawson's complaint for the purpose of supporting their motion to dismiss.

It is also the case that, even if one accepts as true Defendants' suggestion that Lawson's complaint demonstrates that she sought to invoke the second path to judicial review (i.e., that she sued after filing an EEO complaint regarding her ADEA claims), Defendants do not automatically prevail with respect to their argument that these claims are unexhausted and/or untimely. The D.C. Circuit has expressly avoided deciding whether, having filed an EEOC complaint, an ADEA plaintiff "must reasonably *pursue* the process, as an exhaustion requirement would ordinarily entail." *Rann*, 346 F.3d at 195 (emphasis in original); *see also id.* (remarking that the ADEA is silent on this question). The Supreme Court has similarly declined *134 to reach the issue of whether an ADEA plaintiff like Lawson is required to press on with respect to any EEO complaint she has filed and avail herself of all potential administrative remedies before bringing a lawsuit. *See Stevens*, 500 U.S. at 9–10, 111 S.Ct. 1562 (1991) (explaining that this issue was not properly before the Court in light of the Solicitor General's position that "a federal employee who elects agency review of an age discrimination claim need not exhaust his administrative remedies").

In the absence of any argument from Defendants regarding these significant open legal questions—or, for that matter, any argument tailored to the ADEA framework as opposed to the procedures that Title VII prescribes—this Court declines, at this time, to dismiss Lawson's ADEA claims as unexhausted or untimely. Defendants are, of course, free to reassert these defenses (along with the appropriate legal and factual support) at a later stage in this case.

b. *Defendants Have Not Demonstrated That Lawson's Age Discrimination And Retaliation Claims Must Be Dismissed For Failure To State A Claim*

Defendants have also argued, as an alternative to the untimeliness and exhaustion contentions, that each of Lawson's ADEA failure-to-hire claims fails to state a claim upon which relief can be granted. (*See* Defs.' Mot. at 15–20.) For the reasons explained below, this Court disagrees with Defendants' view of Lawson's failure-to-hire claims that allege discrimination (Count II) and retaliation (Count III) in violation of the ADEA, and it will allow Lawson to amend the complaint to clarify the basis for the ADEA failure-to-hire claim that is set forth in Count I.

(i) Lawson has stated a discrimination claim under the ADEA because she has alleged sufficient facts to support the inference that age was a factor in defendants' refusals to reinstate her.

[16] [17] [18] The ADEA requires that "[a]ll personnel actions affecting employees or applicants for employment who are at least 40 years of age ... in executive agencies ... be made free from any discrimination based on age." 29 U.S.C. § 633a(a). "The Act's protections for employees of the federal government" are "more expansive than those for workers employed in the private sector," *Miller v. Clinton*, 687 F.3d 1332, 1336–37 (D.C. Cir. 2012); that is, while a private sector employee must "show that the challenged personnel action was taken *because* of age," a federal employee can prevail by "show[ing] that the personnel action involved '*any* discrimination based on age[,]' " *Ford v. Mabus*, 629 F.3d 198, 205 (D.C. Cir. 2010) (quoting 29 U.S.C. § 633a) (emphasis added). Accordingly, a federal employee alleging age discrimination must demonstrate "that age was *a* factor in the challenged personnel action." *Ford*, 629 F.3d at 206 (emphasis in original). "[A]t the motion-to-dismiss stage, the

guiding lodestar is whether, assuming the truth of the factual allegations, ... the inferences of discrimination drawn by the plaintiff"—i.e., that age was a factor in the challenged decision—"are reasonable and plausibly supported." *Townsend v. United States*, 236 F.Supp.3d 280, 298 (D.D.C. 2017).*

*135 [19] Notwithstanding Defendants' arguments to the contrary, Lawson's complaint alleges sufficient facts, accepted as true, to state a plausible claim for age discrimination based on a failure to hire in Count II. Lawson identifies herself as an "applicant" to the federal government who "is at least 40 years of age" (Compl. 4); hence, she is part of the class that the ADEA protects, *see* 29 U.S.C. § 633a(a). Moreover, according to her complaint, Lawson previously served for over a decade as an FBI special agent (*see* Compl. 10, 13), and was thus "qualified for the FBI SA position" to which she applied (*id.* 19). Despite her record of service and other qualifications, the FBI allegedly denied Lawson's applications for reinstatement on four separate occasions between May 31, 2007 and March 26, 2010. (*See id.* 20, 25, 28, 32.) And in lieu of reinstating Lawson, "between October 2009 and June 2010[,]" the FBI purportedly "hired seven applicants who sought reinstatement in the FBI SA position[,]" all of whom "were age 39 and younger[,]" and several of whom did not possess "a unique or special skill, or ability." (*Id.* 39–41.) Lawson's complaint further alleges that, in a denial "letter dated January 7, 2009, [Unit Chief] Carrico advised [Lawson] that she had 'reached the age' where she could no longer be reinstated in the FBI SA position." (*Id.* 28.) This Court agrees with Lawson's contention that these facts together are more than sufficient to support a reasonable inference that she "applied for and was not hired for the FBI SA position because of her age" in violation of the ADEA. (*Id.* 84.)

Defendants' response is to insist that Lawson's complaint "fails to state any disparate treatment claim because it does not show that the FBI treated Lawson's request for reinstatement differently *because of* her age[.]" (Defs.' Mot. at 17.) This is so, Defendants argue, because under the FBI reinstatement policy as described

---

* To be sure, in order to be entitled to judgment on her failure-to-hire age discrimination claims ultimately, Lawson might need to resort to the familiar three-part burden-shifting framework of *McDonnell Douglas Corp. v. Green*, 411 U.S. 792, 793, 93 S.Ct. 1817, 36 L.Ed.2d 668 (1973), which requires, among other things, that a plaintiff establish a prima facie case of discrimination. *See id.* at 802–05, 93 S.Ct. 1817; *see also Teneyck v. Omni Shoreham Hotel*, 365 F.3d 1139, 1155 (D.C. Cir. 2004) (listing elements to establish a prima facie case of discrimination for failure to hire). But "an employment discrimination plaintiff is not required to plead every fact necessary to establish a *prima facie* case to survive a motion to dismiss[.]" *Jones v. Air Line Pilots Ass'n, Int'l*, 642 F.3d 1100, 1104 (D.C. Cir. 2011) (citing *Swierkiewicz v. Sorema N.A.*, 534 U.S. 506, 511, 122 S.Ct. 992, 152 L.Ed.2d 1 (2002)); *see also Swierkiewicz*, 534 U.S. at 511, 515, 122 S.Ct. 992 (holding that "under a notice pleading system, it is not appropriate to require a plaintiff to plead facts establishing a prima facie case because the *McDonnell Douglas* framework does not apply in every employment discrimination case" and "the Federal Rules do not contain a heightened pleading standard for employment discrimination suits"); *Brown v. Sessoms*, 774 F.3d 1016, 1022–1023 (D.C. Cir. 2014) ("We have been clear, however, that '[a]t the motion to dismiss stage, the district court cannot throw out a complaint even if the plaintiff did not plead the elements of a prima facie case.'" (quoting *Brady v. Office of Sergeant at Arms*, 520 F.3d 490, 493 (D.C. Cir. 2008))).

in Lawson's own complaint, "Lawson's withdrawal from her FERS [retirement] account *automatically* disqualified her from reinstatement." (*Id.* (emphasis in original).) Thus, say Defendants, it was Lawson's retirement account withdrawal, and *not* her age, that was the true cause for the FBI's refusals to re-hire her. What Defendants overlook is the fact that the allegations in Lawson's complaint do not preclude a jury finding that age was "a" factor in the FBI's proffered rationale for refusing to reinstate Lawson, which is all that is required for ADEA claims against federal government employers. *See Ford,* 629 F.3d at 206.

Specifically, as alleged in the complaint, Lawson was told that the FBI's policy regarding FERS withdrawals and reinstatements *turns* on the age of the former agent. (*See* Compl. 21–22 (asserting that the unit chief of the FBI's Human Resources Division told her that "a federal law prohibited [her] from repaying [the *136 FERS] amount" that she had been paid; that "she could not be credited with her prior law enforcement service years" for retirement purposes; and that due to her age, she "could not earn twenty years of federal law enforcement service credit by the mandatory separation age of '57' ").) In addition, Lawson's complaint plainly alleges that the FBI "denied [Lawson] the FBI SA position because she was 41 years old." (*Id.* 20; *see also id.* 16, 28, 34.) Therefore, in contending that Lawson's complaint does not plausibly allege that the FBI rejected her reinstatement because of her age (*see* Defs.' Mot. at 17–18), Defendants refuse to acknowledge the plain text of the very pleading that they purportedly analyze.

[20] [21] This is not to suggest that Lawson's complaint is a model of clarity. The complaint does contain factual allegations that, if true, tend to indicate that Defendants decided to forego reinstating Lawson because she had withdrawn money from her FERS account or because of other reasons that do not relate to discrimination on account of her age (*see, e.g.,* Compl. 21 (discussing Lawson's withdrawal from her FERS account)), and it is true that, "[i]n some cases, it is possible for a plaintiff to plead too much; that is, to plead himself out of court by alleging facts that render success on the merits impossible[,]" *Sparrow v. United Air Lines, Inc.,* 216 F.3d 1111, 1116 (D.C. Cir. 2000). But this Court's "role is not to speculate about which factual allegations are likely to be proved after discovery"; instead, "the only question … is whether [Lawson has] alleged facts that, taken as true, render [her] claim of [discrimination] plausible." *Harris v. D.C. Water & Sewer Auth.,* 791 F.3d 65, 70 (D.C. Cir. 2015). For the reasons explained above, Lawson needs only to include facts that demonstrate that her age was one factor in the FBI's decision-making process, *see Ford,* 629 F.3d at 206, and in this regard, this Court finds that she has "nudged [her] claims across the line from conceivable to plausible," *Twombly,* 550 U.S. at 570, 127 S.Ct. 1955.*

---

* Defendants' contentions that Lawson's complaint references an improper "comparator" with respect to the ADEA discrimination claims (*see* Defs.' Mot. at 17 (arguing that Lawson's "alleged comparator is not a proper comparator")) and that 36 of the 37 employees who were reinstated from 2001 to 2010 "did not take a refund from their FERS pension fund" (*id.* (internal quotation marks and citation omitted))—even if true—do not demand a different result. The pleading standard at this stage is not "onerous[,]" *McManus v. Kelly,* 246 F.Supp.3d 103, 110–12, 2017 WL 1208395, at *5 (D.D.C. 2017),

(ii) Lawson has stated a retaliation claim under the ADEA because she has adequately alleged that defendants failed to rehire her because she engaged in a protected activity.

[22] Defendants next insist that Lawson's complaint fails to state a valid retaliation claim under the ADEA (Count III) "because it fails to show the requisite causal link between Lawson's prior protected activity and the FBI's denial of her reinstatement request." (Defs.' Mot. at 19.) This argument presents a closer question than Defendants' contentions with respect to Lawson's ADEA discrimination claim, but after careful consideration of this contention, this Court concludes that Lawson has alleged a sufficient causal connection between the FBI's refusals to reinstate her and her 2006–2007 EEO activity.

[23] [24] [25] The ADEA prohibits an employer from retaliating against a federal-sector *137 employee because the employee engaged in protected activity by opposing unlawful employment practices or bringing prior charges of age discrimination. See *Gomez-Perez v. Potter*, 553 U.S. 474, 479, 128 S.Ct. 1931, 170 L.Ed.2d 887 (2008). "In the absence of direct evidence of retaliatory intent, to succeed on a claim for retaliation under … the ADEA, [Lawson] must show that [she] '1) engaged in a statutorily protected activity; 2) suffered a materially adverse action by [her] employer; and that 3) a causal connection existed between the two.'" *Townsend*, 236 F.Supp.3d at 315 (quoting *Nurriddin v. Bolden*, 818 F.3d 751, 758 n.6 (D.C. Cir. 2016)). Notably, the requisite causal relationship may be inferred through temporal proximity between the protected act and the adverse employment action. *See id.* at 316. However, where causation is predicated on temporal proximity alone, "the proximity in time must be very close." *Kwon v. Billington*, 370 F.Supp.2d 177, 187 (D.D.C. 2005) (internal quotation marks and citation omitted); *see also Greer v. Bd. of Trustees of Univ. of D.C.*, 113 F.Supp.3d 297, 311 (D.D.C. 2015) ("When relying on temporal proximity alone to demonstrate causation, there is no bright-line rule, although three months is perceived as the outer limit." (citation omitted)). Nevertheless, "[a] large gap between protected activity and retaliation is not necessarily fatal to a claim when the plaintiff can point to other factors leading to an inference of causation." *Greer*, 113 F.Supp.3d at 311 (citation omitted).

In the instant case, Lawson alleges that she participated in the informal and formal EEO complaint process throughout 2006 and 2007, and thereafter, the FBI repeatedly denied her requests for reinstatement. (*See* Compl. 90, 94–96.) Defendants do not dispute that the FBI's refusal to hire Lawson is a materially adverse action, nor do they dispute that Lawson's prior EEO participation constitutes a protected activity. Thus, the only pertinent disputed issue is whether Lawson's complaint alleges sufficient facts related to the causation element to render her claim plausible. This Court concludes that it does, for at least two reasons.

---

and while "[a]llegations regarding comparators … obviously strengthen a discrimination complaint," this evidentiary requirement is "inapplicable at the pleading stage." *Nanko Shipping, USA v. Alcoa, Inc.*, 850 F.3d 461, 467 (D.C. Cir. 2017).

First of all, there is some temporal proximity between Lawson's alleged protected activity and the FBI's decision not to reinstate her. According to the complaint, Lawson "contacted an EEO counselor, initiated the informal discrimination complaint counseling phase, and participated in the formal discrimination complaint process" in "2006–2007[.]" (*Id.* 90.) After resigning from her position in July 2006, and withdrawing her pending EEO complaint at some point in 2007 (*see id.* 13–14), Lawson requested reinstatement (*see id.* 31, 93). Lawson alleges that, at the time she requested reinstatement, human resources personnel "knew of [her] prior protected EEO activity in 2006–2007" (*id.* 93), and conducted "routine EEO database check[s] to determine whether an applicant for the FBI SA position had prior EEO activity involving the FBI" (*id.* 92). The complaint says that, although Lawson "qualified for the FBI SA position" (*id.* 91), the FBI denied her requests for reinstatement in letters dated May 31, 2007, September 2, 2008, January 7, 2009, and March 26, 2010 (*see id.* 20, 25, 28, 32), and that a "causal connection" exists between her 2006–2007 protected activity and the denial letters "based on the close timing between" the two. (*id.* 96.)

This Court finds that temporal proximity between the administrative processes related to Lawson's 2006 EEOC complaint and the FBI's May 31, 2007 denial letter are sufficiently close to support the causal inference. In particular, the complaint alleges that Lawson "initiated the EEOC discrimination complaint process" at some point in 2006 (*id.* 12), and that she continued ***138** to participate in this process through an unspecified period in 2007 (*id.* 90). "[D]raw[ing] all inferences in [Lawson's] favor[,]" *Brown*, 774 F.3d at 1020 (internal quotation marks and citation omitted), the Court will construe the complaint to allege that Lawson was engaged in pursuing her administrative EEO claim into at least the early part of 2007, shortly before she received the May 31, 2007, denial letter. So construed, Lawson has alleged the requisite temporal proximity between her protected activity and the FBI's May 31, 2007, refusal to reinstate her. *See, e.g., Hamilton v. Geithner*, 666 F.3d 1344, 1358 (D.C. Cir. 2012) (holding that a two-month gap between the employee's protected activity and the employer's adverse action was sufficiently brief to support an inference of retaliation).

The Court recognizes that whether or not Lawson has alleged the requisite causal link is a closer question with respect to the final three denial letters, dated September 2, 2008, January 7, 2009, and March 26, 2010, respectively. The FBI issued these letters one to three years after Lawson's latest protected activity, and "[b]ecause of the time lapse, [Lawson] cannot rely solely on the time of [the denial letters] to show causation." *Forman v. Small*, 271 F.3d 285, 301 (D.C. Cir. 2001). However, Lawson offers more, and her additional assertion is the second reason that the Court concludes the complaint states a sufficient ADEA failure-to-hire retaliation claim.

Specifically, Lawson alleges that human resources personnel performed "an EEO database check to determine whether an applicant for [the] FBI SA [position] had prior EEO activity[.]" (Compl. 46.) This database check helps to bridge the one-to three-year gap between Lawson's 2006–2007 EEO activities and her subsequent employment applications, because, if true, such a check would have alerted the employees to Lawson's prior protected activity when they processed her 2008, 2009, and 2010 applications one to three years later. Moreover, this Court is hard-pressed to imagine any *non*-retaliatory justification for such an "EEO database check" as part of a re-hire application review. (*Id.*) The Court notes further that there is nothing in Lawson's complaint regarding the FBI's application process that would render this allegation manifestly implausible. In other words, if it is true that the FBI conducts "database check[s]" to determine whether applicants seeking to be reinstated to SA positions have engaged in prior EEO activity against the FBI, and as a result of this check, the FBI knew about Lawson's 2006–2007 EEO complaint when it processed—and denied—her 2008, 2009, and 2010 reinstatement applications, then Lawson's complaint plausibly suggests that her applications for reinstatement were denied because she engaged in protected activity.

Thus, although it presents a closer question, Defendants' dismissal contentions regarding causation as it pertains to Lawson's ADEA retaliation claims are rejected based on the Court's conclusion that Lawson's allegation regarding the FBI's EEO database check is sufficient to suggest a causal link between her protected activity and the challenged conduct.

(iii) Lawson may amend her complaint to clarify what type of ADEA claim she intends to raise in Count I.

[26] Finally, this Court must evaluate Defendants' dismissal arguments related to Count I, and in doing so, it has observed that, unlike the other ADEA failure-to-hire claims that are brought in the complaint, Lawson's complaint is unclear regarding the nature of the challenge that Lawson is making. That is, at times, Count I appears to object to the FBI's reinstatement policy as discriminatory on its face (*see, e.g.,* ***139** Compl. 57, 58, 64, 67, 69 (making allegations that are in the nature of objections to the policy itself)), but other allegations in Count I suggest that Lawson's Count I challenge is, in fact, a disparate treatment claim (*see, e.g., id.* 65 (alleging that the FBI's "discriminatory age-based policy was not applied to every over age 37 reinstatement applicant who depleted the FERS pension account but was applied to disadvantage Plaintiff because of her age")). To make matters worse, Defendants perceive yet another theory in the interstices of Count I's allegations of fact; they apparently read this cause of action "to allege claims of disparate impact" as well as disparate treatment. (Defs.' Mot. at 8.)

Until this Court has a better understanding of the theory (or theories) of liability that Lawson seeks to advance in Count I, it cannot undertake to analyze the sufficiency of this claim. The three aforementioned theories of liability (i.e., a facial challenge to the FBI's policy, a disparate treatment claim, or a disparate impact claim) represent fundamentally different types of legal claims with different applicable standards. *See, e.g., Ross v. Lockheed Martin Corp.*, No. 16-cv-2508, —— F.Supp.3d ——, —— – ——, 2017 WL 3242237, at *2–3 (D.D.C. July 28, 2017) (describing the differences between disparate treatment and disparate impact claims). Therefore, Lawson will need to clarify the challenge she intends to raise in this cause of action before the Court can adequately entertain Defendants' arguments in support of dismissing Count I. Accordingly, in the accompanying Order, the Court permits Lawson to amend her complaint to clarify the ground (or grounds) on which she seeks relief, and as a result, denies as moot Defendants' current motion pertaining to dismissal of Court I.

B. Defendants Have Not Demonstrated That Lawson's Complaint Fails To State A Title VII and ADEA Claim For Retaliatory Interference With The Administrative Processing Of Her EEO Complaint

[27] In addition to the various failure-to-hire claims discussed above, Lawson has also claimed unlawful retaliation, in violation of the ADEA (Count IV) and Title VII (Count VII), based on her former supervisor's alleged interference with, and improper processing of, Lawson's EEO complaint in 2010. (See Compl. 100–15; 141–55.) As Part I.B (supra) explains, Lawson maintains that Enriquez, the Chief of her former FBI Unit, "knew that [Lawson] filed an EEO complaint against him and other FBI employees in 2006" (id. 107), and that when she "participated in the formal discrimination complaint process in 2010" in connection with the FBI's failure to reinstate her as an SA (id.), Enriquez "appeared in" the office tasked with handling Lawson's complaint (id. 108), "supervised [Lawson]'s pending EEO case" (id.), and "intentionally interfered with [Lawson] in the pending EEO discrimination complaint process" (id. 109; see also id. 148–50). The complaint specifically asserts that Enriquez interfered with the processing of Lawson's 2010 EEO complaint by means of "improper complaint processing, approv[ing] the incomplete investigation of Plaintiff's claims of discrimination and retaliation, and approv[ing] the omission of any investigation of her claims of retaliation." (Id. 109, 150.) And Lawson asserts that these acts of intentional interference with the EEO-complaint process "establish retaliation in violation of [Title VII and] the ADEA." (Id. 113; see also id. 154.)

[28] This Court concludes that Defendants have failed to demonstrate that Lawson's *140 allegations are deficient in any manner that requires dismissal under Rule 12(b)(6). As discussed above, to state a claim for retaliation under Title VII and the ADEA, a plaintiff must plausibly allege that she "1) engaged in a statutorily protected activity; [and] 2) suffered a materially adverse action by [her] employer; and that 3) a causal connection existed between the two." Townsend, 236 F.Supp.3d at 315 (internal quotation marks and citation omitted). Defendants do not attack Lawson's retaliatory interference allegations by reference to any of these three elements; instead, Defendants' sole argument is that Lawson's allegations "are not actionable under Title VII or the ADEA" (Defs.' Mot at 21), because "Title VII and the ADEA create a cause of action for discrimination, not 'an independent cause of action for the mishandling of an employee's discrimination complaints'" (id. at 20 (quoting Douglas-Slade v. LaHood, 793 F.Supp.2d 82, 96 (D.D.C. 2011))). But Defendants' argument misperceives the retaliatory interference claims that Lawson makes in this complaint. It is true enough that the mishandling of an EEO complaint, on its own, does not give rise to an independent cause of action under Title VII or the ADEA. See Douglas-Slade, 793 F.Supp.2d at 96 (dismissing Title VII claim because plaintiff had merely alleged "errors and irregularities with respect to the investigation

of her discrimination claims at the administrative level"). But unlike the plaintiff in *Douglas-Slade*, Lawson alleges that a supervisor intentionally interfered with the processing of her EEO complaint *in retaliation* for prior protected activity. (*See* Compl. 109–10, 113–14, 150–51; 154.) In other words, Lawson has not alleged an independent cause of action for the improper processing of her EEO complaint, as Defendants suggest; rather, she has alleged a claim of retaliation, which is of course an actionable species of discrimination under both Title VII and the ADEA. *See* 42 U.S.C. § 2000e–3(a); *see also Gomez-Perez*, 553 U.S. at 479, 128 S.Ct. 1931.

[29] To the extent that Defendants' assertion that Lawson's retaliatory interference allegations "are not actionable" (*see* Defs.' Mot at 21) is actually intended to argue that interference with the processing of an EEO complaint is not the sort of "materially adverse action" that can support a retaliation claim, *see Townsend*, 236 F.Supp.3d at 315, this argument fares no better. It is well-established that the anti-retaliation provisions of Title VII and the ADEA "extend[ ] beyond workplace-related or employment-related retaliatory acts and harm[,]" and encompass *any* retaliatory acts that "might have dissuaded a reasonable worker from making or supporting a charge of discrimination." *Burlington N. & Santa Fe Ry. Co. v. White*, 548 U.S. 53, 67–68, 126 S.Ct. 2405, 165 L.Ed.2d 345 (2006) (internal quotation marks and citation omitted). The standard for identifying actionable adverse actions is objective, and it requires courts to "focus[ ] on the materiality of the challenged action and the perspective of a reasonable person in the plaintiff's position" in order to identify "those acts that are likely to dissuade employees from complaining or assisting in complaints about discrimination." *Id.* at 69–70, 126 S.Ct. 2405; *see also id.* at 68, 126 S.Ct. 2405.

This Court has no doubt that, when properly understood, the adverse action standard is satisfied under the circumstances presented in this case. Just as a reasonable employee might refrain from filing a discrimination complaint out of fear that a supervisor would retaliate by reassigning her to different duties or suspending her without pay, *see id.* at 70–73, 126 S.Ct. 2405, so too might a reasonable employee refrain from engaging the EEO-complaint *141 process out of a concern that a supervisor's interference with the EEO process would render her efforts futile. Thus, in both situations, the employer's conduct reasonably might "discourage an employee … from bringing [EEO] discrimination charges" that she otherwise would bring, *id.* at 70–71, 126 S.Ct. 2405; *cf. Mogenhan v. Napolitano*, 613 F.3d 1162, 1166 (D.C. Cir. 2010) (holding that a jury could find that adverse action occurred where a supervisor increased the plaintiff's workload in order "to keep [her] too busy to file complaints").

Notably, although some prior decisions in this district have suggested that the improper processing of an administrative complaint cannot constitute a materially adverse action sufficient to sustain a claim of retaliation, *see Briscoe v. Kerry*, 111 F.Supp.3d 46, 59 (D.D.C. 2015); *Diggs v. Potter*, 700 F.Supp.2d 20, 46 (D.D.C. 2010), support for that proposition ultimately stems from *Keeley v. Small*, 391 F.Supp.2d 30, 45 (D.D.C. 2005), which is a case that was decided *prior* to the view of the scope of actionable retaliation that the Supreme Court clarified in *Burlington Northern*. *See Diggs*, 700 F.Supp.2d at 46 (relying on *Keeley* without addressing the

Supreme Court's intervening decision in *Burlington Northern*); *see also Briscoe*, 111 F.Supp.3d at 59 (relying on *Diggs*). In *Keeley*, the judge reasoned that complaints regarding interference with an EEOC investigation could not form the basis of a Title VII retaliation claim because an EEOC investigation does not relate to "a condition of employment[.]" 391 F.Supp.2d at 45 (citation omitted). But the Supreme Court's 2006 *Burlington Northern* opinion spoke definitively to that issue, holding that the scope of "Title VII's substantive provision and its antiretaliation provision are not coterminous[,]" and the latter "is not limited to discriminatory actions that affect the terms and conditions of employment[,]" *id.*, 548 U.S. at 64, 67, 126 S.Ct. 2405.

[30] For the reasons explained above, and in light of *Burlington Northern*, this Court concludes that interference with the processing of an EEO complaint plausibly constitutes the sort of materially adverse action that can support a retaliation claim, and thus Defendants' motion to dismiss Counts IV and VII on the grounds that such alleged retaliation is not actionable must be **DENIED**.

IV. CONCLUSION

This Court has carefully reviewed Lawson's complaint and the arguments that Defendants have raised in their motion to dismiss. While Lawson has failed to exhaust her administrative remedies fully with respect to most of her failure-to-hire claims under Title VII, and it is far from clear that Lawson will ultimately be able to prove the remaining discrimination and retaliation claims she seeks to advance in this action, the Court finds that Lawson's complaint alleges sufficient facts to state a claim under the ADEA for discrimination and retaliation based on the agency's refusal to reinstate her. In addition, the Court concludes that Lawson's complaint states a claim for retaliation under the ADEA and Title VII based on Defendants' purported interference with the processing of her administrative complaint. Finally, the Court will permit Lawson to amend her complaint in three limited respects: *first*, Lawson may clarify the claim she intends to raise in Count I; *second*, she may plead the date that she purportedly received the right-to-sue letter that preceded the filing of the instant action; and *third*, she is permitted to supplement the factual allegations that support her surviving Title VII claims. Accordingly, as set forth in the Order accompanying this ***142** Memorandum Opinion, Defendants' Motion to Dismiss is **GRANTED IN PART AND DENIED IN PART**.

ALL CITATIONS

271 F.Supp.3d 119, 2017 Fair Empl.Prac.Cas. (BNA) 336,584

3 Job Descriptions
The Key to Hiring Success

Shelby L. Schneid

******

Chapter Objectives:

1. Understand the importance of a written job description.
2. Understand the applicable laws and governmental agencies.
3. Understand the legal issues involved in written job descriptions.

******

A well-written job description is essential for all positions no matter the function, level, or practice. Job descriptions are the foundation of all organizations. When you applied for your current role, you applied because you were attracted to the description of the job. Your role was outlined so that you could understand what your realm of responsibilities were and what qualified you for the role. Did you meet all the desired characteristics?

A good job description will allow your candidate to understand what is needed out of them to be successful in the role. The pre-employment process for candidates should consist of some company research, understanding what the company does, what market that they are in, and any big issues that may be in the media spotlight right now. Through the application process, they should get a general understanding of the job in which they are applying for and the scope of the responsibility they would be taking on if they were to get an offer. After applying, if the candidate lands an interview, this is where they get an in-depth introduction to the position, team, site, and business.

The interview phase is where you as the interviewer probe deeper than the resumé. Candidates that move into the interview phase should already meet the basic qualifications, so you will use this time to see how/if their experience will allow the candidate to be successful and take on the designated responsibilities, and if they have any of the desired characteristics outlined in the job description. Therefore, the development of the job description is very important. It is your opportunity as a hiring manager to create your ideal candidate on paper, so when you are filtering through resumés and interviewing candidates all you must do is find the best fit to your job description.

To create a job description, you must start with a job analysis. Is this job description for a position you currently have, or is it for a new role? If it is for a current role, then you will want to interview all employees who currently hold that position. What do they see as their responsibilities? What makes them qualified for the position? What characteristics do they feel are essential for the position? This can be done on a one-to-one basis or in a group setting, depending on the number of employees.

If the position is new to the business, interview those who will be interacting with the role. What will the reporting structure look like? Who will the employee be reporting to? Will they have anyone reporting to them? What team are they part of? By getting 360-degree feedback from all directions which this role will touch, it will help you understand the potential needs of others that this role can assist or potentially take over from.

Once the analysis is complete, you should have an overview of what the person in the position will be responsible for, the qualifications needed to do the job, and the desired characteristics of your ideal candidate. Now it is time to put it on paper using the following sections:

1. General Overview
2. Role Responsibilities
3. Basic Qualifications
4. Desired Characteristics
5. Equal Employment Opportunity Commission (EEOC) Statement

1. GENERAL OVERVIEW

This is an opening statement that gives your candidates a preview of the position's fit within the organization. Just like an opening paragraph of a paper or book, this is what you need to use to attract your candidate into wanting to keep reading. Will this position be responsible for a $1.2 million revenue generating business? Will this position be developing and implementing improved manufacturing processes? Will this position be working with international clients? What is cool about this position? Put it in there. This is the high-level overview of the position.

For example, the below is an overview for a safety and health manager:

The safety and health professional will demonstrate leadership in communicating EHS business goals, programs, and processes for the manufacturing facility generating over $2.2 million dollars of annual revenue. This role will utilize experience and expertise to problem solve, develop, and execute objectives to meet quarterly and annual goals.

2. ROLE RESPONSIBILITIES

This is where you take your general overview and break it down a little further. For example, if you said the position is responsible for developing and implementing improved manufacturing process in the general overview, you will now break down the responsibilities for doing so, such as leading root cause analysis processes and implementing corrective action.

For example:

- *Lead HSE culture defined by leadership engagement, employee ownership, and learning.*
- *Develop and implement an HSE strategy and drive sustainable outcomes.*
- *Lead and develop a team of HSE professionals.*
- *Perform necessary duties for the safe operations of the manufacturing facility.*
- *Develop and implement site-specific HSE programs and processes.*
- *Work with business operational managers to confirm all operational activities are clearly identified and risk assessed to ensure a safe working environment.*
- *Drive improvements in compliance processes, employee training, and HSE performance.*
- *Conduct compliance and framework audits and review as necessary to meet all compliance requirements.*
- *Drive injury prevention initiatives.*

3. BASIC QUALIFICATIONS

This is one of the most important pieces of your job description. This is the first candidate filtering piece of information that determines if the candidate will move on to the next stage of the process. The basic qualifications are your minimum qualifications needed for the position. These are the "must haves" for the position. This section normally outlines the following:

- *Educational Requirements.*
- *Experience Requirements.*
- *Direct Experience.*

For example:

- *Bachelor's degree from an accredited university or college.*
- *Minimum five years of safety and health management experience in a manufacturing environment.*

4. DESIRED CHARACTERISTICS

This section allows you to outline skills you would like to see in the candidate that are not required for the position, but are "nice to haves." Think of this section like your Christmas list: what are the characteristics of your ideal, picture-perfect candidate.

For example:

- *Effective communication skills.*
- *Strong interpersonal skills.*

- *Strong organizational skills.*
- *Ability to effectively multi-task.*
- *OSHA certified.*

5. EEOC STATEMENT

Employment decisions are made without regard to race, color, religion, national or ethnic origin, sex, sexual orientation, gender identity or expression, age, disability, protected veteran status, or other characteristics protected by law.

You created the job description: now what? As with any good process, review the job description prior to posting with the team that you first interviewed for input. This will allow feedback for potential wording issues, additions, or editing. Once you have gone through the review process with the team, review the description with human resources and/or recruiting to take in any feedback from them as well. Once you have reviewed with all parties, you are ready to post and begin filtering resumés to find your next employee.

Case Study: This case may have been modified for the purpose of this text.

2016 WL 4785169
Only the Westlaw citation is currently available.
United States District Court,
S.D. Texas, Corpus Christi Division.

Karen Barsch, Plaintiff,
v.
Nueces County, Defendant.
CIVIL NO: 2:14-CV-435
|
Signed 01/14/2016

MEMORANDUM AND ORDER

Hilda G. Tagle, Senior United States District Judge
1 Effective February 3, 2014, Defendant Nueces County, Texas ("the County"), promoted Plaintiff Karen Barsch ("Barsch") to the position of GIS Engineering Specialist in its Public Works Department and reduced that position's pay group from 32 to 24. Nueces County Personnel Action Request 1, Jan. 27, 2014, Dkt. No. 18 Ex. 10. As a result, Barsch, who is a woman, receives a smaller salary than her male predecessor, Tony Parlamas ("Parlamas"). Orig. Ans. 9, Dkt. No. 3 (so admitting). Barsch alleges that her right to receive equal pay for equal work guaranteed by the Equal Pay Act of 1963 ("Equal Pay Act"), 29 U.S.C. § 206(b), and her right

---

* The reduction from 32 to 24 is sometimes characterized in the summary judgment record as a reduction in pay grade rather than pay group. *E.g.,* Orig. Ans. 23; Dkt. No. 18 at 7. Nothing should be inferred from the Court's use of the phrase "pay group" in this opinion.

to be free from discrimination based on her sex guaranteed by Title VII of the Civil Rights Act of 1964, as amended, 42 U.S.C. § 2000e et seq, was violated. Pl.'s Orig. Compl. 16–22, Dkt. No. 1. The County has filed a motion for summary judgment. Dkt. No. 18. Because genuine disputes over material facts exist on this record, the Court denies the County's motion for summary judgment.

I. BACKGROUND

Except where otherwise noted, the Court recounts the undisputed facts in the record in the light most favorable to Barsch and draws reasonable inferences in her favor because she is the party against whom summary judgment is sought. *See, e.g., Matsushita Elec. Indus. Co., Ltd. v. Zenith Radio Corp.*, 475 U.S. 574, 587–88 (1986) (quoting *United States v. Diebold, Inc.*, 369 U.S. 654, 655 (1962)) ("On summary judgment the inferences to be drawn from the underlying facts ... must be viewed in the light most favorable to the party opposing the motion." (alteration omitted)); *Clift v. Clift*, 210 F.3d 268, 270 (5th Cir. 2000) ("We must view all evidence in the light most favorable to the party opposing the motion and draw all reasonable inferences in that party's favor." (citing *Anderson v. Liberty Lobby, Inc.*, 477 U.S. 242, 255 (1986))). Parlamas began working for the County in 2009. Aff. of Glen R. Sullivan ("Sullivan") 3, Dkt. No. 18 Ex. 2. The job description for GIS Engineering Specialist under which Parlamas was working in 2013 required a four-year degree and five years of experience.* *Id.* Also, "software application support with a County Engineer/Road Department or equivalent public works department was preferred." *Id.*

*2 The County hired Barsch as a Geographic Information System (GIS)† Technician on January 7, 2013. Orig. Ans. 9. Barsch does not possess a four-year degree. Orig. Compl. 9; *see also* Dkt. No. 18 Ex. 8 at 4.

In 2013, the County promoted Parlamas to a role in which he "overs[aw]‡ the entire GIS section of the Nueces County Department of Public Works." Sullivan Aff. 6. By interoffice memorandum dated August 27, 2013, from Sullivan, the director of the County's public works department, directed to "All Department of Public Works Employees," the County solicited applications for the GIS Engineering Specialist

---

* The record contains a Nueces County job description, Dkt. No. 18 Ex. 4, that the County's attorney describes in her affidavit as the job description "prior to reclassification," Dkt. No. 18 Ex. 1 at 1. Viewed in the light most favorable to Plaintiff, however, this document does not establish that it is the job description under which Parlamas was hired in 2009. *See* Fed. R. Evid. 901. The following typewritten text appears near the top of Exhibit Four to the County's motion for summary judgment: "Created: 7/3/09; rev. 08/08/2012." Dkt. No. 18 Ex. 4 at 1. It is reasonable for the Court to infer that the notation "rev. 08/08/2012" means that the document was revised on August 8, 2012, especially in light of the fact that the signature on that document is dated August 13, 2012, *id.*, and the version of the job description for the same position that the parties agree was modified after August 2013 bears a second notation reading "rev. 11/12/2013" next to the first notation "rev. 08/08/2012." Dkt. No. 18 Ex. 6 at 1. There is no dispute that, however, Parlamas worked under the first job description when he was promoted in August 2013 or that the same job description was initially advertised at the same pay group after Parlamas's promotion. *Compare* Dkt. No. 18 Ex. 4 *with* Dkt. No. 21 Ex. A at 2 (same version of job description Barsch alleges was advertised in August 2013).

† *See, e.g.*, Dkt. No. 21 Ex. A at 1 (job description expanding the GIS acronym sometimes used by the parties).

‡ The record does not indicate whether Parlamas remains in this role today.

position. Dkt. No. 18 Ex. 5 at 1 (listing opening and closing dates for applications respectively as September 3 and 10, 2013); Dkt. No. 21 Ex. A (same memorandum). The memorandum listed the job's hourly pay as $27.77, *ibid.* at 1, and the accompanying position description listed a salary in pay group 32, Dkt. 18 Ex. 4 at 1. As it did when Parlamas occupied it, the job description required a four-year degree and at least five years of experience. *Id.* Ex. 4 at 2. The County received no applications for the GIS Engineering Specialist position as advertised on August 27, 2013. Sullivan Aff. 3. Barsch states in her affidavit* that she prepared an application, but did not submit it because she does not have a four-year degree. Dkt. No. 21 Ex. C at 1.

The County revised the position description in the ensuing months "according to the Nueces County Civil Service Rules."† Sullivan Aff. 3; *see also* Memorandum from Sullivan to Julie Guerra ("Guerra"), Director of Human Resources 1, Oct. 29, 2013, Dkt. No. 18 Ex. 3 (memorializing "recent [ ] request" to modify job description). The revision process resulted in three changes. First, the County decreased the position's salary level from 32 to 24. *Compare id.* Ex. 6 at 1 *with id.* Ex. 4 at 4. Sullivan states in his affidavit that "[t]he [new] pay group was based on the State of Texas classifications for Geographic Information Specialists to be consistent with similar positions and qualifications across the State." Sullivan Aff. 4. Second, the revised description eliminated the stated preference for experience with "[s]oftware application support with a County Engineer/Road department or equivalent public works department." *Compare id.* Ex. 6 at 2 *with id.* Ex. 4 at 2. Finally, the County

---

* The fact that Barsch originally attached the affidavit on which she relies in her response to the county's motion for summary judgment to her charge of discrimination filed with the Equal Employment Opportunity Commission ("EEOC") does not prevent Barsch from relying on that affidavit at summary judgment. *See* Dkt. No. 21 Ex. C at 4. A review of Barsch's affidavit confirms that it is sworn and bears a notary public's seal. *Id.* at 1. Therefore, the Court relies on it as competent summary judgment evidence. *See, e.g., Lohn v. Morgan Stanley DW, Inc.,* 652 F.Supp.2d 812, 826 (S.D. Tex. 2009) (overruling objection at summary judgment to sworn document described as "Examination Under Oath" taken during EEOC investigation); *Emanuel v. Mo. Pac. R.R. Co.,* No. H-02-4851, 2004 WL 3249248, at *1 & n.3 (S.D. Tex. Mar. 22, 2004) (relying on affidavit attached to EEOC charging document at summary judgment).

† The Texas legislature has authorized certain counties to establish a general civil service system under Chapter 158, Subchapter A, of the Local Government Code, §§ 158.001–158.015, as well as a sheriff's department system under Subchapter B, §§ 158.031-158.040. *See County of Dallas v. Wiland,* 216 S.W.3d 344, 348 & nn. 11–12 (Tex. 2009) (noting that general civil service system limited to counties with a population of more than 200,000 and sheriff's department civil service system to counties with a population of more than 500,000). The record does not disclose which kind of system the County had in 2013 and 2014. Nueces County apparently had a civil service commission in 2005 that heard appeals from, at least, sheriff's office employees. *See Soliz v. Nueces County Sheriff,* No. 13-06-00243-CV, 2007 WL 2429044, at *1–*2 (Tex. App. —Corpus Christi Aug. 29, 2007) (considering appeal of commission's 2005 decision of sheriff's employee's grievance and discussing commission's powers under § 158.009); *see also* Tex. Gov't Code § 158.031 (Vernon 2015). That commission appears to have been a general civil service commission. *See Soliz,* 2007 WL 2429044, at 2 & n.2 (citing § 158.009 and other provisions of subchapter A when considering appeal of sheriff's department employees); *see also Wiland,* 216 S.W.3d at 348 ("In 1989, ... the Legislature expanded the group of employees that a county could include in its general civil service system, thereby allowing a county to choose to extend civil service protection to deputy constables." (internal footnotes and citation omitted)). Nor have the rules of the County's civil service commission been made known to the Court. A civil service commission has the power, inter alia, to "adopt, publish, and enforce rules regarding ... selection and classification of county employees." Tex. Loc. Gov't Code § 159.009(a)(2) (Vernon 2015); *see also* § 158.035(a)(2) (granting same power to sheriff's department civil service commission).

added the following text under the "Education and/or Experience" heading of the revised job description:

> *3 Associates degree from an accredited college with a major in geographic information systems (GIS), engineering, geography, computer science or a related field plus a minimum of 7 years professional experience supporting and developing software applications and supporting geographical information systems or; [sic] High School or GED plus a minimum of 9 years professional experience supporting and developing software applications and supporting geographical information systems.

Id. Ex. 6 at 2.

In his affidavit, Sullivan states that "[s]ome of the job duties associated with the position were transferred to … Parlamas when he was promoted in 2013 …."* Sullivan Aff. 6. A list of six categories of transferred duties follows, ranging from "cataloging the County's master drainage plan and providing information regarding the plan's contents" to "assisting the Director of Public Works with presentations." *Id.* The record also includes a memorandum regarding the revision of the GIS Engineering Specialist job description from Sullivan to Guerra, dated October 29, 2013. *Id.* Ex. 3. The memorandum responds to Guerra's proposal to lower the position's pay group to 17 in light of the proposed change to the education and experience required. *Id.* Sullivan's memorandum proposes pay group 24. *Id.* Sullivan's memorandum further reads as follows:

> We note that the State of Texas has five titles for Geographic Information Specialist (GIS I–V) for use by the various state agencies. The salary levels for these five positions ranges from the GIS I position with a salary level from $38,746 to $61,664 to the GIS V position with salary level from $68,054 to $112,288.
> *4 The Education and Experience requirements for all five GIS positions is identical and reads as follows: "Experience in geographic information systems analysis and design work. Graduation from an accredited four-year college or university with major course work in computer science, computer information systems geography,

---

* The Court does not consider any factual assertions in Barsch's response to the County's motion for summary judgment that are unsupported by competent summary judgment evidence, including factual assertions not present in her affidavit. *See, e.g., Hockaday v. Tex. Dep't of Criminal Justice, Pardons & Paroles Div.,* 914 F.Supp. 1439, 1446 (S.D. Tex. 1996) ("It is … well settled that arguments set forth in a brief in response to a motion for summary judgment, unsupported by affidavits, deposition testimony, or other adequate summary judgment evidence, are insufficient to raise genuine issues of material fact to defeat a properly supported motion for summary judgment.") (citing *Schroeder v. Copley Newspaper,* 879 F.2d 266, 271 (7th Cir. 1989)) (other citations omitted). For example, Barsch asserts that Parlamas heard in July 2013 that the GIS Engineering Specialist position would soon be available. Resp. Mot. Summ. J. 1, Dkt. No. 21. Barsh also describes her alleged interactions with a supervisor, Eddie Eubanks ("Eubanks"), in her response. Resp. Mot. Summ. J. 1–2. She represents that Eubanks reviewed her initial application for the GIS Engineering Specialist position advertised in August 2013, that she pointed out to him that she did not have the four-year degree the position then required, and that he later made statements to Barsch that left her with the impression that "the position description was rewritten specifically to allow [her] to qualify …." Resp. Mot. Summ. J. 2. Barsch's affidavit makes no mention of a conversation with Parlamas in July 2013; nor does it mention Eubanks by name or role. *See* Dkt. No. 21 Ex. C. Likewise, Barsch cites no evidence to support her assertion that the County transferred some duties to Parlamas "to justify its discriminatory acts." Resp. Mot. Summ. J. 13.

geographic information systems, or a related field is generally preferred. Education and experience may be substituted for one another."

In looking at the salary ranges for the current County positions of Engineering Technician (GP 34), GIS Engineering Specialist (GP 32), and GIS Data Technician (GP 15) we agree that the salary group for GIS Engineering Specialist could be changed to better reflect the pay ranges used by both the State and this Department.

Id.

Sullivan sent a memorandum to the County's public works employees advertising a vacancy under the revised job description on January 7, 2014. Dkt. No. 18 Ex. 7 at 1 (listing position opening and closing dates respectively as January 10 and 17, 2014); Dkt. No. 21 Ex. B (same memorandum). Consistent with the reclassification of the position as pay group 24, the advertisement listed $18.91 as the hourly pay rate. *Id.* at 1. The County received two letters of interest: one from Barsch and one from Michael Flores, a male "Vector Control Technician." Dkt. No. 18 Ex. 8. On January 27, 2014, Sullivan signed a personnel action request promoting Barsch to the GIS Engineering Specialist position. *Id.* Ex. 10 at 1.

II. SUMMARY JUDGMENT STANDARD

"Summary judgment is appropriate where the competent summary judgment evidence demonstrates that there is no genuine issue of material fact and the moving party is entitled to judgment as a matter of law." *Brumfield v. Hollins*, 551 F.3d 322, 326 (5th Cir. 2008) (citing *Bolton v. City of Dall.*, 472 F.3d 261, 263 (5th Cir. 2006)); *accord.* Fed. R. Civ. P. 56(c). "A genuine issue of material fact exists when the evidence is such that a reasonable jury could return a verdict for the nonmovant." *Piazza's Seafood World, LLC v. Odom*, 448 F.3d 744, 752 (5th Cir. 2006) (citing *Anderson v. Liberty Lobby, Inc.*, 477 U.S. 242, 248 (1986)). The Court must view all evidence in the light most favorable to the non-moving party. *Brumfield*, 551 F.3d at 326 (citing *Matsushita Elec. Indus. Co. v. Zenith Radio*, 475 U.S. 574, 587 (1986)); *Piazza's Seafood World*, 448 F.3d at 752; *Lockett v. Wal-Mart Stores, Inc.*, 337 F. Supp. 2d 887, 891 (E.D. Tex. 2004). Factual controversies must be resolved in favor of the non-movant, "but only when there is an actual controversy, that is, when both parties have submitted evidence of contradictory facts." *Little v. Liquid Air Corp.*, 37 F.3d 1069, 1075 (5th Cir. 1994) (en banc, per curiam).

*5 The party moving for summary judgment bears the "burden of showing this Court that summary judgment is appropriate." *Brumfield*, 551 F.3d at 326 (citing *Celotex Corp. v. Catrett*, 477 U.S. 317, 323 (1986)). The burden of production a party must initially carry depends upon the allocation of the burden of proof at trial. *See Shanze Enters., Inc. v. Am. Cas. Co. of Reading*, ___ F. Supp. 3d ____, 2015 WL 8773629, at *3 (N.D. Tex. Dec. 15, 2015) ("Each party's summary judgment burden depends on whether it is addressing a claim or defense for which it will have the burden of proof at trial."). "[I]f the movant bears the burden of proof on an issue, either because he is the plaintiff or as a defendant he is asserting an affirmative defense, he must establish beyond peradventure *all* of the essential elements of the claim or defense to warrant judgment in his favor." *Tesoros Trading Co. v. Tesoros Misticos, Inc.*, 10 F. Supp. 3d 701, 709 (N.D. Tex. 2014) (quoting *Fontenot v. Upjohn Co.*, 780 F.2d 1190,

1194 (5th Cir. 1986) (emphasis in *Fontenot*); *accord. Shanze Enters.*, ___ F. Supp. 3d
____, 2015 WL 8773629, at *3. On the other hand, when the nonmovant will bear
the burden of proof at trial, the movant may discharge its initial burden at summary
judgment by "merely point[ing] to the absence of evidence and thereby shift to the non-
movant the burden of demonstrating by competent summary judgment proof that there
is an issue of material fact warranting trial." *Transamerica Ins. Co. v. Avenell*, 66 F.3d
715, 718–19 (5th Cir. 1995) (*per curiam*); *see also Celotex*, 477 U.S. at 323–25. Once
the party seeking summary judgment has discharged its initial burden, the nonmovant
must come forward with specific evidence to show that there is a genuine issue of fact.
Lockett, 337 F. Supp. 2d at 891; *see also Ashe v. Corley*, 992 F.2d 540, 543 (5th Cir.
1993). The non-movant may not merely rely on conclusory allegations or the pleadings.
Lockett, 337 F. Supp. 2d at 891. Rather, it must cite specific facts identifying a genuine
issue to be tried in order to avoid summary judgment. *See* Fed. R. Civ. P. 56(e); *Piazza's
Seafood World*, 448 F.3d at 752; *Lockett*, 337 F. Supp. 2d at 891. "Rule 56 does not
impose upon the district court a duty to sift through the record in search of evidence to
support a party's opposition to summary judgment." *Ragas v. Tenn. Gas Pipeline Co.*,
136 F.3d 455, 458 (5th Cir. 1998) (quoting *Skotak v. Tenneco Resins, Inc.*, 953 F.2d 909,
915–16 & n.7 (5th Cir. 1992)). Thus, once it is shown that a genuine issue of material
fact does not exist, "[s]ummary judgment is appropriate ... if the non-movant 'fails
to make a showing sufficient to establish the existence of an element essential to that
party's case.'" *Arbaugh v. Y&H Corp.*, 380 F.3d 219, 222–23 (5th Cir. 2004) (quoting
Celotex Corp. v. Catrett, 477 U.S. 317, 322 (1986)).

III. EQUAL PAY ACT

The Equal Pay Act guarantees "equal pay for equal work regardless of sex." *Corning
Glass Works v. Brennan*, 417 U.S. 188, 190 (1974) (citing Pub. L. No. 88–38, 77 Stat.
56§ 3 (Sept. 23, 1963)) (other citations omitted); *accord. Siler-Khodr v. Univ. of
Tex. Health Sci. Ctr. San Antonio*, 261 F.3d 542, 546 (5th Cir. 2001) (citing *Corning
Glass*, 417 U.S. at 195) ("In short, [the Equal Pay Act] demands that equal wages
reward equal work."). It provides:

> No employer having employees subject to any provisions of this section shall dis-
> criminate, within any establishment in which such employees are employed, between
> employees on the basis of sex by paying wages to employees in such establishment at a
> rate less than the rate at which he pays wages to employees of the opposite sex in such
> establishment for equal work on jobs the performance of which requires equal skill,
> effort, and responsibility, and which are performed under similar working conditions,
> except where such payment is made pursuant to (i) a seniority system; (ii) a merit sys-
> tem; (iii) a system which measures earnings by quantity or quality of production; or
> (iv) a differential based on any other factor other than sex: *Provided*, [t]hat an employer
> who is paying a wage rate differential in violation of this subsection shall not, in order
> to comply with the provisions of this subsection, reduce the wage rate of any employee.

29 U.S.C. § 206(d)(1) (2012) (emphasis in original). The plaintiff bears the initial
burden of making out a prima facie case of discrimination under the Equal Pay Act
by showing that "1. her employer is subject to the Act; 2. she performed work in a

position requiring equal skill, effort, and responsibility under similar working conditions; and 3. she was paid less than the employee of the opposite sex providing the basis of comparison." *Chance v. Rice Univ.*, 984 F.2d 151, 153 (5th Cir. 1993) (citation omitted); *see also Thibodeaux-Woody v. Hous. Cmty. Coll.*, 593 Fed.Appx. 280, 283 (5th Cir. 2014) (unpublished) (quoting 29 U.S.C. § 206(d)(1)) (stating elements of prima facie case in slightly different language: "the employer pays different wages to men and women, the employees perform 'equal work on jobs the performance of which requires equal skill, effort, and responsibility,' and the employees perform their jobs 'under similar working conditions'"). In contrast to a Title VII claim,* "'[o]nce a plaintiff has made her prima facie case by showing that an employer compensates employees differently for equal work, the burden shifts to the defendant to' show by a preponderance of the evidence that the differential in pay was made pursuant to one of the [Equal Pay Act's] four enumerated exceptions," which are affirmative defenses on which the employer bears the burden of proof. *King v. Univ. Healthcare Sys., L.C.*, 645 F.3d 713, 723 (5th Cir. 2011) (quoting *Siler-Khodr*, 261 F.3d at 546); *see also Corning Glass*, 417 U.S. at 196–97 (citations omitted) (reasoning that shifting burden to employer "to show that the differential is justified under one of the Act's four exceptions … is consistent with the general rule that the application of an exemption under the Fair Labor Standards Act is a matter of affirmative defense on which the employer has the burden of proof"); *Jones v. Flagship Int'l.*, 793 F.2d 714, 722 (5th Cir. 1986) (citing *Corning Glass*, 417 U.S. at 497) (holding that the Equal Pay Act's four exceptions "are affirmative defenses on which the employer has the burden both of production and of persuasion").

A. Prima Facie Case

*6 Relying on Sullivan's affidavit and a comparison of the job description under which Parlamas worked and the revised description under which Barsch was hired, the County argues that the GIS Engineering Specialist job as Barsch performs it requires substantially less skill and responsibility than it did when Parlamas performed it. Mot. Summ. J. 5. The parties draw competing inferences from the revision of the job description for the GIS Engineering Specialist position between August 2013 and January 2014. *Compare* Dkt. No. 18 Ex. 4 *with id.* Ex. 6. The County argues that the removal of the requirement of a four-year degree demonstrates conclusively that the job Barsch performs does not require the same skill as it did when Parlamas performed it. Mot. Summ. J. 5. The County also maintains that Barsch has fewer responsibilities than [Parlamas] based on Sullivan's statement that Parlamas retained six categories of duties he was performing when he was promoted in 2013. *See* Sullivan Aff. 6–7. Barsch counters that a reasonable jury could find that she has made out a prima facie case because the County did not revise the job description's listed duties and responsibilities when it recategorized the position

---

* *Contra Mathis v. FedEx Corp. Servs., Inc.*, Civ. A. No. H-12-1871, 2014 WL 1278182, at *10 (S.D. Tex. Mar. 27, 2014) (relying on *Browning u. Sw. Research Inst.*, 288 Fed.Appx. 170, 173, 174 (5th Cir. 2008) (unpublished) for the proposition that "Equal Pay Act claims are also analyzed using the burden-shifting framework set forth in *McDonnell Douglas Corp. v. Green*, 411 U.S. 792 (1973)").

between August 2013 and January 2014. *Compare* Dkt. No. 18 Ex. 4 at 1–2 *with id.* Ex. 6 at 1–2.

On this record, a genuine fact issue exists on Barsch's prima facie case. Job descriptions can shed light on the question, but they "should not be accorded as much weight as that given to the duties actually performed by employees in determining whether jobs are substantially equal …." *Brennan v. Owensboro-Daviess Cty. Hosp.*, 523 F.2d 1013, 1017 (6th Cir. 1975), *quoted in E.E.O.C. v. TXI Operations, L.P.*, 394 F. Supp. 2d 868, 877 (N.D. Tex. 2005); *see also* 29 C.F.R. § 1620.13(e) (2015) ("Application of the equal pay standard is not dependent on job classifications or titles but depends rather on actual job requirements and performance."); *see also, e.g., Georgen-Saad v. Texas Mut. Ins. Co.*, 195 F. Supp. 2d 853, 857 (W.D. Tex. 2002) (relying on comparison of job description to decide motion for summary judgment on prima facie Equal Pay Act case). Accordingly, though it can be considered, a fact finder need not treat a difference in the education required as the *sine qua non* of substantially equivalent skill, effort, and responsibility. *See Vasquez v. El Paso Cty. Cmty. Coll. Dist.*, 177 Fed. Appx. 422, 425 (5th Cir. 2006) (per curiam, unpublished) (considering differences in titles, stated responsibilities, and required education for positions to decide that plaintiff did not demonstrate that he performed work substantially equal to comparators); *Romo v. Tex. Dep't of Transp.*, 48 S.W.3d 266, 272 (Tex. App. —San Antonio 2001) (considering type of degree as well as experience of comparators when deciding whether plaintiff made out prima facie retaliation case of pay discrimination); *Schulte v. Wilson Indus., Inc.*, 547 F. Supp. 324, 333 (S.D. Tex. 1982) (sitting as fact finder and rejecting argument that paying male account manager with college degree more than female plaintiff was justified by factor other than sex because, inter alia, "the evidence revealed that since [the comparator with a college degree] was so inexperienced when he first started working for [the employer], plaintiff performed more assistant account managerial duties than he did"). Indeed, differences in employee education sometimes comprise an employer's affirmative defense. *See Schulte*, 547 F. Supp. at 333 (analyzing education claim as affirmative defense); 29 C.F.R.§ 1620.13(c) (2015) (listing education as one factor that may justify pay differential: "When factors such as seniority, education, or experience are used to determine the rate of pay, then those standards must be applied on a sex neutral basis."). Undisputedly, the County added seven years of experience and a two-year degree as alternative qualifications for the GIS Engineering Specialist position between August 2013 and January 2014. Sullivan Aff. 3–5. The County left the five years of experience with a four-year degree alternative unchanged, and the County did not alter any of the listed duties or responsibilities. *Id.* A reasonable jury viewing this evidence in the light most favorable to Barsch could conclude that the alternative stated combinations of qualifying education and experience are equivalent and could, therefore, find that Barsch performs work substantially equal to the work Parlamas performed in 2013 when he held the GIS Engineering Specialist title.

7 As an alternative to its contentions regarding Barsch's prima facie case, the County argues that the difference between Barsch and Parlamas's salary is justified under the Equal Pay Act's catch-all exception for "any other factor other than sex." 29 U.S.C. § 206(d)(1) (2012). That is, the County does not attempt to establish that the salary differential between Barsch and Parlamas results from a seniority system, a merit system, or a system "which measures earnings by quantity or quality of production." *Id.* "So long as it is consistent with the purposes of the Equal Pay Act ... an employer may base a salary differential on a factor other than sex, including: (1) 'different job levels;' (2) 'different skill levels;' (3) 'previous training;' (4) 'experience;' and (5) 'prior salary history, performance, and other factors.'" *Sauceda v. Univ. of Tex. at Brownsville*, 958 F. Supp. 2d 761, 776 (S.D. Tex. 2013) (quoting *TXI Operations, L.P.*, 394 F. Supp. 2d at 879 & n.11); *see also Lenihan v. Boeing Co.*, 994 F. Supp. 776, 798 (S.D. Tex. 1998) (quoting *Irby v. Bittick*, 44 F.3d 949, 955 (11th Cir. 1995)) (listing comparable, nonexhaustive list of factors other than sex, including "special exigent circumstances connected to the business"). An "employer must show that sex provided no part of the basis for the wage differential" to succeed on the Equal Pay Act's catch-all defense. *Lenihan*, 994 F. Supp. at 798 (quoting *Peters v. City of Shreveport*, 818 F.2d 1148, 1154 (5th Cir. 1987)) (internal quotation omitted); *see also Wojciechowski v. Nat'l Oilwell Varco, L.P.*, 763 F. Supp. 2d 832, 854 (S.D. Tex. 2011) (denying summary judgment because the "record len[t] some support to Plaintiff's position" that Defendant's decision to pay her male counterparts more was not pretextual). That is, the difference in sex must be a but-for cause of the differential, and "the factor of sex is not a 'but for' cause of the differential if, without the employer's consideration of the factor of sex, the wage differential would have been the same in character and extent as that which [the employee] receives." *Peters*, 818 F.2d at 1161 (internal quotation omitted) (explaining that sex need not be substantial factor or proximate cause of disparity). "However, even a legitimate, nondiscriminatory reason for the pay differential will not justify summary judgment where plaintiff is able to show that defendant's proffered reasons are a pretext for gender discrimination." *Wojciechowski*, 763 F. Supp. 2d at 853 (quoting *Perales v. Am. Ret. Corp.*, No. SA-04-CA-0928, 2005 WL 2367772, at *6 (W.D. Tex. Sept. 26, 2005)); *accord. Sauceda*, 958 F. Supp. 2d at 776 (citing *TXI Operations*, 394 F. Supp. 2d at 879 & n.11 and *Siler-Khodr*, 261 F.3d at 548–49) ("Summary judgment is not appropriate, however, if a genuine fact issue on pretext exists").

The County characterizes the disparity between Barsch and Parlamas's salaries as the result of "a business decision based on the lowering of the educational qualification" for the GIS Engineering Specialist position. Mot. Summ. J. 7. Sullivan elaborates in his affidavit that "[t]he pay group was based on the State of Texas classifications for Geographic Information Specialists to be consistent with similar positions and qualifications across the State." Sullivan Aff. 4; *see also* Dkt. No. 18 Ex. 3 at 1. To show that a difference in pay for equal work is justified by a difference in employees' education, an employer must show at a minimum that the education possessed by the better-paid comparator relates to the skills actually used to perform the job's duties.† *See E.E.O.C. v. Cash & Go., Ltd.*, No. C-04-416, 2005 WL 1527663, at *7 (S.D. Tex.

---

* The Fifth Circuit has at least tacitly endorsed the proposition that "'compliance with civil service requirements is not sufficient in and of itself to establish a merit system defense under the [Equal Pay] Act.'" *Peters v. City of Shreveport*, 818 F.2d 1148, 1162 (5th Cir. 1987) (quoting *Maxwell v. City of Tuscon*, 803 F.2d 444, 447 (9th Cir. 1986)).

† The Court need not consider whether the Equal Pay Act's but-for causation standard requires the employer to show that the difference in education is proportionate to the difference in salary. *See Peters*, 818 F.2d at 1161.

June 28, 2005) (Ellington, Mag. J.), report & recommendation not adopted as moot Order 1, Dkt. No. 30 (July 28, 2005) (citing *Salazar v. Marathon Oil Co.*, 502 F. Supp. 631, 636 (S.D. Tex. 1980)) (denying summary judgment based on fact question of whether comparator hired to do marketing and stating "whether Davila's college degree is relevant depends on whether the nature of his duties required the use of his marketing skills"). As an example, the *Salazar* court rejected the argument that a difference between the educational attainment of two employees constituted a factor other than sex, stating "the crucial question is not if Mr. Clements possessed additional skill or training, but whether the nature of his duties as maintenance dispatcher required those additional skills." *Salazar*, F. Supp. at 636. Here, the only change other than removing preferred qualifications between the two job descriptions allowed substitution of an additional two years of experience and a two-year degree for a four-year degree. *Compare* Dkt. No. 18 Ex. 4 at 2 *with id.* Ex. 6 at 2. The salient question on the County's affirmative defense therefore is whether the difference between two additional years of experience and a four-year degree is related to the skills actually used to perform the GIS Engineering Specialist job.

8 The County offers no evidence or argument comparing the skills an employee acquires when obtaining a four-year degree to those acquired when earning a two-year degree and two years of additional experience or purporting to show that it considered the demand for employees with the two sets of qualifications to be different. *See* Mot. Summ. J. 6–7; *see also Reznick v. Assoc'd Orthopedics & Sports Med., P.A.*, 104 Fed.Appx. 387, 390–91 (5th Cir. 2004) (unpublished) (holding fact that two orthopedic surgeons who were in different subspecialties coupled with evidence that higher-paid doctor specialized in practice that garnered greater revenue demonstrated that jobs were not comparable and supported factor-other-than-sex defense); *Hofmister v. Miss. Dep't of Health*, 53 F. Supp. 2d 884, 894 (S.D. Miss. Mar. 9, 1999)) (holding education, background, and experience of registered nurses constituted factor other than sex where summary judgment evidence showed that employer believed it needed to increase pay to attract and retain registered nurses and the record included no evidence of pretext); *but see Sauceda*, 958 F. Supp. 2d at 779–80 (discussing limitation of market-forces defense when it would frustrate Congress's purpose in enacting the Equal Pay Act as discussed in *Siler-Khodr*, 261 F.3d at 546 and *Brennan v. City Stores, Inc.*, 479 F.2d 235, 241 n. 12 (5th Cir. 1973)). Nor does the County link the skills needed to perform the GIS Engineering Specialist job to its education requirements beyond labelling the change in salary a business decision. *See* Mot. Summ. J. 7. Decreasing a position's salary always benefits the employer, but the Equal Pay Act nonetheless forbids employers to take advantage of a superior bargaining position as to women performing equal work

---

* In support of its affirmative defense, the County relies on the Fifth Circuit's general pronouncement in *Hodgson v. Golden Isles Convalescent Homes, Inc.* that "Congress intended to permit employers wide discretion in evaluating work for pay purposes." 468 F.2d 1256, 1258 (5th Cir. 1972), *cited in* Mot. Summ. J. 6. The decision in *Golden Isles* has limited applicability here because the general statement on which the County relies was made not when considering an employer's affirmative defense but instead when deciding that the evidence supported the district court's factual finding that two groups of employees did not perform equal work. *See id.* at 1257–59 ("We decide only that the evidence supports the finding that the work done by the orderly at Golden Isles was sufficiently different from the work done by the aide to justify different pay scales.").

when that position results from "subjective assumptions and traditional stereotyped misconceptions regarding the value of women's work." *Sauceda*, 958 F. Supp. 2d at 779 (quoting *Shultz v. First Victoria Nat'l Bank*, 420 F.2d 648, 656 (5th Cir. 1969)); *see also id.* at 779–80 (quoting statement in *City Stores*, 479 F.2d at 241 n. 12, that "this is no excuse for hiring saleswomen and seamstresses at lesser rates simply because the market will bear it"). As the Court has already noted, the job description under which Barsch was hired in January 2014 states that experience and education may be substituted for one another. Dkt. No. 18 Ex. 6 at 2. Sullivan's memorandum to the County's Director of Human Resources states that all five of the State of Texas's GIS specialist job descriptions, which span a salary range encompassing the salary the County paid Parlamas, also allow substitution of experience for education. Dkt. No. 18 Ex. 3 at 1. A reasonable jury could view this evidence in the light most favorable to Barsch as establishing that the required education and experience with which Barsch and Parlamas qualified are interchangeable insofar as they relate to the skill required to perform the GIS Engineering Specialist job. Because that inference would in turn permit the reasonable conclusion that the County's stated reason for reclassifying the job is unrelated to the skill needed to perform the job, a fact issue precludes summary judgment on the County's factor-other-than-sex affirmative defense. *See Sauceda*, 958 F.Supp.2d at 778 (denying summary judgment on employer's affirmative defense where employer stated that salary differential resulted from comparators holding different degrees because record permitted reasonable finding that difference in education was unrelated to skill needed to perform teaching jobs); *Cash & Go*, 2005 WL 1527663, at *7 (denying summary judgment because of fact issues on relationship between education requirements for position and skills used to perform job); *Schulte*, 547 F. Supp. at 332–33 (relying in part on evidence that comparator's degree, which purportedly justified pay differential, was unrelated to job duties actually performed to reject affirmative defense as pretextual and noting that education and experience were subjectively used in discriminatory fashion to set salaries).

IV. TITLE VII

The parties agree that the familiar burden-shifting framework of *McDonnell Douglas Corp. v. Green* should be used to evaluate the County's motion for summary judgment. *See Goudeau v. Nat'l Oilwell Varco, L.P.*, 793 F.3d 470, 474 (5th Cir. 2015) ("Because there will seldom be eyewitness testimony as to the employer's mental processes, claims brought under these laws typically rely on circumstantial evidence that is evaluated under the burden-shifting framework first articulated in *McDonnell Douglas* for Title VII claims of employment discrimination." (citations and internal quotations omitted). In the *McDonnell Douglas* framework, the plaintiff "must first make a prima facie showing of employment discrimination." *Burton v. Freescale Semiconductor, Inc.*, 798 F.3d 222, 227 (5th Cir. 2015) (citing *E.E.O.C. v. Chevron Phillips Chem. Co., LP*, 570 F.3d 606, 615 (5th Cir. 2009)). The burden of production then shifts to the employer to "articulate a legitimate non-discriminatory reason for the adverse employment action." *Id.* (quoting *Chevron Phillips Chem.*, 570 F.3d at 615). At the last step, the burden returns to the employee who must "show the articulated reason is pretextual." *Id.* (citing *Chevron Phillips Chem.*, 570 F.3d at 615).

The County challenges the sufficiency of Barsch's evidence at each of the three *McDonnell Douglas* steps. "Generally, … a Title VII claim of wage discrimination parallels that of an EPA violation," *Siler-Khodr*, 261 F.3d at 546 (citing *Kovacevich v. Kent State Univ.*, 224 F.3d 806, 828 (6th Cir. 2000)), but "[t]he allocation of burdens in EPA and Title VII claims, however, differ in a way that may, in some cases, result in an employee prevailing on her [Equal Pay Act] claim but not her Title VII claim," *King*, 645 F.3d at 724. "[A] Title VII claim alleges 'individual, disparate treatment,'" *Siler-Khodr*, 231 F.3d at 545 (quoting *Plemer v. Parsons-Gilbane*, 713 F.2d 1127, 1135 (5th Cir. 1983)); the Equal Pay Act "has a higher threshold," *id.* at 546.

*9 Specifically, where a plaintiff makes an adequate prima facie case for both an EPA and a Title VII claim, the defendant bears the burden of persuasion to prove a defense under the EPA, whereas it has only a burden of production to show a legitimate nondiscriminatory reason for its actions under Title VII, with the ultimate burden of persuasion remaining with the plaintiff.

King, 645 F.3d at 724 (citing *Plemer*, 713 F.2d at 1136–37); *accord. Peters*, 818 F.2d at 1154–55 and n.3.

A. Prima Facie Case

In their briefing on the instant motion, the parties list the generalized elements of a prima facie case based on the denial of a job application: (1) she was a member of a protected class; (2) she was qualified for the position she was denied; (3) she suffered an adverse employment action; and (4) she was replaced by or treated differently than someone outside the protected class. McDonnell Douglas, 411 U.S. at 802, *cited in* Resp. Mot. Summ. J. 10, Dkt. No. 21; *see also* Mot. Summ. J. 8. Under the third element, the County argues, without citing authority, that Barsch cannot complain that she suffered an adverse employment action because her salary was increased when she was promoted in February 2014, albeit not to the level Parlamas was paid when he held her title. *See* Dkt. No. 18 Ex. 10. Also, the County maintains that Barsch cannot carry her burden on the fourth element because she and Parlamas are not similarly situated.

The parties misapprehend the elements of Barsch's prima facie case. The phrase "adverse employment action" is a "judicially-coined term referring to an employment decision that affects the terms and conditions of employment" as proscribed in Title VII. *Thompson v. City of Waco*, 764 F.3d 500, 503 (5th Cir. 2014) (citing *Pegram v. Honeywell, Inc.*, 361 F.3d 272, 281–82 (5th Cir. 2004)) (other citation omitted). As applicable here, Title VII forbids an employer, among other things, "to fail or refuse to hire or to discharge any individual, or otherwise to discriminate against any individual with respect to his *compensation*, terms, conditions, or privileges of employment, because of such individual's … sex …" 42 U.S.C. § 2000e-2(a)(1) (2012) (emphasis added). Accordingly, the Fifth Circuit has "held that adverse employment actions consist of 'ultimate employment decisions' such as hiring, firing, demoting, promoting, granting leave, and compensating." *Thomson*, 764 F.3d at 503 (citing *McCoy v. City of Shreveport*, 492 F.3d 551, 560 (5th Cir. 2007)) (other citations omitted). Conversely, "an employment action that 'does not affect job duties, compensation, or benefits' is not an adverse employment action." *Id.* (quoting *Pegram*, 361 F.3d at 282). After articulating the elements of the plaintiff's prima facie case in *McDonnell Douglas*, the Supreme Court "added ... that this standard is not inflexible, as '[t]he facts necessarily will vary in Title VII cases, and the specification above of the prima facie proof required from respondent is not necessarily applicable in every respect in differing factual situations.'" *Tex. Dep't of Cmty. Affairs v. Burdine*, 450 U.S. 248, 253 n.6 (1981) (quoting *McDonnell Douglas*, 411 U.S. at 802 n.13) (second alteration in original); *accord. Reed v. Neopost USA, Inc.*, 701 F.3d 434, 439 (5th Cir. 2012) ("[T]he precise elements of [the prima facie] showing will vary depending on the circumstances ...." (quoting *Mission Consol. Indep. Sch. Dist. v. Garcia*, 372 S.W.3d 629, 634 (Tex. 2012))); *Siler-Khodr*, 231 F.3d at 545–46 ("Because the facts in a particular Title VII case will differ, the evidence necessary to prove a prima facie case of discrimination under Title VII will vary." (citation omitted)); *LaPierre v. Benson Nissan, Inc.*, 86 F.3d 444, 448 n.3 (5th Cir. 1996); *Lavigne v. Cajun Deep Founds., LLC*, 86 F. Supp. 3d 524, 532–33 (M.D. La. 2015) (citation omitted). The Fifth Circuit has held that "[t]o establish a prima facie case of discrimination regarding compensation, a plaintiff must prove that (1) he is a member of a protected class and (2) he is paid less than a nonmember for work requiring substantially the same responsibility." *Runnels v. Tex. Children's Hosp. Select Plan*, 167 Fed.Appx. 377, 384 (5th Cir. 2006) (per curiam, unpublished) (citing *Uviedo v. Steves Sash& Door Co.*, 738 F.2d 1425, 1431 (5th Cir. 1984)); *accord. Taylor v. United Parcel Serv., Inc.*, 554 F.3d 510, 522 (5th Cir. 2008) (citing *Uviedo*, 738 F.2d at 1431).

*10 Regarding the elements of Barsch's prima facie case, the parties do not dispute that Barsch is a member of a protected class. *See* Dkt. No. 21 Ex. C at 2. Barsch seeks to compare her salary to the salary Parlamas earned when he held the GIS Engineering Specialist title in 2013. *Id.* at 4. The County's proposed analysis of the adverse-employment-action element ignores Parlamas's salary and instead compares Barsch's salary before her promotion in February 2014 to her salary after that promotion. *See* Mot. Summ. J. 10. The Fifth Circuit's holdings that the analysis of a Title VII plaintiffs prima facie case of pay discrimination "is the same even where the two employees whose salaries are being compared are employed at different times in the same position" allows Barsch to compare her salary to Parlamas's former salary at the prima facie stage. *Uviedo*, 738 F.2d at 1431 (citing *Pittman v. Hattiesburg Mun. Separate Sch. Dist.*, 644 F.2d 1071, 1074 (5th Cir. 1981), and *Bourque v. Powell Elec. Mfg. Co.*, 617 F.2d 61, 64 (5th Cir. 1980)) (affirming comparison of plaintiff's salary with that of employee she "succeeded" in personnel department).

Because there is no dispute that Parlamas is male, only the question whether Barsch and Parlamas performed work requiring substantially the same responsibility remains. *See Minnis v. Bd. of Supervisors of La. State Univ. & Agric. and Mech. Coll.*, 620 Fed.Appx. 215, 218 (5th Cir. 2015) (per curiam, unpublished) ("No one disputes that Minnis is a member of a protected class. Thus ... the only issue at the prima facie stage is whether Minnis was paid less than white employees for substantially the same job responsibilities."). Barsch's burden to make out a prima facie case under Title VII obliges her to "show that [her] circumstances are 'nearly identical' to those of a better-paid employee who is not a member of the protected class." *Taylor*,

554 F.3d at 523 (citing *Little v. Republic Ref. Co.*, 924 F.2d 93, 97 (5th Cir. 1991)). Compared to the Equal Pay Act, "Title VII incorporates a more relaxed standard of similarity between male and female-occupied jobs." *Wojciechowski*, 763 F. Supp. 2d at 854 n.5 (quoting *Miranda v. B& B Cash Grocery Store, Inc.*, 975 F.2d 1518, 1529 n.11 (11th Cir. 1992)); *Lenihan v. Boeing Co.*, 994 F. Supp. 776, 799 (S.D. Tex. 1998) ("[U]nder Title VII, the 'plaintiff is not required to meet the exacting standard of substantial equality of positions set forth in the Equal Pay Act.' " (quoting *Miranda*, supra, 975 F.2d at 1529 n.11)). Therefore, the conclusion, supra, that genuine fact issues exist on the substantially similar element of Barsch's prima facie case under the Equal Pay Act precludes summary judgment on the substantially similar element of Barsch's prima facie case under Title VII. *See Plemer*, 713 F.2d at 1137 (analyzing prima facie case under the Equal Pay Act and then concluding the "evidence also sufficed to make out a prima facie compensation case under Title VII" without further analysis); *Wojciechowski*, 763 F. Supp. 2d at 854 (denying summary judgment on prima facie Title VII case for same reason summary judgment denied on Equal Pay Act claim); *Lenihan*, 994 F.Supp. at 799 (analyzing prima facie Title VII and Equal Pay Act elements together).

B. Employer's Non-Discriminatory Reason

As its reason for paying Barsch less than Parlamas, the County states that "the only relevant evidence shows that even though Plaintiff is being paid less than her predecessor as a GIS Engineering Specialist, the reevaluation and reclassification of the position occurred before the position was posted, before she applied, and before she was hired ...." Mot. Summ. J. 10 (citing *id.* Ex. 3 and 6). The County's summary judgment evidence shows that the GIS Engineering Specialist position's pay grade was reclassified after a review of the GIS Specialist job descriptions promulgated by the State of Texas and the County's elimination of the mandatory requirement of a four-year degree. *See* Sullivan Aff. 4; *id.* Ex. 3. According to Sullivan, the change in pay grade was "to be consistent with similar positions and qualifications across the State." Sullivan Aff. 4. The Fifth Circuit has held in at least one unpublished case that an employer's calculation of the market rate at the time of hiring constitutes a legitimate, non-discriminatory reason for a pay disparity under Title VII, and the County has accordingly met its burden at the second *McDonnell Douglas* step. *See Minnis*, 620 Fed.Appx. at 219 (holding employer's statement that the "[plaintiff]'s salary was calculated based on the market for the position" carried employer's burden at second *McDonnell Douglas* step).

C. Pretext

*11 At the third *McDonnell Douglas* step, the "plaintiff may attempt to establish that he was the victim of intentional discrimination 'by showing that the employer's proffered explanation is unworthy of credence.'" *Goudeau*, 793 F.3d at 476 (quoting *Reeves v. Sanderson Plumbing Prods., Inc.*, 530 U.S. 133, 143 (2000)). The employer's articulated reason for its action need not have been correct; instead, "[t]he issue at the pretext stage is whether [the employer's] reason, even if incorrect, was the real reason for" the employer's action. *Id.* (quoting *Sandstad v. CB Richard Ellis, Inc.*, 309 F.3d 893, 899 (5th Cir. 2002)) (alterations in original).

Barsch has carried her burden to show that genuine fact issues exist at the pretext stage on her Title VII claim. In this regard, "a plaintiffs prima facie case, combined with sufficient evidence to find that the employer's asserted justification is false, may permit the trier of fact to conclude that the employer unlawfully discriminated." *Id.* (quoting *Reeves*, 530 U.S. at 147–48) (first alteration omitted). Regarding pretext, Barsch emphasizes the fact that the required duties and responsibilities did not change and that the second job description states that seven years of experience and a two-year degree may be substituted for five years of experience and a four-year degree. *Compare* Dkt. No. 18 Ex. 4 *with id.* Ex. 6. In addition, Sullivan's affidavit and his memorandum to Guerra can be reasonably viewed as inconsistent with one another on what salary factors were sought to be equalized and the role of stated qualifying education and experience in the salary realignment. *See Evans v. City of Houston*, 246 F.3d 344, 355 (5th Cir. 2001) (noting that evaluation of plaintiff in letter in record stood "in stark contrast" to later-dated memorandum supporting employer's articulated reason for firing employee). As noted, Sullivan's affidavit states that he believed the new pay grade was "consistent with similar positions and qualifications across the state." Sullivan Aff. 4. Sullivan's memorandum to Guerra, however, states that the qualifications for all five state GIS Specialist positions are identical and allow for the substitution of experience for education. Dkt. No. 18 Ex. 3 at 1. Additionally, Sullivan does not mention equalizing internal salaries in his affidavit, Sullivan Aff. 4, but he compares the GIS Engineering Specialist salary to the salary levels of other job titles in the public works department in his memorandum dated October 29, 2013, *see id.* Ex. 3 at 1 ("In looking at the salary ranges for the current County positions of Engineering Technician (GP 34), GIS Engineering Specialist (GP 32), and GIS Data Technician (GP 15) we agree that the salary group for GIS Engineering Specialist could be changed to better reflect the pay ranges used by both the State and this Department.").

The difference in the reasons given by Sullivan in his affidavit and his memorandum dated October 29, 2013, sufficiently raises an issue of material fact about the County's articulated reason, considering the competent summary judgment evidence which support a finding that Sullivan revised the job description with Barsch in mind. Unlike the record that supported summary judgment for the employer in *Stewart v. Work Connection (JTPA)*, Civ. A. No. 90-2239, 1991 WL 111019, at *7 (E.D. La. June 10, 1991), the summary judgment record here permits the reasonable inference that the GIS Engineering Specialist job description was revised with Barsch in mind. Undisputedly, the County revised the GIS Engineering Specialist job description knowing that Parlamas was being promoted, *see* Sullivan Aff. 3. The only evidence in the record shows that the County advertised the GIS Engineering Specialist position internally. *See* Dkt. No. 18 Ex. 5 at 1; *id.* Ex. 7 at 1. While the record does not show how many people Sullivan supervises, it includes evidence that he considered the salary levels of public works department employees with GIS-related qualifications when revising the pay grade. *See* Dkt. No. 18 Ex. 3 at 1 (listing positions consulted). If the inference that Sullivan knows his employees' educational background and relevant experience is drawn, a jury could reasonably conclude on this record that he revised the required education and experience in the GIS Engineering Specialist job description with Barsch in mind. *See Stewart*, 1991 WL 111019, at *7 (granting employer summary judgment at pretext stage because plaintiff did not create fact issue on employer's evidence that it revised job description to add education requirements before it knew incumbent would resign or that plaintiff, who did not possess degree, would apply). Finally, nothing in the record suggests that Sullivan, or anyone else, interviewed either applicant before promoting Barsch—something the jury could reasonably view as unusual and indicating that Barsch's promotion was a foregone conclusion in the eyes of management. *See Evans*, 246 F.3d at 355–56 (holding evidence of "four dates on one demotion memorandum and two dates on the reassignment form is rather peculiar" and that peculiarity supported inference of pretextual backdating that was sufficient along with other evidence to create fact issue for trial). Viewing all of the foregoing evidence together and in the light most favorable to Barsch, a reasonable jury could therefore conclude that the County's articulated reason for reclassifying Barsch's salary is unworthy of credence. *See id.*

V. CONCLUSION

*12 For the foregoing reasons, the Court DENIES the County's motion for summary judgment, Dkt. No. 18.

It is so ORDERED.

SIGNED this 14 day of January, 2016.

ALL CITATIONS

Not Reported in F.Supp.3d, 2016 WL 4785169

4 Employee Motivation and Progressive Discipline

******

Chapter Objectives:

1. Acquire an understanding of what motivates employees to work safely.
2. Appraise the positive reinforcement elements of progressive discipline.
3. Appraise the negative reinforcement elements of progressive discipline.
4. Understand the need for progressive discipline within the safety function.

******

One of the foundational questions which all safety and health professionals ask is "How do I motivate ALL employees to work safely ALL of the time?" Safety and health professionals establish programs, analyze trends and data, train and educate employees, and ensure all equipment and operations are safe; however, some employees still find a way to injure themselves on the job. Is it a lack of focus? Baggage from outside of the workplace? Inattention to the job activities? Broken equipment? The work functions themselves? Safety and health professionals have tried a multitude of different programs and techniques; however, the myriad of variables which cannot be controlled by the safety and health professional often leads to minimal or no change in the workplace.

Since the inception of the Occupational Safety and Health Act (hereinafter OSHA) in 1970, achieving and maintaining compliance with the proscribed safety standards has been a mainstay of most safety and health professionals' portfolios. However, compliance with the OSHA standards is the minimum level under which a safety and health professional cannot permit his/her program to fall, without the potential of penalty. However, meeting or even exceeding compliance with the OSHA standards does not produce a safety and health program which can eliminates or minimizes the potential of workplace injuries and illnesses. The OSHA standards are the foundation upon which safety and health professionals must build in order to achieve the results in reduction of workplace injuries and illnesses, as well as the related costs.

Another of the foundational questions for safety and health professionals is "Who owns the safety and health program?" Prior to 1970 and the enactment of the OSHA, on-the-job safety programs were virtually non-existent at many companies, and employees tended to fend for themselves and their crews when it came to safety and health. With the OSHA, the federal government (and state plan states) became actively involved in the regulation of safety and health in the workplace and, to a great part, the position of "safety and health professional" was born. In the early years of the safety and health profession, the professional was often the "safety cop," ensuring compliance with the standards being virtually the entire job responsibility. Throughout the past 40+ years since the enactment of the OSHA, the focus and responsibilities of the safety and health professional have vastly expanded and the focus for safety and health professionals has morphed and multiplied from basic compliance to a monetary emphasis resulting from increased workers' compensation and related insurance, operational, turnover, and correlating cost increases. The job duties and responsibilities of today's safety and health professional still includes compliance with the OSHA standards; however, this responsibility has become a far smaller percentage of the safety and health professional's overall job duties and responsibilities. Although compliance is essential, safety and health professionals focus far more of their time and efforts in the areas of injury/illness reduction, employee training and education, cost controls, trend and data analysis, and related areas.

So, is the safety and health professional solely responsible for the safety and health performance within the operations? Is the safety and health function the responsibility of all employees as part of their job duties? With the myriad of potential variables impacting individual employee's safety and health performance, as well as production, quality, and other workplace stressors, can and should the safety and health professional be held solely responsible for safety and health performance? Today's safety and health professional's responsibilities and duties should be more aligned with that of a "coach" or a "leader" who designs, directs, manages, motivates, directs, and controls the overall safety and health function for the organization, with managerial team members and employees possessing an ownership interest in the success or failure of the safety and health program.

From the chief executive officer to the newest hired employee of the company, safety and health should be a component of each employee's job function. Just as shareholders own a piece of publicly held companies, each and every employee should have an ownership interest in the safety and health program. The safety and health professional should not have sole responsibility for the safety and health function and most definitely should not be tasked with being the "safety police" to ensure employees are wearing their personal protective equipment (PPE) and performing their jobs in a safe and healthful manner. Although different job functions may have different levels of responsibility for the safety and health function, all owners should play an active role making safety and health a priority on a daily basis.

In general, every level of management, as well as hourly personnel, has their role within the safety and health function; however, the motivational factors may be significantly different. The executive level of most companies is motivated by financial aspect and performance. The mid or supervisory level is often motivated by financial aspects, but also by time. The hourly or non-exempt level employees are often motivated by factors outside of the workplace (such as family), as well as internal recognition and relationships (e.g., award for welding team). Safety and health professionals should be cognizant of different motivators for different levels, and among and between levels, when designing as well as leading safety and health activities.

Employees must trust their safety and health professional to "do the right thing." Conversely, the safety and health professional must trust his/her employees to make sound safety and health judgements. Employees should be empowered to take responsibility for their own safety and health, as well as the safety and health of others within the company. From the CEO to the new hire, the safety and health professional should train, educate, and prepare each employee, thus providing the "tools" to successfully perform their responsibilities as an owner of the safety and health function. However, in addition to providing the education and guidance, the safety and health professional must also hold each employee accountable for his/her safety and health performance. With empowerment comes responsibility. With ownership comes duties. With leadership comes trust.

Motivation can be positive or negative. Most studies identify that positive reinforcement is generally more successful than negative reinforcement. However, in the safety and health arena, both positive and negative reinforcement is needed. From a positive reinforcement prospective, the gambit runs from telling an employee he/she did a good job to sophisticated safety performance incentive programs. However, from a performance and legal perspective, most, if not all companies, utilize varying forms of a progressive disciplinary system. In general, a progressive disciplinary system is a policy wherein the levels of disciplinary action increase in severity up to the final step of involuntary termination from the company. Although there are varying forms and levels, generally the initial steps involve coaching and counseling of the employee, followed by a verbal warning (which is documented in writing), a written warning, suspension without pay from work, and involuntary termination. This type of progressive discipline policy is usually broad in nature, encompassing all areas from absenteeism through safety violations, with certain offenses resulting in immediate suspension or involuntary termination.

EXAMPLE OF SAFETY-RELATED PROGRESSIVE DISCIPLINE

Failure to Wear Required PPE:

First Offense – Coaching by immediate supervisor on the importance of properly wearing PPE.

(Positive reinforcement and education of the employee regarding the need and importance of properly wearing his/her PPE).

Second Offense – Counseling by the Safety Director.

(Positive reinforcement and further education. Any further violation will impact employee's continued employment with the company).

Third Offense – Verbal warning in the employee's personnel file.

Fourth Offense – Formal written warning in the employee's personnel file.

Fifth Offense – Three (3) day suspension from work WITHOUT PAY.

Sixth Offense – Involuntary termination from employment.
Note: With most company policies, the time period for accumulated offenses is one year.

As noted previously in this chapter, employee turnover is very expensive for both the employee as well as the company. A progressive disciplinary policy is designed to provide the employee multiple chances to correct his/her inappropriate or unsafe behavior in the workplace. However, in certain extreme circumstances, the discipline provided for the offense can be "up to and including discharge." Within the safety and health function, these extreme circumstances resulting in immediate suspension or involuntary termination from employment involve circumstances where the employee's actions or inactions could or did result in a severe injury or even a workplace fatality. As an example, under many Control of Hazardous Energy programs (also known as "Lockout and Tagout" or "LOTO"), special emphasis is placed on the removal of a protective lock which prevents re-energization of equipment. Under many policies, if an employee cuts a secure lock from the equipment and energizes the equipment, disciplinary action may begin at the suspension level or result in immediate involuntary termination.

Safety and health professionals should also be aware that the documentation generated through the progressive disciplinary system can be utilized as evidence, not only in legal proceedings, such as unemployment hearings, but also as evidence of the enforcement of safety and health compliance program requirements, as well as in some defenses to OSHA violations. Prudent safety and health professionals should always consider including a reference to the company's progressive disciplinary policy within their compliance programs (or separate safety-specific progressive disciplinary policy) in order to prove to the compliance officer during the site visit that safety and health related requirements are being enforced. Additionally, in the event of an OSHA violation, if the facts permit, the use of "Lack of Employee Knowledge," "Employee was Experienced," or "Isolated Incident" defenses, proof of enforcement can be exhibited through the employee's written disciplinary documents, as well as training and education documents.

Motivating employees to work safely each and every day is a daunting task. Employees bring with them to work each day "baggage" in the form of personal, relationship, and other issues which can impact their focus on the job. Internal factors, such as a rumor of a layoff or a supervisor increasing the production speed, can also allow the employee to lose focus on their safety responsibilities. It is the duty and responsibility of the safety and health professional, as the leader of the safety and health program, to ensure all employees are educated and "provided with the tools" to work safely, be empowered to address issues which can impact the safety of the environment and fellow employees, and are motivated to create and maintain a safe and healthful work environment each and every day. This can be a delicate balancing act with unforeseen and uncontrolled factors impacting the equilibrium.

Case Study: This case may have been modified for the purposes of this text.

2010 WL 199268
United States District Court,
E.D. Pennsylvania.

CONOCOPHILLIPS CO., Plaintiff
v.
UNITED STEELWORKERS, LOCAL, Defendant.
Civil Action No. 09–3842.
|
Jan. 19, 2010.

West Key Summary

1   Labor and Employment⊷Discipline
Labor and Employment⊷Determination

Arbitrator was not authorized to make a particular progressive discipline structure of his own creation a part of the collective bargaining agreement (CBA) he was interpreting. Although the arbitrator appropriately determined that the CBA's "just cause" language included progressive discipline, it was possible to read the arbitrator's opinion as improperly incorporating into the CBA a five-step progressive discipline structure. To eliminate any uncertainty or ambiguity, the court modified the arbitrator's opinion to clarify the non-binding nature of his discussion of the specific progressive discipline regime.

Cases that cite this headnote

MEMORANDUM

PAUL S. DIAMOND, District Judge.

*1 In this labor dispute, Plaintiff ConocoPhillips has moved for summary judgment, asking me to vacate an arbitrator's decision to reduce the level of discipline Conoco had imposed on one of its employees. Defendant, United Steelworkers, Local 10–234, has cross-moved for summary judgment, asking me to uphold the arbitrator's decision. Although I agree with the Union that the arbitrator could properly reduce the level of discipline, I also agree with the Company that in making his decision, the arbitrator was not entitled to rewrite the Parties' Collective Bargaining Agreement. Accordingly, I will modify the Arbitrator's Opinion.

I. JURISDICTION

I have jurisdiction under Section 301(a) of the Labor Management Relations Act to vacate or to modify an arbitration award. 29 U.S.C. § 185; *see Newark Morning Ledger Co. v. Newark Typographical Union, Local 103*, 797 F.2d 162, 167 (3d Cir.1986) (agreeing with the district court's decision that it was "compelled to modify the award").

II. LEGAL STANDARDS

A. Summary Judgment

Upon motion of any party, summary judgment is appropriate "if there is no genuine issue as to any material fact and the moving party is entitled to judgment as a matter of law." Fed.R.Civ.P. 56(c). The moving party must initially show the absence of any genuine issues of material fact. *See Celotex Corp. v. Catrett*, 477 U.S. 317, 106 S.Ct. 2548, 91 L.Ed.2d 265 (1986). An issue is material only if it could affect the result of the suit under governing law. *See Anderson v. Liberty Lobby, Inc.*, 477 U.S. 242, 106 S.Ct. 2505, 91 L.Ed.2d 202 (1986).

In deciding whether to grant summary judgment, the district court "must view the facts in the light most favorable to the non-moving party," and take every reasonable inference in that party's favor. *Hugh v. Butler County Family YMCA*, 418 F.3d 265 (3d Cir.2005). If, after viewing all reasonable inferences in favor of the non-moving party, "the record taken as a whole could not lead a rational trier of fact to find for the non-moving party, there is no genuine issue for trial," and summary judgment is appropriate. *Matsushita Elec. Indus. Co. v. Zenith Radio Corp.*, 475 U.S. 574, 586, 106 S.Ct. 1348, 89 L.Ed.2d 538 (1986); *Delande v. ING Employee Benefits*, 112 Fed. Appx. 199, 200 (3d Cir.2004). Granting summary judgment thus "avoid[s] a point-less trial in cases where it is unnecessary and would only cause delay and expense." *Walden v. Saint Gobain Corp.*, 323 F.Supp.2d 637, 641 (E.D.Pa.2004) (quoting *Goodman v. Mead Johnson & Co.*, 534 F.2d 566, 573 (3d Cir.1976)).

B. Judicial Review Under the LMRA
District courts have a limited and deferential role in reviewing arbitration awards arising from labor disputes. *Pa. Power Co. v. Local Union No. 272, Int'l Bhd. of Elec. Workers, AFL–CIO*, 276 F.3d 174, 178 (3d Cir.2001) (citing *United Paperworkers Int'l Union, AFL–CIO v. Misco, Inc.*, 484 U.S. 29, 36, 108 S.Ct. 364, 98 L.Ed.2d 286 (1987)). The district court must uphold an arbitration award that "draws its essence" from the parties' collective bargaining agreement "and is not merely [the arbitrator's] own brand of industrial justice." *Misco*, 484 U.S. at 36.

*2 An award draws its essence from a collective bargaining agreement "if its interpretation can in any rational way be derived from the agreement, viewed in light of its language, its context, and any other indicia of the parties' intention." *United Transp. Union Local 1589 v. Suburban Transit Corp.*, 51 F.3d 376, 379–80 (3d Cir.1995). *See also Exxon Shipping Co. v. Exxon Seamen's Union*, 993 F.2d 357, 360 (3d Cir.1993) ("As a general rule, we must enforce an arbitration award if it was based on an arguable interpretation and/or application of the collective bargaining agreement, and may only vacate if there is no support in the record for its determination or if it reflects manifest disregard of the agreement, totally unsupported by principles of contract construction.") Thus, I must uphold an arbitration award that was derived from the collective bargaining agreement, even if I disagree with the arbitrator's conclusions. *Brentwood Med. Assoc. v. United Mine Workers of America*, 396 F.3d 237, 241 (3d Cir.2005).

I am bound by the arbitrator's factual findings, provided that they have evidentiary support. *See Misco*, 484 U.S. at 36 ("[I]t is the arbitrator's view of the facts and of the meaning of the contract that [the Parties] have agreed to accept. Courts thus do not sit to hear claims of factual or legal error by an arbitrator as an appellate court does in reviewing decisions of lower courts.").

The Third Circuit has cautioned, however, that the district court is neither "entitled nor encouraged simply to 'rubber stamp' the interpretations and decisions of arbitrators." *Matteson v. Ryder Sys., Inc.*, 99 F.3d 108, 113 (3d Cir.1996) (citations omitted). The arbitrator's authority to resolve a dispute is bound by the terms of the parties' collective bargaining agreement. *See Barrentine v. Arkansas–Best Freight Sys., Inc.*, 450 U.S. 728, 744, 101 S.Ct. 1437, 67 L.Ed.2d 641 (1981) ("An arbitrator's power is both derived from, and limited by, the collective-bargaining agreement.")

Section 301 of the LMRA confers on the district court the authority to modify an arbitration award that is unfaithful to the collective bargaining agreement. *Newark Morning Ledger,* 797 F.2d at 167. The Federal Arbitration Act also confers on the court authority to modify a labor arbitration award where, as here, the arbitrator "awarded on a matter not submitted to [him]." 9 U.S.C. § 11(b); *see Misco,* 484 U.S. at 41 n. 9 (acknowledging "federal courts have often looked to the [FAA] for guidance in labor arbitration cases," particularly because Section 301(a) "empowers the federal courts to fashion rules of federal common law to govern [s]uits for violation of contracts between an employer and a labor organization" under the federal labor laws"). Under the FAA, any modification is entirely discretionary and should "promote justice between the parties." *Andorra Servs. v. Venfleet, Ltd.,* 2009 U. S.App. LEXIS 26969, 2009 WL 4691635 (3d Cir.2009).

In sum, although my review of an arbitrator's decision must be deferential, I may vacate or modify the decision if I determine that he exceeded his authority under the parties' collective bargaining agreement. *See Matteson,* 99 F.3d at 113–14; *Newark Morning Ledger,* 797 F.2d at 167.

III. BACKGROUND

*3 ConocoPhillips is a Delaware corporation with its headquarters in Houston, Texas. *(Compl. & Answer 9.)* The Company operates an oil refinery in Trainer, Pennsylvania. *(Id.)* United Steelworkers, Local 10–234 is the exclusive collective bargaining representative for the employees' bargaining unit at the Trainer refinery. *(Compl. & Answer 11.)* At all relevant times, relations between the Parties were governed by a Collective Bargaining Agreement (CBA) effective since February 1, 2002. *(Doc. No. 13.)*

A. Mr. Bishop's Employment History

The Company hired Richard Bishop in September 2004 as a "B Pumper Operator" at the Trainer refinery. *(Compl. & Answer, 23, 24.)* After a six-month training program comprising basic safety classes and a series of written and field tests, Bishop was qualified as a "B Pumper Operator." *(Id. 26.)*

On November 9, 2005, Bishop issued safe work permits without reviewing the accompanying paperwork, necessitating a work stoppage until the permits were properly completed. *(Compl. & Answer, 27, 28.)* He received a verbal warning for this work rules violation. *(Id. 29.)*

In a subsequent disciplinary proceeding, Conoco determined that Bishop also violated work rules on October 26, 2005 by opening the wrong valve for a shipment of butane, creating a safety risk. *(Compl. 31–33.)* With evidentiary support, the Arbitrator found as a fact that Bishop and the Union did not receive the written warning issued to Bishop for this second work rules violation. *(Compl. Ex. B, Arb. Opinion and Award at 8.)*

On September 4, 2007, Bishop committed a third work rules infraction that resulted in the discipline challenged in the Arbitration. As found by the Arbitrator, Bishop failed to follow the proper three-part procedure for sampling an oil tank for certification. *(Compl. Ex. B, Arb. Opinion and Award at 8.)* When he met with his supervisor to discuss the incident, he did not deny taking the samples in violation of the prescribed procedure. *(Compl. 39.)* As a result of this failure to follow work rules, on September 18, 2007, the Company suspended Bishop for two days. *(Compl. 40.)* On September 21, 2007, the Union filed a grievance on behalf of Bishop, urging that the suspension was "too harsh." *(Compl. & Answer, 41.)*

B. The Parties' Collective Bargaining Agreement

CBA Article 24 (entitled "Discipline") provides that Conoco may discipline employees for "just cause." *(Id., Art.24.1.)* "Just cause" is not defined; rather, Article 24 states that "[c]ommitting a posted offense, failing to obey working rules, or unsatisfactory work performance may be cause for discipline." *(Id. Art. 24.3.)* The CBA requires Conoco to "issue to the Union a copy of any disciplinary letter placed in an employer's file." *(Id. Art. 24.4)* The CBA does not provide for or include any mention of "progressive discipline."

*4 Article 25 sets out the "Grievance and Arbitration Procedure": a four-part process for hearing employee complaints that Conoco violated the CBA. *(Id. Art. 25.6.)* After three informal hearings, the dispute is submitted to arbitration before a member of the National Academy of Arbitrators. *(Id. Art. 25.8.)* Significantly, Article 25.8 provides that "[i]n reaching a decision, the arbitrator shall be restricted to *the specific terms and provisions of this agreement, and shall not add to, subtract from, or in any way alter any of the provisions of the Agreement." (Id.)* (emphasis supplied).

C. The Arbitration Award and the Instant Litigation

On May 7, 2009, Arbitrator Robert E. Light conducted a hearing. *(Compl. & Answer, 43.)* The Parties stipulated that Mr. Light should decide the following questions: "Was there just cause for the two (2) day suspension imposed on Richard Bishop? *If not, what shall be the remedy?" (Id. 44)* (emphasis supplied).

Mr. Light issued his Arbitration Opinion and Award on July 22, 2009. *(Id. 46.)* He found that Bishop "violated the written and required operating procedures for obtaining certification samples," and that "[h]e knew what those procedures were and he simply did not follow them, offering no legitimate excuse for not doing so." *(Compl. Ex. B, Arb. Opinion and Award at 8.)* In his Opinion, the Arbitrator stated that

> while argument was made to the contrary by the Company, it is clear to this Arbitrator, that this Company follows the dictates of progressive discipline as respects discipline meted out to employees. That is to say, as was the case here, normally the sequence is an oral warning, a written warning, a penalty of some duration, perhaps a penalty of longer duration and subsequently, if no improvement is made, then termination.
>
> *(Id.)*

Once again, Mr. Light found that Bishop received a verbal warning for the November 5, 2005 violation. Having also found that Bishop never received a written warning for the October 26, 2005 offense, Mr. Light ruled that the "two (2) day disciplinary suspension [for the 2007 offense] was not the next 'building block' in the progressive disciplinary scheme." *(Id. at 9.)* Thus, the Arbitrator issued the following Award:

> There was not just cause for the two-day disciplinary suspension imposed on Richard Bishop. There was just cause for a written warning for the incident which occurred on September 4, 2007. Mr. Bishop shall receive two (2) days' pay, but his record shall reflect the fact that he received a disciplinary written warning for the incident on September 4, 2007.
>
> ***(Id. at 10.)***

On August 21, 2009, Conoco filed its Complaint seeking to vacate the Arbitrator's Opinion and Award. Both sides have cross-moved for summary judgment. (Doc. Nos.12, 14.)

IV. DISCUSSION

A. The Relief the Company Seeks

Conoco asks me to vacate Mr. Light's "Award," arguing that he did not have authority under the CBA to reduce Mr. Bishop's discipline. Yet, at the arbitration's outset, Conoco and the Union jointly asked Mr. Light to determine whether or not to reduce Mr. Bishop's two-day suspension ("If not, what shall be the remedy?"). *(Compl. & Answer, 43.)* The Company thus acknowledged that Mr. Light had the authority to make this determination.

*5 In its Complaint, the Company alleged that the Arbitrator improperly "concluded that ConocoPhillips was obliged to adhere to a five-part progressive discipline structure created by the Arbitrator himself …. The parties, however, never agreed to any such progressive discipline system." *(Compl. 50, 52.)* In asking me to vacate, Conoco argues as follows:

> Here, Arbitrator Light went well beyond interpreting the [CBA] to include some general term of progressive discipline. Instead, he prescribed a specific, five-step system …. The mere fact that progressive discipline schemes similar to the five-step system espoused by Arbitrator Light have been recognized [elsewhere] does not mean that [the CBA] … includes the imposition of a five-part progressive discipline scheme.
>
> ***(Doc. No. 15 at 8–9.)***

It is apparent that the Company's true objection here is not to the Award itself—reducing Bishop's discipline to a written warning—but to the Arbitrator's means of reaching that decision: by suggesting in his Opinion that the CBA includes the five-step structure of progressive discipline "espoused by Arbitrator Light." *(Id.)* Accordingly, I will construe Conoco's Complaint as challenging that part of the Arbitrator's Opinion in which he conceives and implements the five-step structure. *Pantry Pride, Inc. v. Retail Clerks Tri-State Pension Fund,* 747 F.2d 169,

171 (3d Cir.1984) (district courts may liberally construe complaints). This is, in effect, a request that I modify the Arbitrator's decision.

B. Employee Discipline Under the CBA

There is tension between two CBA provisions. Article 25.8 provides that the arbitrator is "restricted to the specific terms and provisions of this agreement, and shall not add to, subtract from, or in any way alter any of the provisions of the Agreement." *(CBA Art. 25.8.)* The Third Circuit has called for strict enforcement of such provisions, and has vacated awards made in contravention of those provisions. *See*, e.g., *Pa. Power Co. v. Local Union No. 272 of the Int'l Bhd. of Elec. Workers, AFL–CIO*, 886 F.3d 46, 48–49 (3d Cir.1989) ("[T]he arbitrator exceeds her authority if she deems arbitrable those issues whose resolution calls for the addition of new terms or conditions to the agreement.")

Labor arbitrators have authority, however, under "just cause" provisions like Article 24.1 to reduce discipline they believe is too harsh. *Exxon Shipping Co. v. Exxon Seaman's Union*, 73 F.3d 1287, 1297 (3d Cir.1996). Moreover, the Third Circuit has held that an arbitrator may construe an undefined "just cause" provision—such as Article 24.1—"to include a progressive discipline requirement." *Arco–Polymers, Inc., v. Local 8–74*, 671 F.2d 752, 756 (3d Cir.1982). *See also Suburban Transit Corp.*, 51 F.3d at 381 ("To the extent the arbitrator's award was based upon a theory that the parties intended 'proper cause' to incorporate some form of progressive discipline, that interpretation has some basis in the CBA."). In both decisions, the Court held that when a collective bargaining agreement does not include a definition of "just cause," the parties "bargained for contractual ambiguity." *Id.* at 380.

*6 In construing Articles 24.1 and 25.8, an arbitrator may not read either provision out of the CBA. *Barrentine*, 450 U.S. at 744. Rather, the arbitrator must seek to reconcile these provisions in a manner consistent with the Parties' intent. *Tanoma Min. Co., Inc. v. Local Union No. 1269, United Mine Workers*, 896 F.2d 745, 747–48 (3d Cir.1990).

The *Suburban Transit* Court—reviewing a collective bargaining agreement similar to that presented here—held that the provision limiting the arbitrator's authority did not preclude the arbitrator from construing the "just cause" provision as including progressive discipline. *See Suburban Transit*, 51 F.3d at 381. Accordingly, like the arbitrator in *Suburban Transit*, Mr. Light appropriately determined that Article 24.1's "just cause" language included progressive discipline. Unlike the *Suburban Transit* arbitrator, however, Mr. Light went on to conceive and set out a five-step program of progressive discipline. *See Compl. Ex. B, Arb. Opinion and Award at 8–9* (the two-day suspension "was not the next 'building block' in the progressive disciplinary scheme," which should consist of "an oral warning, a written warning, a penalty of some duration, perhaps a penalty of longer duration and subsequently, if no improvement is made, then termination.")

No court has allowed a labor arbitrator to re-write the collective bargaining agreement he is charged with interpreting. In *Pennsylvania Power Company*, when the labor arbitrator disregarded a provision similar to Article 25.8 and thus amended the Parties' collective bargaining agreement to extend benefits to a class of employees not previously covered, the Third Circuit reversed, holding that the arbitrator "altered the Agreement in direct violation of its provision that he had no power to do so." *Pa. Power Co.*, 276 F.3d at 179. *See also United Steelworkers of America v. Warrior & Gulf Navigation Co.*, 363 U.S. 574, 80 S.Ct. 1347, 4 L.Ed.2d 1409 (1960) ("The arbitrator created a limitation on [the company's] management rights that is not found in the Agreement. In so doing, the arbitrator's decision failed to draw its essence from the Agreement."); *Appalachian Regional Health Care, Inc. v. United Steelworkers of America, AFL–CIO, Local 14398*, 245 F.3d 601, 605 (6th Cir.2001), *cert. denied*, 534 U.S. 952, 122 S.Ct. 350, 151 L.Ed.2d 264 (2001) ("Even if we were to credit the arbitrator's construction of the Agreement ..., we would still have to vacate the award as it imports notions not found in the Agreement itself.")

Had Mr. Light re-written the CBA to include the five-step progressive discipline structure he favors, this would have been an impermissible attempt to afford the Parties Mr. Light's "own brand of industrial justice." *Misco*, 484 U.S. at 36. Whether the CBA should include a three, five, or ten-step structure of progressive discipline— and the definition of each step—should be determined by collective bargaining, not by arbitrator fiat. It does not appear, however, that Mr. Light intended to re-write the CBA or to make the five-step structure a part of the CBA. Rather, the language Mr. Light employed ("*normally* progressive discipline would include ... a penalty of some duration, *perhaps* a penalty of longer duration ...") suggests that he intended his five-step structure to be merely illustrative. *(Compl. Ex. B, Arb. Opinion and Award at 8.)* Indeed, the Union describes that part of the Opinion as "dicta." *(Doc. No. 12 at 7.)* It is, nevertheless, possible to read Mr. Light's Opinion as incorporating into the CBA the five-step progressive discipline structure favored by Mr. Light. To eliminate any uncertainty or ambiguity in this respect, I will modify Mr. Light's Opinion to clarify the non-binding nature of his discussion of a specific progressive discipline regime.

V. CONCLUSION

*7 Mr. Light had the authority to conclude that Article 24.1's undefined "just cause" provision includes the concept of progressive discipline. Moreover, in light of the Parties' stipulation that the Arbitrator determine if the two-day suspension should be reduced, and Mr. Light's factual finding that Bishop had not previously received a written warning, Mr. Light also had the authority to reduce the two-day suspension imposed on Bishop to a written warning. Nothing in the CBA or the law, however, authorized Mr. Light to make a particular progressive discipline structure of his own creation a part of the CBA. Accordingly, I will confirm Mr. Light's Award and modify his Opinion to make clear that the five-step progressive discipline structure he describes is merely illustrative and is not required by or part of the CBA.

An appropriate Order follows.

ORDER

AND NOW, this 19th day of January, 2010, it is **ORDERED** that the Motion for Summary Judgment by Plaintiff ConocoPhillips (Doc. No. 14) and the Cross-Motion for Summary Judgment by Defendant United Steelworkers, Local 234 (Doc. No. 12) are GRANTED IN PART AND DENIED IN PART. The Arbitration Opinion and Award shall be MODIFIED as follows: that portion of the Arbitrator's Opinion describing a five-step progressive discipline structure is merely illustrative and that structure is not part of the Parties' Collective Bargaining Agreement and is not required by the Collective Bargaining Agreement. The Opinion and Award are otherwise CONFIRMED.

IT IS SO ORDERED.

5 Safety and Employee Behavioral Issues in the Workplace

******

Chapter Objectives:

1. Why develop written safety and health programs to achieve Occupational Safety and Health Act (OSHA) compliance.
2. Employee behavior issues and safety impacts.
3. Why do employees work unsafely (when they know better)?
4. Concepts of safety and health management.

******

Why do safety and health professionals develop and codify their compliance programs as well as other directives into written policies, procedures, and programs? Written compliance programs provide steady and consistent guidance as to the prescribed activities, as well as proof as to the directives, training, and enforcement during an Occupational Safety and Health Act (OSHA) compliance inspection, in grievance and other internal disputes and in a court of law. Written compliance programs serve as the foundational elements or base from which safety and health professionals build and expand their safety and health programs focused on the achievement of specified goals and objectives.

Historically, safety and health professionals utilized lagging indicators, such as the OSHA recordkeeping data, from which to design and implement their structure and programs to achieve the company's specified goals or objectives. This data, although important, provided a view of past activities and programs; however, it did not often address the potential risk within the workplace. In recent years, safety and health professionals have been more frequently relying on leading indicators, or a combination of leading and lagging indicators, in order to surgically focus their efforts on the current or potential risks rather than the historical data provided by lagging indicators.

Today's safety and health professional faces new and difficult issues as well as correlating laws and regulations which have emerged within our society. Thorny issues, such as employee psychological and stress-related claims, prescription medication, alcohol, and controlled substance addiction, off-duty behaviors by employees, social media utilization, and others have created unforeseen challenges for safety and health professionals. New laws and employee protections, along with our litigious society and societal issues, have placed increasing pressures on safety and health professionals to broaden their prospective when developing programs and policies to ensure laws and regulations which could impact the function are adequately addressed.

Safety and health professionals should be aware that the Americans with Disabilities Act (ADA) protects any qualified person with a permanent mental or physical disability.

Historically, the safety and health programs conducted alcohol and controlled substance testing, including for THC (marijuana), to ensure that employees were not working under the influence. Some state workers' compensation laws attempted to address work-related injuries by employees under the influence by reducing or denying benefits. Most company progressive disciplinary policies addressed the possession and use of controlled substances on the job as a condition of employment. Empirical data over the years has found that employees working under the influence tend to have a higher injury rate in most industries. However, many state laws have changed, permitting the use of medical marijuana or recreational use marijuana. (Note: Federal law still classified marijuana as a controlled substance). For safety and health professionals with operations in states which have legalized the use of marijuana, this can blur the historical "lines" and create new issues and potential conflicts between employee and employer rights. For safety and health professionals, this can often create policy and program conflicts which require careful analysis to ensure compliance, not only with state laws, but also with federal laws and OSHA regulations.

For safety and health professionals with operations in states which have legalized marijuana use, safety and health and company policies prohibiting use on the job are usually enforceable. However, correlating issues impacting the safety and health function such as off-the-job usage or lingering effects when returning to work can create issues in enforcement. As more states legalize marijuana use, whether medical only, recreational use, or different combinations and levels, safety and health professionals should be cognizant of the potential impacts on the safety and health of their workforce, as well as the potential legal conflicts with safety and health programs, policies, and procedures.

In the area of controlled substances and rehabilitation, safety and health professionals should be aware of the protections afforded to individual applicants and employees following completion of alcohol or controlled substance rehabilitation. The ADA provides no protection for employees who are actively using alcohol and controlled substances on the job. Most companies possess strict policies prohibiting the use of alcohol and controlled substances, and enforcement under many progressive disciplinary policies can include discipline up to and including discharge from employment. However, when an employee returns from rehabilitation and is not actively using alcohol or a controlled substance, the employee is protected from discrimination or harassment, as well as any aspect of his/her employment. However, if the employee relapses and again uses alcohol or a controlled substance, he/she loses the proscribed protection and is subject to disciplinary actions up to and including discharge. Many companies provide rehabilitation serves to employees within their health care plans and through other sources.

Safety and health professionals should be aware of that, in general, employers may require an employee to submit to a controlled substance test (a "drug test"). Generally, controlled substance testing for current employees requires cause exemplified by physical evidence, behavior, or high-risk job functions. Additionally, specific to the safety and health function is the required testing for controlled substances for work-related injuries where controlled substance use is suspected. Prudent safety and health professionals should be aware that many states have enacted laws limiting the circumstances when an employer can test for controlled substances, as well as the methods which can be used for such testing.

DISABILITY DISCRIMINATION

Disability discrimination occurs when an employer or other entity covered by the Americans with Disabilities Act, as amended, or the Rehabilitation Act, as amended, treats a qualified individual with a disability who is an employee or applicant unfavorably because she has a disability. Learn more about the Act at *ADA at 25*.

Disability discrimination also occurs when a covered employer or other entity treats an applicant or employee less favorably because she has a history of a disability (such as cancer that is controlled or in remission) or because she is believed to have a physical or mental impairment that is not transitory (lasting or expected to last six months or less) and minor (even if she does not have such an impairment).

The law requires an employer to provide reasonable accommodation to an employee or job applicant with a disability, unless doing so would cause significant difficulty or expense for the employer ("undue hardship").

The law also protects people from discrimination based on their relationship with a person with a disability (even if they do not themselves have a disability). For example, it is illegal to discriminate against an employee because her husband has a disability.

Note: Federal employees and applicants are covered by the Rehabilitation Act of 1973, instead of the Americans with Disabilities Act. The protections are mostly the same.

DISABILITY DISCRIMINATION AND WORK SITUATIONS

The law forbids discrimination when it comes to any aspect of employment, including hiring, firing, pay, job assignments, promotions, layoff, training, fringe benefits, and any other term or condition of employment.

DISABILITY DISCRIMINATION AND HARASSMENT

It is illegal to harass an applicant or employee because he has a disability, had a disability in the past, or is believed to have a physical or mental impairment that is not transitory (lasting or expected to last six months or less) and minor (even if he does not have such an impairment).

Harassment can include, for example, offensive remarks about a person's disability. Although the law doesn't prohibit simple teasing, offhand comments, or isolated incidents that aren't very serious, harassment is illegal when it is so frequent or severe that it creates a hostile or offensive work environment or when it results in an adverse employment decision (such as the victim being fired or demoted).

The harasser can be the victim's supervisor, a supervisor in another area, a co-worker, or someone who is not an employee of the employer, such as a client or customer.

DISABILITY DISCRIMINATION AND REASONABLE ACCOMMODATION

The law requires an employer to provide reasonable accommodation to an employee or job applicant with a disability, unless doing so would cause significant difficulty or expense for the employer.

A reasonable accommodation is any change in the work environment (or in the way things are usually done) to help a person with a disability apply for a job, perform the duties of a job, or enjoy the benefits and privileges of employment.

Reasonable accommodation might include, for example, making the workplace accessible for wheelchair users or providing a reader or interpreter for someone who is blind or hearing impaired.

While the federal anti-discrimination laws don't require an employer to accommodate an employee who must care for a disabled family member, the Family and Medical Leave Act (FMLA) may require an employer to take such steps. The Department of Labor enforces the FMLA. For more information, call: 1-866-487-9243.

DISABILITY DISCRIMINATION AND REASONABLE ACCOMMODATION AND UNDUE HARDSHIP

An employer doesn't have to provide an accommodation if doing so would cause undue hardship to the employer.

Undue hardship means that the accommodation would be too difficult or too expensive to provide, in light of the employer's size, financial resources, and the needs of the business. An employer may not refuse to provide an accommodation just because it involves some cost. An employer does not have to provide the exact accommodation the employee or job applicant wants. If more than one accommodation works, the employer may choose which one to provide.

DEFINITION OF DISABILITY

Not everyone with a medical condition is protected by the law. In order to be protected, a person must be qualified for the job and have a disability as defined by the law.

A person can show that he or she has a disability in one of three ways:

- A person may be disabled if he or she has a physical or mental condition that substantially limits a major life activity (such as walking, talking, seeing, hearing, or learning).
- A person may be disabled if he or she has a history of a disability (such as cancer that is in remission).
- A person may be disabled if he is believed to have a physical or mental impairment that is not transitory (lasting or expected to last six months or less) and minor (even if he does not have such an impairment).

DISABILITY AND MEDICAL EXAMS DURING EMPLOYMENT APPLICATION AND INTERVIEW STAGE

The law places strict limits on employers when it comes to asking job applicants to answer medical questions, take a medical exam, or identify a disability.

For example, an employer may not ask a job applicant to answer medical questions or take a medical exam before extending a job offer. An employer also may not ask job applicants if they have a disability (or about the nature of an obvious disability). An employer may ask job applicants whether they can perform the job and how they would perform the job, with or without a reasonable accommodation.

DISABILITY AND MEDICAL EXAMS AFTER A JOB OFFER FOR EMPLOYMENT

After a job is offered to an applicant, the law allows an employer to condition the job offer on the applicant answering certain medical questions or successfully passing a medical exam, but only if all new employees in the same type of job have to answer the questions or take the exam.

DISABILITY AND MEDICAL EXAMS FOR PERSONS WHO HAVE STARTED WORKING AS EMPLOYEES

Once a person is hired and has started work, an employer generally can only ask medical questions or require a medical exam if the employer needs medical documentation to support an employee's request for an accommodation or if the employer believes that an employee is not able to perform a job successfully or safely because of a medical condition.

The law also requires that employers keep all medical records and information confidential and in separate medical files.

AVAILABLE RESOURCES

In addition to a variety of formal guidance documents, EEOC has developed a wide range of fact sheets, question and answer documents, and other publications to help employees and employers understand the complex issues surrounding disability discrimination.

- Employer-Provided Leave and the Americans with Disabilities Act
- Recruiting, Hiring, Retaining, and Promoting People with Disabilities
- Pandemic Preparedness in the Workplace and the Americans with Disabilities Act
- Your Employment Rights as an Individual with a Disability
- Job Applicants and the ADA
- Understanding Your Employment Rights Under the ADA: A Guide for Veterans
- Questions and Answers: Promoting Employment of Individuals with Disabilities in the Federal Workforce
- The Family and Medical Leave Act, the ADA, and Title VII of the Civil Rights Act of 1964
- The ADA: A Primer for Small Business
- Your Responsibilities as an Employer
- Small Employers and Reasonable Accommodation
- Work at Home/Telework as a Reasonable Accommodation
- Applying Performance and Conduct Standards to Employees With Disabilities
- Obtaining and Using Employee Medical Information as Part of Emergency Evacuation Procedures
- The Mental Health Provider's Role in a Client's Request for a Reasonable Accommodation at Work
- Employer Best Practices for Workers with Caregiving Responsibilities
- Reasonable Accommodations for Attorneys with Disabilities
- How to Comply with the Americans with Disabilities Act: A Guide for Restaurants and Other Food Service Employers
- Final Report on Best Practices for the Employment of People with Disabilities in State Government
- Depression, PTSD, and Other Mental Health Conditions in the Workplace: Your Legal Rights
- *ABCs* of Schedule A Documents

The ADA Amendments Act

- Final Regulations Implementing the ADAAA
- Questions and Answers on the Final Rule Implementing the ADA Amendments Act of 2008
- Questions and Answers for Small Businesses: The Final Rule Implementing the ADA Amendments Act of 2008
- Fact Sheet on the EEOC's Final Regulations Implementing the ADAAA*

Any qualified applicant or employee with a permanent mental or physical disability should qualify for protection under the ADA. Safety and health professionals

---

* Equal Employment Opportunity Commission website located at www.eeoc.gov.

confronted with psychological or mental health issues with employees should be aware of any requests for accommodation by the employee and the protections afforded to the employee under the ADA. Prudent safety and health professionals should review the protections and accommodations as well as the guidance provided by the Equal Employment Opportunity Commission (EEOC) located at www. EEOC.gov.

The issues of workplace violence, workplace bullying, and workplace harassment have become major issues for most safety and health professionals. Most companies have adopted a "zero tolerance" policy with enforcement through the negative reinforcement provided under the progressive disciplinary systems. Some companies have developed specific workplace violence prevention programs which often address "red flags" in employee behavior and other proactive approaches. However, safety and health professionals should be aware of the delicate balance between employee rights and restrictive actions structured to reduce the potential of workplace violence. The development of a working environment where employees are empowered, safety and health are the highest priority, employees and management treat everyone with respect, and individuals in need of help are provided with help will go a long way in reducing the potential risks of violence, bullying, and harassment in the workplace.

OVERVIEW–WORKPLACE VIOLENCE

HIGHLIGHTS

What is workplace violence?

Workplace violence is any act or threat of physical violence, harassment, intimidation, or other threatening disruptive behavior that occurs at the work site. It ranges from threats and verbal abuse to physical assaults, and even homicide. It can affect and involve employees, clients, customers, and visitors. Homicide is currently the fourth-leading cause of fatal occupational injuries in the United States. According to the Bureau of Labor Statistics Census of Fatal Occupational Injuries (CFOI), of the 4,679 fatal workplace injuries that occurred in the United States in 2014, 403 were workplace homicides. [More ...] However it manifests itself, workplace violence is a major concern for employers and employees nationwide.

Who is at risk of workplace violence?

Nearly two million American workers report having been victims of workplace violence each year. Unfortunately, many more cases go unreported. Research has identified factors that may increase the risk of violence for some workers at certain worksites. Such factors include exchanging money with the public and working with volatile, unstable people. Working alone or in isolated areas may also contribute to the potential for violence. Providing services and care, and working where alcohol is served may also impact the likelihood of violence. Additionally, time of day and location of work, such as working late at night or in areas with high crime rates, are also risk factors that should be considered when addressing issues of workplace violence. Among those with higher risk are workers who exchange money with the public, delivery drivers, healthcare professionals, public service workers, customer service agents, law enforcement personnel, and those who work alone or in small groups.

How can workplace violence hazards be reduced?

In most workplaces where risk factors can be identified, the risk of assault can be prevented or minimized if employers take appropriate precautions. One of the best protections employers can offer their workers is to establish a zero tolerance policy toward workplace violence. This policy should cover all workers, patients, clients, visitors, contractors, and anyone else who may come in contact with company personnel.

By assessing their worksites, employers can identify methods for reducing the likelihood of incidents occurring. OSHA believes that a well-written and implemented workplace violence prevention program, combined with engineering controls, administrative controls, and training can reduce the incidence of workplace violence in both the private sector and federal workplaces.

This can be a separate workplace violence prevention program or can be incorporated into a safety and health program, employee handbook, or manual of standard operating procedures. It is critical to ensure that all workers know the policy and understand that all claims of workplace violence will be investigated and remedied promptly. In addition, OSHA encourages employers to develop additional methods as necessary to protect employees in high-risk industries.*

As safety and health off-the-job becomes an emerging portion of the safety and health professionals' areas of responsibility, correlating issues regarding off-duty activities and behaviors has moved to the forefront. Given the rising cost of healthcare, more companies have moved to self-insurance and/or addressing employees' potentially risky activities or behaviors outside of the workplace which may result increased claims and increased costs. For example, many companies have adopted policies to not only prohibit smoking in their workplace, but have also adopted being tobacco free as a condition of employment. Companies offer "quit smoking" classes as well as aid to support the employee in quitting the smoking habit. For safety and health professionals, this often changes the focus from preventing just work-related accidents to preventing all accidents by employees. Simple activities such as providing protective safety glasses to employees and requiring their use on the job has expanded to providing protective safety glasses to any and all employees for use when performing activities at home or in their outside activities.

---

* OSHA website located at www.osha.gov.

Correlating with safety and health professional's expanding venue is the issue of workplace privacy and monitoring/surveillance. This often-controversial issue attempts to balance the employee right to privacy in the workplace with the employer's right to monitor their business-related operations. With employers monitoring for security and operation reasons, safety and health professionals are often placed in the bullseye of this issue. Conversely, in today's society with cellphones with high-quality camera and video, employers have often established policies prohibiting employees from using such devices within the workplace. In general, safety and health professionals should consider employee privacy (bathrooms, showers, etc.) when considering surveillance or monitoring systems while ensuring employee safety against outside risks and workplace violence through the use of such monitoring systems.

Issues involving the monitoring of internet usage, email, texting, phonecalls, voicemail, and related electronic communications have quickly moved to the forefront for safety and health professionals. In general, employees using computers or other electronic equipment provided by the employer have few, if any, rights under the current law. These emails and other communications are considered the property of the employer, since the employees are using employer-provided equipment. Employers generally have a right to monitor email and other electronic communications if they possess a valid business purpose. For safety and health professionals, this monitoring of emails, website use, and other electronic communications through employer-owned computers and phones has been used as evidence in state and federal courts. The Electronic Communications Privacy Act (ECPA)* and state laws have placed some restrictions on employers' monitoring of phonecalls at work. Under this law, employers are prohibited from monitoring employees' personal calls, even if made on company phones or on company property. The employee can consent to such monitoring by the employer and voicemail is also considered protected. Safety and health professionals should check with counsel before establishing monitoring or surveillance within the workplace.

Lastly, the "hot" issue of guns in the workplace has become an issue for safety and health professionals. Historically, safety and health professionals have developed policies and procedures to create a safe and healthful workplace as required under OSHA's General Duty clause, prevent workplace violence, and limit the employer's potential liability under workers' compensation and tort liability theories. Part and parcel of these policies and programs was the prevention of guns entering the workplace or being on company property (e.g., parking lots). However, with recent changes in the law, a substantial number of individuals have acquired concealed carry weapon permits and several states have further enacted laws specifically addressing the carrying of guns within the workplace.

In general, safety and health professionals can prohibit the carrying of firearms on the job by policy and posting. However, in approximately 15 states, specific individual state laws have been enacted addressing the carrying of guns on the job. These laws, in general, protect employees' rights to maintain a gun in their

---

* 18 U.S.C. § 2510–22.

vehicle in the parking lot, limit the employer's right to search a vehicle on company property, prohibit discrimination against gun owners, and specifically permit employees to carry guns in the workplace without violating the OSHA General Duty clause. However, some of these laws specifically permit the employer to prohibit guns in the workplace if the employer posts the required notice, and offer protection to the employer through immunity for gun-related injuries if complying with the law.

The safety and health function has come a long way since 1970. Safety and health professionals today are challenged by a wide array of issues, protections, and requirements, many of which require a balance between employee rights and employer requirements. The safety and health function no longer works in a "silo," but is impacted by a myriad of different laws, regulations, requirements, and interests. Safeguarding your company's human investment has become a 24–7 responsibility with new and emerging issues arising constantly. Safety and health professionals must always be at the "top of their game" and consider the correlating and impacting issues and law in every decision.

6 Preparing for OSHA
Standards, Violations, and Actions

******

Chapter Objectives:

1. Understand the Occupational Safety and Health Act (OSH Act) and rights and responsibilities under it.
2. Acquire an understanding of the OSHA standards, promulgation to enforcement.
3. Understand the OSHA penalty schedule.
4. Acquire an understanding of an OSHA inspection.

******

Preparation for an inspection by the Occupational Safety and Health Administration (OSHA) should start the first day on the job by the safety and health professional. As part of the overall safety and health efforts, achieving and maintaining compliance with the OSHA standards is a foundational component for any successful safety and health program. Conversely, failure to achieve and maintain compliance with the OSHA standards places the company, as well as the safety and health professional, at risk for monetary as well as criminal sanctions in severe cases. Achieving and maintaining compliance is not a voluntary safety and health activity ... it is mandatory for all companies within the jurisdictional coverage of the Occupational Safety and Health Act.

As defined in Section 3(8) of the OSH Act, an occupational safety and health standard is "a standard which requires conditions, or the adoption or use of one or more practices, means, methods, operations, or processes reasonably necessary or appropriate to provide safe or healthful employment or places of employment."* The OSH Act requires an employer to comply with specific occupational safety and health standards. Section 6(b) of the OSH Act authorizes the Secretary of Labor to "promulgate, modify or revoke any occupational safety or health standard," provides procedures for doing so, and establishes criteria for those standards.† Section 6(f) establishes the standard of review of OSHA promulgated standards by the federal courts of appeal.

---

* 29 U.S.C. § 3(8).
† 29 U.S.C. § 6(b).

When combining the sections of the OSH Act, as well as the interpretations of a number of court decisions, these provisions establish the legal framework and the following requirements for an OSHA standard:

1. The standard must substantially reduce a significant risk of material harm;
2. Compliance must be technologically feasible, in the sense that the protective measures being required by the standard already exist, can be brought into existence with available technology, or can be created with technology that can reasonably be developed;
3. Compliance with the OSHA standard must be economically feasible, in the sense that the standard will not threaten the industry's long-term profitability or substantially alter its competitive structure;
4. Health standards must eliminate significant risk or reduce a significant risk to the extent feasible, and safety standards must be highly protective;
5. OSHA standards must employ the most cost-effective protective measures capable of reducing or eliminating significant risk; and
6. OSHA standards must be supported by substantial evidence in the rule-making record and be consistent with prior agency practice or supported by some justification for departing from that practice.

OSHA standards are categorized by General Industry, Construction, Maritime, and Agriculture. For most safety professionals, General Industry and Construction are the most widely utilized. OSHA standards can be narrowly focused on a specific industry, known as vertical standards, or broadly focused for the vast majority of the industries, known as horizontal standards. The early OSHA standards, or National Consensus Standards, often parallel other voluntary codes (such as NFPA and NEC); however, when adopted by OSHA, these voluntary requirements became mandatory. Many of the early standards were design or specification standards providing minimum criteria to achieve compliance. In recent years, OSHA has moved to more performance-based standards, providing more employer flexibility to achieve compliance.

Within the General Industry standards (29 CFR 1910), every industry or operation will be unique; however, there are several "horizontal" standards common among and between most industries including:

1. Control of Hazardous Energy (Lockout/Tagout)
2. Emergency and Disaster Preparedness
3. Hazard Communications Standard (Haz Com)
4. Respiratory Protection
5. Fall Protection
6. Personal Protective Equipment (PPE)
7. Bloodborne Pathogens Standard.

Compliance with these types of horizontal standards, as well as other vertical standards, must be identified by the safety and health professional, and compliance must be achieved and maintained. OSHA does not advise the safety and health professional what to do, but simply provides standards identifying the minimum requirements to achieve compliance. And, as most safety professionals will identify, simply achieving and maintaining compliance with the OSHA standards does not make a successful safety program. Compliance with the OSHA standards is the "bare bones" minimum necessary to avoid the potential penalties under the OSH Act.

For most safety and health professionals, the probability of a compliance inspection is fairly small when comparing the number of compliance officers with the number of employers and workplaces in the United States. However, if the safety and health professional is in a targeted industry, earned a high injury and illness rate or other factors, the probability of inspection increases. And remember, the safety and health professional must contact OSHA if a workplace fatality occurs, or three or more employees are transported to the hospital. These types of incidents usually mandate an investigation or inspection. So, what generally happens during a typical compliance inspection?

DEVELOPMENT OF YOUR SAFETY AND HEALTH PROGRAMS

When entering the safety and health arena, it is essential that the safety and health professional have, at a minimum, a working knowledge of the operations in order to begin the assessment and analysis as to the specific OSHA standards which are applicable to the operations and how to achieve compliance with the specific applicable standard. One important component of this assessment is to identify any past violations within the operations in order to provide specific emphasis and priority in this area to avoid any potential for a repeat violation. The safety and health professional should become familiar with the OSHA standards located at www.osha.gov and determine whether the general industry or construction standards, or both, are applicable to the operations.

Utilizing past programs, if available, and the safety and health professional's knowledge of the operations, as well as knowledge of the OSHA standards, a careful analysis should be conducted to assess whether or not a particular standard is applicable to the specific circumstance within the operations. Remember, OSHA promulgates a large number of standards covering virtually every industry; however, it is up to the safety and health professional to determine whether or not the particular standard is applicable to the situation and operation. Prudent safety and health professionals may want to encompass all potential applications in a conservative manner. Given the fact that the safety and health professional will not know if he/she has missed a hazard until after an inspection, it is often prudent to address all recognized hazards and applicable standards within the specific workplace. When the hazard is recognized by the compliance officer during an inspection and the safety and health professional has not addressed or developed a program, it is going to cost your company money, at a minimum, and substantial embarrassment or even more for the safety and health professional.

Although OSHA does not mandate written compliance programs, the vast majority of safety and health professionals utilize written compliance programs to ensure compliance and consistency and as an offer of proof during a compliance inspection or on appeal. In preparing to write a compliance program, safety and health professionals generally adopt a particular format or utilize the company's current format. In general, written compliance programs are developed and codified in the following manner:

1. Safety and health professional identifies the applicable standard.
2. Safety and health professional reads and re-reads the standard.
3. Safety and health professional identifies the key elements within the standard.
4. Safety and health professional "visualizes" the program in action and inserts additional elements necessary for effective and efficient operations.

Once the safety and health professional has the concept and elements in his/her mind, it is time to codify this concept on paper. Although some companies possess a specific format, most companies follow some form of the following model:

1. PURPOSE of the written program.
2. SCOPE of the written program.
3. SPECIFIC elements from the standard.
4. APPLICATION of the standard (specific protocol or procedures).
5. TRAINING requirements for the program and standard.
6. ENFORCEMENT of the program.

Safety and health professionals should be aware that there may be several drafts of the codified program to ensure that each and every potential issue has been addressed. Most companies possess a review process through the human resource, legal, and other departments for review and approval before launching the compliance program. Additionally, once the written program is approved, acquisition of the funding, equipment, and other components may be required. Specific training for supervisors or instructors in order that they can train the employees may be required. In general, if a safety and health professional is starting with no programs, it may take between 3–5 years to develop, codify, and implement all required compliance programs within a standard size operation.

When developing compliance programs, it is important for safety and health professionals with operations in state plan states to identify any and all nuisances between the federal and state standards. State plan states usually must adopt any new federal standards within a specified period of time and may modify the federal standard to be more stringent. For safety and health professionals with multiple operations in state plan as well as federal states, it is essential to write a company-wide compliance program at the highest level to ensure that all states plan regulations and requirements, as well as the federal standard requirements, are addressed within the company-wide program.

Implementation of a compliance program, after approvals, usually includes, but is not limited to, equipment acquisition, training and education, and enforcement. Safety and health professionals should be very specific when ordering equipment, such as protective safety glasses, to ensure that the appropriate level of protection required by the standard is acquired, e.g., if safety and health professionals notify the purchasing department of other acquisition unit to simply purchase "safety glasses," they will be provided with the most inexpensive safety glasses which can be purchased. Additionally, the method utilized by the safety and health professional in "rolling out" this new program is also very important. Safety and health professionals may consider discussing equipment selections with employees (who will be required to wear the PPE) and permit selection to be driven by the employees.

In the area of training, there is a spectrum of methods through which employees and management can be trained and educated on the compliance program requirements, ranging from individual face-to-face meetings to traditional classroom training to computerized self-paced training. Safety and health professionals should ensure that the training is efficient and effective and show a verifiable acquisition of the knowledge through testing or other method and documented participation or attendance. Safety and health professionals should maintain this training documentation for a substantial period of time.

For many compliance programs, implementation is date-specific following the completion of training and having all equipment issued or installed. Enforcement is essential, and this enforcement should not be by the safety and health professional. Supervisors, team leaders, or other managerial positions which interact and supervise employees on a daily basis for production and quality are often the appropriate level for enforcement to achieve compliance. If the safety and health professional is considered the "safety cop," he/she will not have sufficient hours in a given day to enforce all of the various compliance requirements. Safety and health professionals should provide all of the tools and skills necessary to the first line supervisors or managers and hold them accountable for enforcement of the compliance program requirements.

PRIOR TO A COMPLIANCE INSPECTION

Safety and health professionals should be aware that OSHA provide no notice prior to the inspection. In fact, it is a crime punishable by a fine of up to $1000.00 and/or imprisonment for up to six months if advance notice is provided to an employer without the authorization of the Secretary of Labor or designee.* However, there are times when an inspection can be scheduled, namely voluntary inspection through state plan education and training divisions, voluntary inspections for VPP or SHARPS programs, and other voluntary situations. Additionally, the OSHA compliance officer must come to the operation during working hours; however, if the shift or location operates outside of traditional working hours, the safety and health professional can contact the area or regional director to possibly schedule other times to inspect the specific location. These are exceptions rather than the rule.

As noted, compliance inspections are virtually always conducted without advance notice. In special circumstances, OSHA can provide notice to the employer; however, such notice is normally less than 24 hours. These circumstances include:

- Imminent danger circumstances requiring immediate correction;
- Accident investigations where the safety professional notified OSHA of a fatality or catastrophe;

---

* 29 U.S.C.A. § 1903.6.

- Inspection after regular business hours or requiring special preparation;
- Situations where notice is required to ensure that the employer and employee representatives or other personnel will be present;
- Situations where the inspection must be delayed more than five days where there is good cause;
- And situations where the OSHA Area Director determines that advance notice would produce a more thorough or effective investigation.

Additionally, safety and health professionals should be aware of the programmed inspections which include:

- Site-Specific Targeting (SST) Program Inspections;
- Construction Inspections;
- Special Emphasis Inspections (SEP); and
- Severe Violator Enforcement Program (SVEP) Inspections.*

As a result of the Supreme Court decision in *Marshall v. Barlow's, Inc.*,[†] employers possess the right to require a search warrant prior to entry into the operation and conducting a compliance inspection. In *Barlow's*, the Supreme Court held that Section 8(a) of the OSH Act, which empowered OSHA compliance officers to search the work area of any employment facility within OSHA's jurisdiction without a search warrant, was unconstitutional. The Court concluded that "the concerns expressed by the Secretary (of Labor) do not suffice to justify warrantless inspections under OSHA or vitiate the general constitutional requirement that for a search to be reasonable a warrant must be obtained."[‡] Safety professionals should be aware that there are exceptions to the search warrant requirement for OSHA, including the consent exception, the plain view exception, and the emergency exception. A fourth and controversial exception is the licensure or *Colonnade–Biswell* exception which has been applied to the OSH Act.[§]

OSHA'S ARRIVAL

With no prior knowledge or notice, the safety and health professional will receive the proverbial "knock at the door" and the OSHA compliance officer(s) will present himself/herself to the company at the security entrance or front desk. The reason for the inspection could be varied, from a random inspection to a targeted inspection to a complaint inspection. Complaint inspections have historically been the top category through which to generate a compliance inspection. The compliance officer

---

* Note: OSHA's priorities for inspection usually provide imminent danger situations top priority, followed by fatality/catastrophe, complaint/referral, and programmed inspection.
[†] 436 U.S. 307 (1978).
[‡] *Id.*
[§] *Colonnade Catering Corp. v. United States*, 397 U.S. 72 (1970) (warrantless nonconsensual searches of liquor stores) and *United States v. Biswell*, 406 U.S. 311 (1972) (nonconsensual warrantless searches of pawn shops).

will produce their credentials and often inform the safety professional of the nature, purpose, and scope of the inspection or the specific location in which he/she would like to conduct the inspection.

■■

The OSH Act specifies that a representative of the employer (e.g., the safety professional) and a representative of the employees (e.g., the labor organization representative or selected employee in non-union operation) must be given an opportunity to accompany the OSHA compliance officer on the inspection. The employer is not required to pay the employee their regular wages for the time spent during the inspection (although most companies do pay the employee's wages for the time spent during the inspection).* The employer representative must be an employee and not a third party.

COMPLIANCE INSPECTION

With the consent of the safety professional or other agent of the employer, the inspection usually starts with a review of the OSHA recordkeeping documents and specific requested compliance programs, records, or documents. The compliance officer often inspects equipment, views processes, and talks with employees during the inspection. The compliance officer has the right to consult with employees concerning safety and health matters and also provides that employees shall have the right to bring OSHA violations to the attention of the compliance officer during the inspection.†

OSHA compliance officers are required to conduct their inspections in such a manner as to avoid disrupting the employer's business operations. Compliance officers are also required to wear and use appropriate protective equipment and clothing and must comply with the employer's safety and health rules while in the workplace. Of particular importance when compliance officers take photographs and samples, compliance officers must take precautions not to create or cause hazardous conditions in the workplace. Additionally, compliance officers must treat trade secrets as confidential. This is especially important with photographs or other company documents or information which may be available from OSHA under the Freedom of Information Act (FOIA).

It is important that the safety professional and members of the safety team accompany the compliance officer(s) and document the same equipment, items, documents, etc. and photograph the same items as the compliance officer. In short, OSHA is collecting evidence for possible use if violations are identified, and OSHA is not required to provide the employer with this information. Safety professionals should be collecting the same evidence during the inspection to prepare to defend against the violation if a citation is issued. As noted later in this chapter, safety professionals will only be provided 15 working days to appeal and may have less time if an informal conference

---

* *Chamber of Commerce v. OSHA*, 636 F.2d 464 (CA-DC, 1980).
† 29 U.S.C.A. § 657(a)(2).

is requested. Every aspect of the compliance officer's inspection should be well documented for possible utilization in the event of a challenge to a violation. It is better to have the documentation and not need it rather than needing the documentation and not having it!

CLOSING CONFERENCE

When the compliance officer(s) determine the compliance inspection is completed, OSHA regulations provide that the compliance officer should confer with the safety professional or employer representatives and inform of any apparent safety and health violations identified during the inspection. Safety professionals should be aware that the compliance officer(s) are required to present their findings to the Area Director for review and approval before the identified violations would be included in the citation. When the closing conference is completed and the compliance officer(s) depart the property, the time starts for OSHA to issue the citation. OSHA is provided with six (6) months to issue the written citation.

CITATION

Under Section 9(a) of the OSH Act, if the Secretary of Labor believes that an employer *has violated a requirement under Section 5 of this Act, of any standard, rule or order promulgated pursuant to Section 6 of this Act, or of any regulation prescribed pursuant to this Act, he shall within reasonable promptness issue a citation to the employer.** Safety professionals should be aware that "reasonable promptness" has been determined as within six months from the occurrence of the violation.[†]

Safety and health professionals should be aware that Section 9 of the OSH Act requires the citation to be in writing and "describe with particularity the nature of the violation, including a reference to the provision of the Act, standard, rule regulation or order alleged to have been violated."[‡] Most citation documents are four columns which include the standard allegedly violated, a description of the violation, the category of the citation, and the proposed monetary penalty. There is no specific method of service of the citations required by the OSH Act; however, most citations are sent to the employer via certified mail with return verification to OSHA. Safety and health professionals should be aware that the 15 working days for appeal starts when a representative of the employer receives and signs for the certified mail.

Under Section 10 of the Act, after the citation is issued, the employer, any employee, and any authorized union representative has 15 working days to file a Notice of Contest.[§] Safety and health professionals should be aware that the

---

* 29 U.S.C. § 651 *et seq.*
[†] *Id.* Note: the statute of limitations contained in Section 9(c) will not be vacated on reasonable promptness grounds unless the employer was prejudiced by the delay.
[‡] *Id.*
[§] *Id.*

citation document must be posted for employee review. If the employer does not contest the violation, abatement date, or proposed penalties within the 15 working days, the citation becomes final and is therefore not subject to review by any court or agency. If a timely Notice of Contest is filed in good faith, the abatement requirement is met and a hearing is scheduled with the Occupational Safety and Health Review Commission (OSHRC). An employer may contest any part or all of the citations, proposed penalties, or abatement dates. Employees have the right to elect party status after an employer has filed a notice of contest. Safety and health professionals should be aware that the Notice of Contest also requires posting for employees.

Of particular importance for safety and health professionals is the format for a Notice of Contest. The Notice of Contest does not have to be in any particular form or format. It must be sent to the OSHA Area Director who issued the citation. The OSHA Area Director is required to forward the Notice of Contest to the OSHRC, which is required to docket the case for hearing. It should be noted that several state plan programs offer fill-in-the-blank forms to assist employers in filing a Notice of Contest.

Safety and health professionals should carefully review each and every word and aspect of the citation and ensure the correctness of the alleged violation, categorization, and penalty. One avenue through which a safety professional can check the appropriateness of the citation information is through review of the *OSHA Field Manual* which is available on the OSHA website. This manual is utilized by the compliance officers as a directive when addressing the appropriate categorization and requirements for alleged violations. After review, appropriate defenses can be developed, with supporting evidence acquired and prepared for hearing.

INFORMAL CONFERENCE

Although discussed in depth later in this text, safety and health professionals should be aware that within 15 working days from receipt of the citation, the safety and health professional may request an informal conference with the area of the regional director in an attempt to settle the identified violations. Safety and health professionals should note that this is a request, not a requirement, and it must be permitted by the area of regional director for OSHA. Safety and health professionals selecting to pursue an informal conference must possess the appropriate authority from the company to settle any/all violations. Additionally, safety and health professionals should prepare their defenses and arguments for an informal conference and be prepared to provide evidence to support any defenses.

Safety and health professionals should be prepared for a compliance inspection before, during, and after the inspection. Before an inspection, safety professionals should prepare their teams and methods to acquire and document evidence, and have the equipment to document the same evidence, such as photographs, as the compliance officer. At the time of the inspection, safety professionals should know the type of requirements of inspections and how the inspection and closing conference is conducted. After the inspection, time is of the essence with the 15 working day time limitation. Safety professionals should be prepared if requesting an informal conference, pursuing a settlement agreement, or appealing all or part of the citation. Remember, if the 15 working day limitation for appeal is not addressed, your company will, in essence, lose all appeal rights. Be prepared and ALWAYS be professional!

Case Study: This case may have been modified for the purposes of this text.

United States of America

OCCUPATIONAL SAFETY AND HEALTH REVIEW COMMISSION
1924 Building – Room 2R90, 100 Alabama Street SW Atlanta,
Georgia 30303-3104

Secretary of Labor,	
Complainant,	
v.	OSHRC Docket No. **15-1928**
Sanderson Farms, Inc.,	
Respondent.	

Appearances:

> Jean C. Abreu, Esquire, U.S. Department of Labor, Office of the Solicitor, Dallas, Texas
>> For Complainant

> Darren Harrington, Esquire, Key Harrington Barnes, PC, Dallas, Texas
>> For Respondent

Before: Administrative Law Judge Sharon D. Calhoun

DECISION AND ORDER

Sanderson Farms, Inc. (Sanderson) operates a poultry processing plant located at 1111 North Fir, Collins, Mississippi. Following a reported injury, the Occupational Safety and Health Administration (OSHA) commenced an inspection of the work-site on July 1, 2015, with an additional site visit on July 29, 2015. As a result of the inspection, a Citation and Notification of Penalty (Citation) was issued to Sanderson on October 5, 2015. As amended, the Citation alleged a violation of 29 C.F.R. § 1910.212(a)(1) and proposed a $7,000.00 penalty.

For the following reasons, the Citation is vacated and no penalty is assessed.

JURISDICTION

The parties stipulated the Occupational Safety and Health Review Commission (Commission) has jurisdiction over this matter and Sanderson is an employer engaged in a business affecting interstate commerce within the meaning of Section 3(5) of the Occupational Safety and Health Act (the Act) (Stip. 1–2). Sanderson's principle place of business is in Laurel, Mississippi (Stip. 3). In its Answer, Sanderson admits the Commission has jurisdiction over this proceeding and acknowledges: "it is an employer engaged in a business affecting commerce as these terms are defined in the Act." (Answer at 1). Based upon the parties' stipulations and the record, Sanderson is a covered business and the Commission has jurisdiction under the Act.

PROCEDURAL BACKGROUND

On October 5, 2015, the Secretary issued one serious Citation for a violation of 29 C.F.R. § 1910.212(a)(3) with a $7,000.00 penalty. OSHA received Sanderson's Notice of Contest on November 2, 2015.* The Secretary filed a timely Complaint and a Motion to Amend the Citation to allege a violation of 29 C.F.R. § 1910.212(a) (1), a different subpart of the same machine guarding standard initially cited.

Sanderson first filed a timely Answer and then filed a Motion for Summary Judgment. The court issued an Order Denying Respondent's Motion for Summary Judgment and Granting the Secretary's Motion to Amend the Citation on June 8, 2016. A hearing was held from July 18, 2016 to July 19, 2016, in Jackson, Mississippi. Both parties filed post-hearing briefs.

FACTUAL BACKGROUND

Sanderson operates a poultry processing plant where employees debone chickens on approximately eight "cone lines" (Stip. 5; Tr. 154, 162–63). Each cone line has a continuously moving conveyor with several "cones" approximately twelve inches apart (Tr. 39, 56; Exhs. C2f, R-9). Employees stand on one side of the cone line making cuts on chicken carcasses as they come down the line (Exh. R-9; Tr. 56, 162, 261–62, 288). Opposite from where the employees stand, there is a splashguard running the length of the cone line (Exhs. C-2e, R-9; Tr. 55, 71, 232, 252, 255–56, 279). The splashguard had been installed to prevent chicken from falling on the floor (Tr. 72–73, 232, 251, 279). Prior to OSHA's first site visit, there was a gap of between one-quarter and three-quarters of an inch between the bottom of the splashguard and the conveyor belt (Tr. 47, 52–53, 252; Exhs. C-2c, C-2d).

Around 9:00 a.m., on June 29, 2015, most of the employees on one of the cone lines took a ten minute break (Tr. 163, 165). One employee declined to take her break and instead worked to catch up on chicken carcasses she did not have time

---

* The Secretary does not allege that Sanderson's Notice of Contest was untimely.

to complete when the line was running (Tr. 163–65, 228; Stip. 6-7). After completing the catch-up work, the employee saw a piece of chicken stuck between the splashguard and the conveyor belt and attempted to remove it with her fingers (Tr. 155, 165, 167; Exhs. C-1, C-2a). As the conveyor belt moved along, one of her fingers became trapped (Tr. 39, 155, 167; Exhs. C-1, C-2a). She screamed and another employee pushed an emergency button to stop the conveyor belt (Tr. 167, 226). At that point, the trapped employee pulled back her hand and realized her left index finger had been amputated (Tr. 167; Stip. 8). She promptly sought medical care with the aid of her co-workers* (Tr. 167; Stip. 8; Exh. C-1).

A few hours after this incident, Sanderson's Corporate Safety and Health Coordinator, Richard Ward, reported the accident to OSHA (Tr. 29; Exh. C-1). Two days later, on July 1, 2015, Albert Smith, an OSHA compliance officer CSHO, inspected the worksite (Tr. 31). During the course of OSHA's inspection, on July 25 and 26, 2016, Sanderson lowered the splashguard to eliminate the gap between the splashguard and the conveyor belt (Tr. 53–54, 235).

On July 29, 2016, the CSHO again visited the worksite and viewed the cone line as modified (Tr.31, 53; Exh. C-2e).

DISCUSSION

A. APPLICABLE LAW

For the Secretary to establish a violation of any OSHA standard, he must prove that: (1) the cited standard applies; (2) its terms were violated; (3) employees were exposed to the violative condition; and (4) the employer knew or could have known with the exercise of reasonable diligence of the violative condition. *See Astra Pharm. Prods., Inc.*, 9 BNA OSHC 2126, 2129 (No. 78-6247, 1981), *aff'd in pertinent part*, 681 F.2d 69 (1st Cir. 1982). The Secretary has the burden of proving each of these elements by a preponderance of the evidence.
Id.

B. APPLICABILITY

The cited standard requires the employer to provide one or more methods of machine guarding: "to protect the operator and other employees in the machine area from hazards such as those created by point of operation, ingoing nip points, rotating parts, flying chips and sparks." 29 C.F.R. § 1910.212(a)(1). The guarding requirements apply to the moving parts of all types of industrial machinery and regardless of whether the hazards are created during production or nonproduction operations. *Ladish Co.*, 10 BNA OSHC 1235, 1237 (No. 78-1384, 1981); *Gen. Elec.*
Co., 10 BNA OSHC 1687, 1690 (No. 77-4476, 1982).

---

* The employee's finger was unable to be reattached by medical professionals.

Because § 1910.212(a) requires employers to guard against "hazards," without providing specific ways to do so, the Secretary must prove operating the machinery presented a hazard. *Landish*, 10 BNA OSHC at 1237; *Buffets, Inc.*, 21 BNA OSHC 1065, 1066 (No. 03-2097, 2005) (Secretary must establish the rotating parts presented a hazard). The fact it is possible for an employee to come into contact with a machine's moving parts is insufficient to establish a violation. *Id*. The Secretary must show employees are exposed to a hazard because of the way the machine functions and how it is operated.* *Id. See also Jefferson Smurfit Corp.*, 15 BNA OSHC 1419, 1421 (No. 89-0553, 1991) (possibility of exposure to an unguarded nip point insufficient to sustain a violation).

The Citation, as amended, alleges Sanderson violated 29 C.F.R. § 1910.212(a)(1) because: "[o]ne or more methods of machine guarding was not provided to protect employees on the cone line from hazards created by the opening between the splash guard and the cone line." The Secretary contends in the Citation, this failure to guard resulted in an employee being "Exposed to an amputation hazard when performing tasks including, but not limited to, cleaning debris jammed between the splashguard and the cone line" on or about July 1, 2015.[4]

Sanderson implores the court to read the cited standard as not requiring guarding of the gap between the splashguard and the rest of the cone line (Sanderson's Brief at 8). Citing two machine specific guarding standards and a 2007 OSHA publication titled *Safeguarding Equipment and Protecting Employees from Amputations* (Safeguarding Publication), Sanderson argues OSHA has concluded gaps of the size present on the cone line are safe[†] (Sanderson's Brief at 8–9; Exh. R-8). It asserts the Secretary failed to establish it had noticed the small gap was a hazard and therefore the Citation must be vacated. *Id*. at 9.

As noted above, the cited standard requires employers to protect employees from hazards in machine areas "such as those created by points of operation, ingoing nip points, rotating parts, flying chips and sparks." 29 C.F.R. § 1910.212(a)(1). The CSHO testified there were no hazards related to flying chips or sparks. He did not indicate whether a point of operation, ingoing nip point, or rotating part created any hazards[‡] (Tr. 103). Although the video of the cone line appears to show rotating parts, the citation does not refer to these and the CSHO did not testify they should

---

* Although this case can only ultimately be appealed to the D.C. or Fifth Circuit, Sanderson urges the Court to follow the approach in *Perez v. Loren Cook Co.*, 803 F.3d 935 (8th Cir. 2015) (Sanderson's Brief at 7–8). The Court finds the case neither applicable nor persuasive. In *Loren Cook*, the Eighth Circuit determined because 29 C.F.R. § 1910.212(a)(1) identifies specific hazards related to the regular use of the machinery (such as rotating parts), it was not applicable to a hazard stemming from the catastrophic failure of a machine. 803 F.3d at 940–41. In the instant matter, there is no allegation the hazard presented by the cone line was a result of a catastrophic failure. As initially issued, the Citation alleged this conduct violated 29 C.F.R. § 1910.212(a)(3), which requires point of operation guarding. Both the CSHO and Ward testified the gap between the splashguard and the rest of the cone line was not a point of operation (Tr. 102, 279–82). Thus, it appears the parties agree the initially cited standard is not applicable.

† The Safeguarding Publication was not timely identified as an exhibit by Sanderson but was admitted over the Secretary's objection (Ex. R-8; Tr. 98–99).

‡ Sanderson argues the CSHO indicated there was no point of operation hazard (Sanderson's Brief at 7). The CSHO's actual testimony was the gap itself was not a point of operation, ingoing nip point, or rotating part (Tr. 103).

have been guarded (Exh. R-9). *See MB Consultants, Ltd. d/b/a Murray Chicken*, 25 BNA OSHC 1146, 1161 n. 11 (No. 12-1165, 2014) (Consol.) (ALJ) (declining to *sua sponte* amend citation to add struck by and entanglement hazards on a poultry cone line). In fact, the CSHO offered no explanation for what needed to be guarded on the cone line and did not understand why the Secretary amended the citation to allege a violation of § 1910.212(a)(1) (Tr. 101–3).

While the sources of possible hazards set forth in the cited standard is not exhaustive, the Secretary nonetheless must present evidence supporting his view of what the hazard was Sanderson needed to guard. Here, the Secretary merely alleges employees working on the cone line were exposed to a caught by hazard, without identifying if that hazard related to the conveyor, the splashguard, the gap, or something else (Secretary's Brief at 7.) When asked at the hearing what hazards he considered employees working on the cone line to be exposed to, the CSHO's response was unclear (Tr. 73). The transcript reads "[t]o a (unintelligible) hazard." *Id.* Neither party sought to correct the transcript nor did the Secretary present other evidence clarifying the testimony. The occurrence of the accident alone is insufficient to establish there was a hazard to which the cited standard applied. *See Ormet Corp.*, 9 BNA OSHC 1055, 1058 (No. 76-530, 1980) (vacating a citation of § 1910.212(a)(1) due to lack of evidence that a hand or arm could be pulled or caught in an unguarded area).

The Secretary also failed to establish Sanderson had sufficient notice of the standard's applicability to the gap between the splashguard and the conveyor. With generally worded standards, the Fifth Circuit has made clear: "[d]ue process requires employers be given reasonably clear advance notice of what is required of them." *Owens-Corning Fiberglass Corp. v. Donovan*, 659 F.2d 1285, 1288 (5th Cir 1981). Such notice may be provided by: (1) industry custom and practice; (2) the injury rate for the type of work; (3) interpretations of the regulation by the Commission; or (4) the obviousness of the hazard.* *Corbesco, Inc. v. Dole*, 926 F.2d 422, 427 (5th Cir. 1991).

The record contains no evidence about industry custom or practice. Nor did the Secretary present any evidence of past injuries or accidents which would have put the employer on notice the cone line required additional guarding. Indeed, Sanderson represents there were no injuries or other issues with the gap, such as gloves or fingers becoming stuck, prior to the accident which led to the CSHO's inspection (Tr. 224, 278–79).

As for Commission precedent, the Commission found 29 C.F.R. § 1910.212(a) applicable to conveyors and a judge applied it to a cone line at a poultry plant.

---

* Neither the Fifth Circuit, nor the D.C. Circuit, has applied the rational of *Corbesco* to 29 C.F.R. § 1910.212(a). However, the Commission cited the case when vacating a violation of it. *Martin v. Miami Indus., Inc.*, 15 BNA OSHC 1258, 1263 (No. 88-671, 1991), *aff'd in pertinent part*, 983 F.2d 1067 (6th Cir., 1992) (unpublished). *See also E.I. Du Pont De Nemours & Co.*, 17 BNA OSHC 2110 (No. 96-0354, 1997) (ALJ) (applying *Corbesco* in a case appealable to the Fifth Circuit which involved a violation of 29 C.F.R. § 1926.212(a)(1)); *Wal-Mart Distrib., Ctr. No. 6016*, 25 BNA OSHC 1397, 1400 n.10 (No. 88-1292, 2015) (vacating a violation of a personal protective equipment standard for lack of notice), *aff'd in pertinent part*, 819 F.3d 200 (5th Cir. 2016) (not addressing the citation item the Commission vacated for lack of notice).

See U.S. Steel Corp., 5 BNA OSHC 2063, 2064 (No. 15500, 1977) (finding §
1910.212(a) applicable to a conveyor); *Landish*, 10 BNA OSHC at 1237 (same);
Ormet, 9 BNA OSHC at 1058 (finding the standard applicable to a conveyor but
finding no violation); *Murray Chicken*, 25 BNA OSHC at 1161 (ALJ) (finding
the standard applicable to chicken leg splitter and a cone line at a poultry plant).
However, none of these cases relate to a gap that was not a point of operation,
ingoing nip point, or rotating part. *Id.* (Tr. 103.) Further, in the instant matter, the
Secretary is not arguing the conveyor required a guard. He argues the gap needed
to be guarded without presenting evidence as to what type of hazard the gap created (Tr. 73). Thus, the Court finds Commission precedent did not provide sufficient notice of 29 C.F.R. § 1910.212(a)'s applicability to the condition the Citation
references.*

In terms of the obviousness of the hazard, Sanderson argues machine specific
guard standards and the Safeguarding Publication indicate it did not need to guard
the one-quarter inch gap (Sanderson's Brief at 8–9). The cited standard is contained
within Subpart O-Machinery and Machine Guarding. Some of the standards within
this subpart relate to specific types of machines, while others, including the cited
standard, more generally address "machine areas." *Compare* 29 C.F.R. § 1910.212
(general requirements for all machines) *with* 29 C.F.R. § 1910.213 (woodwork-
ing machinery requirements). Neither the Secretary nor Sanderson alleges any
of the standards in Subpart O relating to specific types of machinery apply here
(Sanderson's Brief at 8–9). Instead, Sanderson argues a standard related to a certain
type of guard on an abrasive wheel and one concerning power presses establishes
a one-quarter inch opening is safe[†] (Sanderson's Brief at 8 *discussing* 29 C.F.R. §§
1910.215(b)(9), 1910.217(f)(4)).

There is no dispute the cone line was not an abrasive wheel or a power press (Tr.
92, 9495). What may be a permissible type of guarding for one type of machine or
one aspect of its use is not acceptable for all machines or for all uses as the hazards
likely differ. OSHA has adopted different standards for different types of machinery
and imposes different requirements depending on how a machine is being used and
what other precautions are taken. For example, in the abrasive wheel guarding standard, a one-quarter inch gap is permissible at the point of operation only when necessary and only if several safety other measures are in place, including other guards. 29
C.F.R. § 1910.217(c)(1). In the same way, the power press standard permits openings
of a one-quarter inch in size only under certain narrow circumstances. 29 C.F.R. §
1910.217(c)(3)(iii)(f) ("Guards shall be used to protect all areas of entry to the point
of operation not protected by a presence sensing device").

---

* Sanderson contends the cited standard does not apply because the hazard could have been remedied by
removing the splashguard altogether (Sanderson's Brief at 8). Whether Sanderson could have complied
with the standard by removing rather than lowering the splashguard is a question not before this Court
and not relevant to whether there was a violation as the cone line existed at the time alleged in the
Citation.

† Sanderson cites 29 C.F.R. § 1910.217(f)(4) in its brief (Sanderson's Brief at 8). That provision addresses
overloading of power presses and does not refer to permissible gaps in guarding. 29 C.F.R. § 1910.217(f)
(4). The court believes Sanderson may have intended to refer to Table O-10, which is referenced in 29
C.F.R. § 1910.217(c)(1), and relates to safeguarding points of operation (Tr. 87–88, 90–91).

Sanderson also cites the Safeguarding Publication, which states for die presses, the employer must ensure the use of point of operation guards for openings of one-quarter inch or greater (Exh. R-8 at 22). In another section of the document concerning power press brakes, the document indicates a one-quarter inch opening can be permissible if the employer also implements other safeguards. *Id.* at 28.

Although not cited by Sanderson, the Safeguarding Publication makes clear OSHA's position that: "[a]ny machine part, function, or process may cause injury must be safeguarded"* (Exh. R-8 at 9). Because the Safeguarding Publication specifies what constitutes appropriate guarding depends on the machine's design (including what types of safeguards are in place) and how it is used, it does not establish OSHA has concluded gaps of one-quarter inch are permissible[†] *Id.* at 9, 22, 28 (Tr. 94–95).

However, while neither the machine specific standards nor the Safeguarding Publication, establish one-quarter inch openings are universally permissible, these things lend credibility to Sanderson's claim under Fifth Circuit precedent it did not have notice the one-quarter inch gap was a hazard to which the cited standard applied. *Corbesco*, 926 F.2d at 427. Melvin Butler, the supervisor of the cone lines, completed daily safety inspections but never identified the gap as a hazard[‡] (Tr. 203–4; 232–33; Stip. 9). He acknowledged reaching over the line and into the gap to remove chicken would be dangerous, but indicated he was unaware employees did this and expressed skepticism they would have time to do so given the rapid pace at which they were required to cut the chickens coming down the cone line[§] (Tr. 215–16, 218; Exh. R-9). *See Armour Food Co.*, 14 BNA OSHC 1817 (No. 86-0247, 1990) (discussing how difficult it would be for employees to come into contact with unguarded parts).

Ward, Sanderson's safety and health coordinator, also testified he was unaware of any hazard on the cone line and expressed similar skepticism employees would have time to reach across the line while it was moving (Tr. 236, 279, 285, 288). He viewed the splashguard not as a machine guard or something related to safety but solely as a

---

* Sanderson also does not discuss the Hazards of Conveyors section, which sets out various hazards related to employees reaching into machine areas to free debris or jammed materials. (Ex. R-8 at 27–30.) This section does not indicate openings of any size are permissible. *Id.*

† In its brief, Sanderson also cites to Exhibit R-7, which was offered but ruled inadmissible (Tr. 86). Exhibit R-7 was not prepared by OSHA, has no bearing on worksites in Mississippi, and accordingly was found inadmissible under Fed. R. Evid. 401 at the hearing. *Id.* Although Sanderson indicates it is "Michigan OSHA's state plan guidance," it was not prepared by the Michigan Occupational Safety and Health Administration, the agency in charge of administering the Act in Michigan. In fact, it appears no governmental authority prepared or authorized the publication. Rather, it was published by a workers' compensation insurance provider. *See* Michigan Municipal League, Workers' Compensation Fund, http://www.mml.org/insurance/fund/, last visited December 8, 2016. Particularly in light of the fact both the worksite and Sanderson's principal place of business are both in Mississippi, and the lack of any authentication of the document, consistent with the ruling at the hearing, the court finds Exhibit R-7 is inadmissible and not relevant to this matter (Tr. 86; Stip. 3–4).

‡ Other Sanderson employees also completed inspections regularly without identifying the gap as a hazard (Tr. 250–51, 307–8, 315).

§ Sanderson also argues chicken in the gap had no value to the company and suggests it would not care if employees did not remove it (Sanderson's Brief at 5). However, although a piece of "chicken scrap" could not be sold, it could cause the line to shut down and thereby delay production (Tr. 158–59, 176, 219–20, 223–24).

means to keep product on the line* (Tr. 251). It was not intended to protect employees from a moving part and was not located at a point of operation (Tr. 251–52). Ward considered the gap to be too small to trap a finger or hand. (Tr. 255–56, 282–83.) He believed the employee was injured because she failed to don her gloves appropriately causing the tip of her glove to become trapped, leading to her injury[†] (Tr. 256).

The Secretary neither disputes this explanation for the injury nor provided evidence the gap was of a sufficient size to trap a finger when gloves were appropriately worn (or without any personal protective equipment). The Secretary does not allege Sanderson's personal protective equipment policies and procedures were deficient such that it could have been anticipated an employee would improperly don a glove. Likewise, the Secretary does not suggest Sanderson knew or could have known an improperly donned glove could become trapped in the gap.[‡] In addition, there is no evidence Sanderson's overall pattern of inspecting for hazards was insufficient or it otherwise was capable of recognizing the hazard before the accident (Tr. 25051, 278). While the Secretary need not prove how an injury occurred, or even that there was one, for the cited standard to apply, he must show the gap presented a hazard the employer was capable of recognizing. *See Corbesco*, 926 F.2d at 427.

The evidence does not show Sanderson had the level of notice the Fifth Circuit requires. There had been no injuries and there was no evidence of industry customs. Nor has the Commission previously applied the standard the way the Secretary attempted to do so here. Finally, the Secretary did not show the hazard was obvious enough such that the employer should have known or inquired about guarding.[§] *Corbesco*, 926 F.2d at 427. *Cf. Ladish*, 10 BNA OSHC at 1238 (finding the hazard readily apparent). The Court therefore finds the Secretary has failed to establish the violation alleged in Citation 1, Item 1.

---

* In her opening statement, the Secretary's counsel alleged Sanderson failed to properly install and maintain the splashguard and this resulted in a hazard (Tr. 19). However, no witness indicated the splashguard was a machine guard or provided evidence supporting the theory it had been improperly installed or maintained.

† Ward also indicated the splashguard was on the opposite side of where the employees stood (Tr. 254). In his view, employees did not have any contact with it (Tr. 255, 285, 288). However, two employees testified about workers routinely using their hands or scissors to remove product from the gap between the splashguard and the rest of the cone line (Tr. 159–60, 190–91, 263–64, 268–69).

‡ The standard takes into account the inability of humans to always follow every safety precaution. *See Brick & Block Co.*, 3 BNA OSHC 1876, 1878 (No. 4859, 1976) (noting, in connection with a violation of § 1910.212(a), that guarding is still required even if policies are set up to limit contact with machines). However, under Fifth Circuit precedent, the Secretary still had to show the employer had actual or constructive notice of applicability. *Corbesco*, 926 F.2d at 427. The Secretary failed to show Sanderson should have known its glove policies were deficient such that the gap presented a hazard. Sanderson was not cited for any training or personal protective equipment violations and the CSHO did not even ask the injured employee if she had been trained on how to wear the required gloves (Tr. 130–31).

§ While the accident provided sufficient notice, the Secretary does not allege there was any violation after the investigation commenced.

FINDINGS OF FACT AND CONCLUSIONS OF LAW

The foregoing decision constitutes the findings of fact and conclusions of law in accordance with Rule 52(a) of the Federal Rules of Civil Procedure.

ORDER

Based upon the foregoing decision, it is ORDERED that:

Citation 1, Item 1, alleging a violation of 29 C.F.R. § 1910.212(a)(1),
is VACATED and

no penalty is assessed.

SO ORDERED.

/s/
_____

Date: December 27, 2016

SHARON D. CALHOUN
Administrative Law Judge
Atlanta, Georgia

7 Violations and Appeal
The Occupational Safety and Health Review Commission

******

Chapter Objectives:

1. Understanding the Occupational Safety and Health Administration (OSHA) citation appeal process.
2. Appraise the Occupational Safety and Health Review Commission (OSHRC) protocol and case structure
3. Understand the defenses to an OSHA violation.
4. Understand the procedures, protocol and processes in an OSHRC appeal hearing

******

Unlike a court of law where there is discovery in civil actions and prosecutors are required to provide evidence in criminal actions, OSHA is not required to provide their evidence to employers and, with a 15 working day deadline for appeal, acquisition of possible evidence through the Freedom of Information Act takes too long, requiring several months at best. To this end, it is important for safety and health professionals be prepared to gather the same information and potential evidence as the compliance officer during the duration of the inspection. It is important for safety and health professionals to begin far in advance of an inspection by OSHA or a state plan agency to prepare to assemble the tools necessary to gather the same potential evidence as the compliance officer during the inspection. In general, this would include selecting a team, training the team, and equipping the team members with the appropriate equipment to be able to document each and every aspect of the compliance inspection. Generally, the equipment necessary would parallel that of a compliance officer, including a writing pad and writing instrument, working camera, calibrated noise dosimeter, air sampling equipment, and other related which may be necessary during the inspection.

For many organizations, the safety and health professionals team meet immediately after the closing conference to assemble the gathered potential evidence and documentation. After review and assembly, the safety and health professional, armed with the identification of possible violations from the closing conference and the gathered information, should begin the process of assessing the information and identifying potential defenses for each potential violation. Safety and health professionals should be aware that the compliance officer(s) conducting the inspection do not issue the citation, violations, and proposed penalties. This is issued by the area or regional director in consultation with the compliance officer(s). However, the safety and health professional should have a "good idea" of the potential violations identified during the inspection at the closing conference.

Given the fact that "time is of the essence," with 15 working days provided to appeal after receipt of the citation, preparation and identification of potential defenses, as well as supporting evidence in advance of receipt of the citation, provides additional time to assess and study the viability and probability of possible defenses. Additionally, if an informal conference is requested, the time period is substantially shorter, given the fact that the informal conference cannot be used to extend the 15 working day limitation for appeal.

As noted, safety and health professionals should begin to prepare their defenses far in advance of an OSHA compliance inspection. Although it is far better not to have any violations, if a violation and proposed penalty should arise, the safety professional should be prepared to search his/her index of possible defenses to ascertain if one or more defenses fit the facts and circumstances of the situation as well as can be supported by the evidence. Safety and health professionals should remember that the burden of proof is on OSHA for any citation or proposed penalty. Thus, the safety and health professional at an informal conference or as part of the legal team on appeal is attempting to disprove what the compliance officer identified as a citable violation. Additionally, it is important to review each and every element of the alleged violation (identified in the OSHA Field Manual) to ensure that OSHA can prove all elements, as well as the categorization and proposed monetary penalties.

It should be noted that safety and health professional, as well as the company, possesses rights under the United States Constitution. Although most, if not all of the constitutional challenges to the Occupational Safety and Health Act (the OSH Act) were addressed in the early days of the Act, if a constitutional right has been infringed upon as part of the compliance inspection, constitutional challenges remain and can be utilized depending on the circumstances.

The other categories of identifiable defenses which have been used before the OSHRC fall into the categories of procedural defenses and factual defenses. Safety and health professionals should be aware that once OSHA has established a prima facie case that the citation issued to the employer is supported by the evidence, the employer has the burden of rebutting OSHA's prima facie case and may present contrary evidence or establish affirmative defenses to justify noncompliance with the particular standard.

Under the category of procedural defenses, these defenses address the method through which OSHA conducted the inspection or required procedures as well as jurisdictional issues. These defenses can include the Statute of Limitations defense, the Lack of Reasonable Promptness defense, the Failure to Timely Forward Notice of Contest Defense, and the Improper Service Defense. These defenses are very specific to a certain time requirement for OSHA to have completed some task. Additionally, under the category of procedural defenses, issues involving OSHA's jurisdiction to conduct the inspection and issue citations are addressed. These defenses include the Lack of Commerce Clause jurisdiction defense, Preemption by another Federal Agency, and State Plan coverage. Under these defenses, another entity possessed jurisdiction over the particular workplace, and thus OSHA did not possess the authority to conduct the inspection and issue citations.

Procedural defenses also encompass the required employer–employee relationship request under the OSH Act, as well as challenges to the standard itself. Defenses challenging the employer–employee relationship addressed construction sites, loaded employees, and related areas where OSHA may assume an employer–employee relationship. Safety professionals should also refer to the Multi-employer Worksite Rule. Challenges to the standard itself can include the standard was not properly promulgated (for new standards), the standard lacks particularity, the standard was improperly amended, and the standard is arbitrarily vague. Correlating defenses in the area of lack of due process may also be available.

Factual defenses are the primary area of defenses utilized by safety professionals today. The availability of the below listed factual defenses will depend solely on the specific facts of the situation. The recognized factual defenses include the following:

1. A Greater Hazard in Compliance Defense

    This defense is based on the theory that compliance with the OSH Act or a specific standard would create a greater hazard for employees than noncompliance with the standard. For example, the ALJ in *Ashland Oil, Inc.* vacated a citation of the standard requiring a guardrail around the open sides of a platform which was four or more feet above the ground. The judge found that the risk of injury increased with the installation of the guardrail and the guardrail created a risk which did not exist.*

2. Lack of Employer's Knowledge Defense

    Although the employer's knowledge or lack thereof is an element which OSHA is required to prove as part of their prima facie case, safety professionals have often utilized this as a defense to an appropriate violation. In essence, when analyzing the definition of a "serious violation," the OSH Act includes the statement "unless the employer did not, or could not with the exercise of reasonable diligence, know of the presence of the violation."†
    Thus, with a serious violation, employer knowledge is a required element of proof. Other categories of violations do not appear to possess this specific requirement for employer knowledge. Safety professionals should note that the Lack of Employer's Knowledge defense is closely related to the Isolated Incident Defense discussed below. However, safety professionals should note that the Isolated Incident defense is an affirmative defense which shifts the burden of proof to the employer, whereas the Lack of Employer's Knowledge defense maintains the burden on OSHA and is an element of the OSHA violation. The absence of prior incidents may bolster this defense;

---

* 1 OSHC 3246 (ALJ, 1973).
† OSH Act Section 17(k); 29 U.S.C. § 666(j).

however, the existence of prior violations or incidents may undermine the use of this defense.

3. The Machine is Not in Use Defense

The Machine is Not in Use defense is also an affirmative defense. This defense is based on the theory that the equipment which the compliance officer cited was not in use at the time viewed by the compliance officer and thus could not create a hazard. Safety professionals should be prepared to prove that the particular machine or operation was not in use with substantial supporting evidence.

4. Isolated Incident Defense Safety and health professionals should be aware that to prove the isolated incident affirmative defense, the safety professional must be able to demonstrate that: (A) the violation identified resulted exclusively from the employee's conduct; (B) the violation was not participated in, observed by, or performed with the knowledge and/or consent of any supervisory or management personnel; (C) that the employee's conduct contravened a well-established company policy or work rule which was in effect at the time, well-published to employees, and actively enforced through disciplinary action or other appropriate procedures.* Additionally, the safety and health professional must have a specific and verifiable safety program in place for instructing employees in safe work practices.†

It should be noted that the Isolated Incident defense is an affirmative defense thus this defense must be raised "affirmatively during the formation of the issues of the case."‡ Safety and health professionals can also raise this defense at informal conference with supporting evidence. However, safety and health professionals should ensure that appropriate and compelling evidence for each of the elements is readily available, as well as a viable and operational safety program with appropriate and documented training provided to employees.

5. Impossibility of Compliance Defense

The Impossibility of Compliance defense involves the theory that compliance with a specific OSHA standard is impossible because of the nature of the specific work being performed. In essence, the safety professional is admitting that the work being performed is not in compliance, but is impossible to achieve or maintain compliance with the standard. Safety professionals should exercise caution in that any admission or implied admission that compliance is possible can result in the defense failing.§

---

* *Bill Turpin Painting, Inc.*, 5 OSHC 1576 (1977)
† OSHA, *Field Operations Manual* at Section X-€(1).
‡ See, for example, *Otis Elevator Co.*, 5 OSHC 1514 (1977).
§ See *Taylor Building Associates*, 5 OSHC 1083 (1977).

6. Employee Has No Exposure to the Hazard

OSHA possesses the burden of proving that at least one employee was exposed to, or had access to, the alleged condition(s) for which OSHA cited the employer. Safety and health professionals should review the facts and evidence and if no employees were exposed to the hazard and can prove this fact, this defense may be available. The safety and health professionals must be able to prove that no employees were exposed to the hazard identified by the compliance officer.

Although safety and health professionals have addressed this defense through several approaches, including demonstration of engineering and administrative controls to show that there is no feasible method through which to achieve compliance with the specific requirements of the standard, and demonstrating the impossibility of performing the necessary work if compliance with the standard is met, safety professionals should be aware that the employer has the burden to prove or demonstrate the applicability of this defense.

7. Employee Misconduct Defense

The theory of this defense is that the employer with a good safety and health program should not be penalized for a condition or hazard created by a trained employee. Additionally, it would be unfair to the employer if the hazard created by the employee through his/her misconduct was not preventable by the employer. Safety and health professionals should be aware that in order to establish this defense, the safety professional must demonstrate that: (A) the violation identified by the compliance officer resulted exclusively from the employee's misconduct; (B) the violation was not participated in, observed by, or performed with the knowledge and/or consent of any supervisory or management personnel; and (C) that the employee's misconduct contravened a well-established company policy or work rule which was in effect at the time, well-publicized to the employees, and actively enforced through disciplinary action or other appropriate procedures. Additionally, the safety and health professional must prove through documentation or other sources that the employee was appropriately trained and instructed in safe work practices. For this defense, the safety professional should be aware that the burden is on the employer to prove and establish that the safety program exists and is uniformly enforced.

8. No Hazard Defense

When OSHA promulgates a standard, there is a presumption that a hazard exists. In situations where the employer has been previously inspected by a compliance officer and the compliance officer indicated that the particular situation or condition was not a violation, safety professionals may be able to argue that the current alleged violation is not a recognizable hazard and only the subjective opinion on the compliance officer. Safety and health professionals should be aware that OSHA compliance officer possess different backgrounds and different opinions with regards to hazards. No opinion by an OSHA compliance officer is binding on OSHA or the OSHRC.

Safety and health professionals should be aware that although the factual defenses identified are recognized defenses, other potential defenses can be argued by creative practitioners. Depending on the facts and circumstances, new and creative challenges can be made by safety professionals at informal conference, as well as before the OSRRC. Safety and health professionals are encouraged to review and utilize the OSHA *Field Operations Manual* to "reverse engineer" the citation and alleged violations, with a special focus on the categorization of the penalties and standard(s) cited. Safety professionals should remember that the burden of proof is on OSHA, and safety and health professionals should "turn over every stone" to find applicable defenses if the facts and circumstances support the defense. In the event that the safety professional possesses no defense(s), good faith and immediately addressing and eliminating the hazard can go a long way in reducing the penalties through settlement.

Upon receipt of the safety and health professional's timely notice of contest or appeal of an OSHA citation (including proposed penalties), the OSHA Area Director is required to notify the Occupational Safety and Health Review Commission (hereinafter "OSHRC") of the notice of contest in order that the OSHRC can properly schedule or docket a hearing. Scheduling or docketing a case would include appointment of a judge, scheduling a courtroom, and scheduling the date and time of hearing.

Within 20 days after receipt of the safety and health professional's Notice of Contest to an OSHA citation, the Secretary of Labor must file a written complaint document (known as a "complaint") with the OSHRC identifying the location, time, and circumstances of the alleged violation and specify the circumstances upon which the proposed penalties are founded.* Upon receipt of the complaint, the safety and health professional, usually through legal counsel, are provided 15 days after receipt to file a written "Answer" to the complaint with the OSHRC. Safety and health professionals should note that these documents usually are required to be posted and employees or union representatives are provided an opportunity to participate in an OSHA hearing if they properly petition the OSHRC to intervene as parties of interest.

Safety and health professionals should be aware that, in most circumstances, the employer will acquire legal counsel for the OSHRC hearing. However, the safety and health professional, who possesses the most knowledge regarding the inspection and alleged violations, will be an important and essential component of the legal defense team. It is essential that the safety professional provide all evidence, including inspection notes, documents, photographs, and other evidence collected during the inspection to legal counsel. Safety and health professionals should also be aware that the OSHRC has authority to issue subpoenas requiring attendance and testimony of witnesses and the production of documents upon application of any party to this OSHA proceeding.

OSHA hearings are conducted by administrative law judges (known as "ALJ") who are assigned by the chairperson of the OSHRC. The ALJ possesses many of the same powers as a federal district court judge and often requires the parties to

---

* 29 C.F.R. § 2200.33.

participate in prehearing conference(s) primarily for the purpose of exploring and facilitating settlement of the issues or simplifying the issues prior to the hearing. The mechanics of the OSHRC hearing parallel the rules of evidence applicable to the federal district court. OSHA, or the Secretary of Labor, has the burden of proof in any hearing involving a citation or proposed penalties. OSHA's attorney(s) as well as the employer's attorney(s) are entitled to a reasonable period of time for oral arguments at the close of the hearing upon proper request. All parties are permitted, upon proper request prior to the close of the hearing, to file briefs in support of their position as well as findings of fact and conclusions of law.

Safety and health professionals should be aware that there is a simplified OSHRC proceeding, as well as mediation, which may be offered by the OSHRC. Simplified proceedings are designed to save time and expense, with many of the formal pleadings, such as complaint and answer, being eliminated and the Federal Rules of Evidence are not be applicable. Interlocutory appeals and discovery are not permitted. The ALJ will conduct the simplified hearing and issue a written decision. In mediation, OSHA and the company are brought together in an attempt to settle the various issues in question. If the parties agree, a settlement document will be drafted for review and approval by the OSHRC.

Safety and health professionals should be aware that the OSHRC hearing parallels the processes in federal district court. After the Complaint and Answer, there may be motions, as well as discovery and other preliminary matters. The hearing is scheduled before the ALJ, usually at the local federal courthouse or other appropriate location. There is no jury. At the hearing, there may be opening statements by the parties, followed by witnesses testifying under oath and being cross-examined by the other party. A verbatim transcript is usually made of the hearing, and the Federal Rules of Evidence apply. The parties often provide a closing statement, and then the ALJ will close the hearing. The parties may submit briefs after the hearing which the ALJ will review. After a period of time, the ALJ will issue a written decision which contains findings of fact and conclusions of law, and either affirms the citation and penalties, vacates the citation, or modifies the citation, proposed penalties, and/or abatement requirements. The ALJ's decision is filed with the OSHRC and may be directed for review by any OSHRC member *sua sponte* or in response to a party's petition for discretionary review. Safety and health professionals should be aware that failure to file a petition for discretionary review precludes subsequent judicial review by the courts. This petition for discretionary review must be filed within 20 days of the ALJ's mailing of a copy of the decision to the parties.*

The OSHRC is empowered to issue an order affirming the ALJ's decision, modifying the ALJ's decision or vacating the citation and/or proposed penalties. Although the OSHRC has the authority to direct "other appropriate relief" upon review of the ALJ's decision, the OSHRC does not have the authority to award attorney fees or assess costs against a party. Safety professionals should be aware that the OSHRC is not bound by the ALJ's findings of fact; however, the OSHRC is required to substantiate its contentions should it reject the ALJ's decision in a case. Parties who were adversely affected by a decision or order from the OSHRC may obtain review of the

---

* 29 C.F.R. § 2200.91.

decision or order in an appropriate U.S. Court of Appeal. As permitted under Section 11(a) of the OSH Act, any person adversely affected by the OSHRC's final order may file, within 60 days of the decision, a petition for review in the U.S. Court of Appeals for the circuit in which the alleged violations occurred or in the U.S. Court of Appeal for the District of Columbia.*

Safety and health professionals should be aware of the rules and requirements in order to provide their companies or organizations the protections afforded to them under the law. Safety professionals should not fear an inspection by OSHA; however, preparation is necessary to ensure that all legal rights are protected. With most citations and penalties, effective preparation and communications lead to a cost-effective and amicable settlement at informal conference. However, it is imperative that safety professionals review each and every detail of the citation and proposed penalties in preparation for any appeal. This preparation starts well before there ever is a knock at the door!

■■

OSHRC – Rules of Procedure

■■

Occupational Safety and Health Review Commission
(Updated as of 01/15/10)

PART 2200 – RULES OF PROCEDURE

SUBPART A – GENERAL PROVISIONS

2200.1 Definitions.
2200.2 Scope of rules; applicability of Federal Rules of Civil Procedure; construction.
2200.3 Use of gender and number.
2200.4 Computation of time.
2200.5 Extensions of time.
2200.6 Record address.
2200.7 Service and notice.
2200.8 Filing.
2200.9 Consolidation.
2200.10 Severance.
2200.11 [Reserved]
2200.12 References to cases.

SUBPART B – PARTIES AND REPRESENTATIVES

2200.20 Party status.
2200.21 Intervention; Appearance by non-parties.

---

* 29 C.F.R. § 651 *et seq.*

2200.73 Interlocutory review.

2200.74 Filing of briefs and proposed findings with the Judge; Oral argument at the hearing.

SUBPART F – POSTHEARING PROCEDURES

2200.90 Decisions of Judges.

2200.91 Discretionary review; Petitions for discretionary review; Statements in opposition to petitions.

2200.92 Review by the Commission.

2200.93 Briefs before the Commission.

2200.94 Stay of final order.

2200.95 Oral argument before the Commission.

2200.96 Commission receipt pursuant to 28 U.S.C. 2112(a)(1) of copies of petitions for judicial review of Commission orders when petitions for review are filed in two or more courts of appeals with respect to the same order.

SUBPART G – MISCELLANEOUS PROVISIONS

2200.100 Settlement.

2200.101 Failure to obey rules.

2200.102 Withdrawal.

2200.103 Expedited proceeding.

2200.104 Standards of conduct.

2200.105 Ex parte communication.

2200.106 Amendment to rules.

2200.107 Special circumstances; waiver of rules.

2200.108 Official Seal of the Occupational Safety and Health Review Commission.

SUBPART H – SETTLEMENT PART

2200.120 Settlement procedure.

Subpart I–L – [Reserved]

SUBPART M – SIMPLIFIED PROCEEDINGS

2200.200 Purpose.

2200.201 Application.

2200.202 Eligibility for Simplified Proceedings.

2200.203 Commencing Simplified Proceedings.

2200.204 Discontinuance of Simplified Proceedings.

2200.205 Filing of pleadings.

2200.206 Disclosure of Information.

2200.207 Pre-hearing conference.

2200.208 Discovery.

2200.209 Hearing.

PART 2204 – RULES IMPLEMENTING THE EQUAL ACCESS TO JUSTICE ACT

SUBPART A – GENERAL PROVISIONS

SUBPART B – INFORMATION REQUIRED FROM APPLICANTS

SUBPART C – PROCEDURES FOR CONSIDERING APPLICATIONS

■■■

(Table 11.1)

■■■

* OSHRC website located at www.oshrc.gov.

Guide to Review Commission Procedures

Occupational Safety and Health Review Commission

November 2007

Table of Contents

SECTION 1 – INTRODUCTION

The Review Commission
Purpose of this Guide
Rules of Procedure
Using This Guide
Parties May Represent Themselves
Time is of the Essence
Sample Legal Documents
Questions Regarding Proper Procedure

SECTION 2 – PRESERVING RIGHTS AND CHOOSING A PROCEEDING

OSHA Citation
Employer's Notice of Contest
Informal Conference with OSHA
Content and Effect of Notice of Contest
Notice of Docketing
Employee Notification
Employees May Contest Abatement Period
Employees May Elect Party Status
Party Requests for Simplified Proceedings
Choosing Simplified Proceedings or Conventional Proceedings

SECTION 3 – AN OVERVIEW OF HEARINGS CONDUCTED UNDER CONVENTIONAL PROCEEDINGS

The Complaint
The Answer
Discovery
Scheduling Order or Conference
Withdrawal of Notice of Contest
Settlement
Hearings
Hearing Transcripts
Post-hearing Briefs
Judge's Decision and Petition for Discretionary Review
Decisions Final in 30 Days

SECTION 4 – SIMPLIFIED PROCEEDINGS – AN OVERVIEW FOR EMPLOYERS AND EMPLOYEES

What are Simplified Proceedings?
Major Features of Simplified Proceedings
Cases Eligible for Simplified Proceedings
Employee or Union Participation
Should You Ask for Simplified Proceedings?
Complaint and Answer
Beginning Simplified Proceedings
Notifying Other Parties
Objections to and Discontinuing Simplified Proceedings
Pre-hearing Conference
Review of the Judge's Decision

SECTION 5 – OTHER IMPORTANT THINGS TO KNOW

Appearances in Commission Procedures
Penalties
Private (Ex Parte) Discussions
Petition for Modification of Abatement
Expedited Proceedings
Maintaining Copies of Papers Filed with the Judge

SECTION 6 – DESCRIPTIVE TABLE OF CONVENTIONAL PROCEEDINGS FOR CONTESTING AN OSHA CITATION

Events Common to All Proceedings
Events Pertaining to Conventional Proceedings

SECTION 7 – DESCRIPTIVE TABLE OF CONTESTING AN OSHA CITATION AND CHOOSING SIMPLIFIED PROCEEDINGS

Events Common to All Proceedings
Events Pertaining to Simplified Proceedings

SECTION 8 – DESCRIPTIVE TABLE OF EVENTS PERTAINING TO REVIEW OF AN ADMINISTRATIVE LAW JUDGE'S DECISION

GLOSSARY

APPENDIXES/SAMPLE LEGAL DOCUMENTS

Notice of Contest
Complaint and Certificate of Service
Answer
Request for Simplified Proceedings

Notice of Decision
Petition for Discretionary Review
Direction for Review
Notice of Withdrawal

SECTION 1 – INTRODUCTION

THE REVIEW COMMISSION

The Occupational Safety and Health Review Commission ("Review Commission") is an independent agency of the U.S. Government that was established by the Occupational Safety and Health Act of 1970 ("Act") to be like a court that resolves certain disputes under the Act. The Review Commission is composed of three members who are appointed by the President of the United States and confirmed by the Senate for six-year terms. It employs Administrative Law Judges to hear cases.

The Act was passed by Congress to "assure safe and healthful working conditions for working men and women." The Act also established another agency, the Occupational Safety and Health Administration ("OSHA"), which is part of the U.S. Department of Labor, to enforce the law. OSHA issues regulations setting occupational safety and health standards that an employer must follow. As part of its enforcement responsibilities, OSHA may also conduct an inspection of a workplace. If OSHA's inspectors find what they believe are unsafe or unhealthy conditions, they may issue a citation to an employer. A citation includes allegations of workplace safety or health violations, proposed penalties, and proposed dates by which an employer must correct the alleged hazardous conditions.

If the cited employer or any of its employees or an employee representative disagrees with the citation, they may then file a timely **notice of contest.** The Review Commission (which is **completely independent** of OSHA) then comes into the picture to resolve the dispute over the citation.

PURPOSE OF THIS GUIDE

This Guide is intended to inform employers, employees, and other interested persons about Review Commission proceedings. It provides an overview of the proceedings conducted before the Administrative Law Judges and the Commission Members and it is primarily intended to assist persons in defending their business or their employer's business after having contested an OSHA citation. It will also be useful to other persons who desire a general overview of the Review Commission and its procedures.

The Review Commission also publishes a Guide to Simplified Proceedings and an Employee Guide to Review Commission Procedures that may be obtained at the Review Commission website, located at http://www.oshrc.gov, or by writing or calling:

Executive Secretary
U.S. Occupational Safety and Health Review Commission
1120 20th Street, N.W., 9th Floor
Washington, D.C. 20036-3457
(202) 606-5400

RULES OF PROCEDURE

The Review Commission's Rules of Procedure are published in Part 2200 of Title 29, Code of Federal Regulations ("C.F.R."). These Rules may be available in a local library. They can also be obtained at the Review Commission Website, http://www.oshrc.gov, or by contacting the Review Commission's Office of The Executive Secretary at the address or telephone number above. References to the Rules in this Guide state "See Rule" and the appropriate number. (For example, "See Rule 4" refers to 29 C.F.R. § 2200.4.)

This guide is intended to provide an overview of the Review Commission's procedures and it is not intended to be a substitute for the Rules of Procedure, which are followed in the Review Commission's proceedings in deciding cases. Parties to cases should review the Rules and follow them in proceedings before judges and the Commission members.

USING THIS GUIDE

This guide describes many of the documents and steps in proceedings before the Commission members and judges. Throughout this Guide, important terms are shown in bold italics and many are included in the **Glossary**.

PARTIES MAY REPRESENT THEMSELVES

The Review Commission's Rules do not require that a **party**—an employer, a union, or affected employee(s)—be represented by a lawyer. However, proceedings before the Review Commission are legal in nature. Certain legal formalities must be followed. OSHA will be represented by lawyers from the **Solicitor of Labor's Office,** the employer may be represented by a lawyer, and the decision in the case may have consequences beyond the amount of the penalty. For example, a decision may require corrective actions at a worksite. Parties to cases should consider carefully whether to hire a lawyer to represent them in their case.

TIME IS OF THE ESSENCE

Many of the documents parties are required to file, such as those needed to disagree with an OSHA citation or proposed penalty, must be filed within a specific time period. Failure to file documents as required could result in a citation becoming a final order without an opportunity to appeal. Therefore, parties to cases must respond promptly to communications received from the judge, the Commission, or any of the other parties to the dispute.

SAMPLE LEGAL DOCUMENTS

The Appendixes contain forms and sample correspondence that may be used or referred to in preparing a case. These are mentioned as appropriate throughout the Guide.

QUESTIONS REGARDING PROPER PROCEDURE

Parties to cases having questions regarding the Commission's procedures in cases pending before a judge should call the judge's office. At other stages of the proceedings, inquiries should be directed to the Executive Secretary's Office at 202-606-5400. Commission employees cannot give legal advice or advise a party how to proceed. However, they can provide information about the Rules of Procedure and the Commission's methods of processing cases.

SECTION 2 – PRESERVING RIGHTS AND CHOOSING A PROCEEDING

OSHA CITATION

Cases that come before a Review Commission judge arise from inspections conducted by OSHA, an agency of the United States Department of Labor. When OSHA finds what it believes to be a **violation** at a worksite, it will notify the employer in writing of the alleged violation and the period of time it thinks reasonable for correction by issuing a written **citation** to the employer.

The period of time stated in the citation for an employer to correct the alleged violation is the **abatement period.** OSHA likely will also propose that the employer pay a monetary penalty.

The Act requires that the employer **immediately post a copy of the citation** in a place where **affected employees** will see it, to have legal notice of it. An affected employee is an employee who has been exposed to or could be exposed to any hazard arising from the cited violations.

EMPLOYER'S NOTICE OF CONTEST

If an employer disagrees with any part of the OSHA citation—the alleged violation, the abatement period, or proposed penalty—**it must notify OSHA in writing of that disagreement within 15 working days** (Mondays through Fridays, excluding Federal holidays) of receiving the citation. This written notification is referred to as a notice of contest, and if it is filed late with OSHA, the employer is not usually entitled to have the dispute resolved by the Commission.

The notice of contest must be **delivered in writing to the Area Director of the OSHA office that mailed the citation.** The Area Director's name and address will be listed on the citation. **A notice of contest must not be sent to the Commission.**

INFORMAL CONFERENCE WITH OSHA

If a citation is issued, an employer may schedule an informal conference or engage in settlement discussions with the OSHA Area Director, **but this does not delay the 15 working day deadline for** filing a notice of contest. Thus, if an informal conference is conducted that does not result in a written settlement agreement, if a notice of contest is not filed within the 15 working day deadline, all citation items must be abated and all penalties must be paid.

CONTENT AND EFFECT OF NOTICE OF CONTEST

The notice of contest is a statement that an employer intends to contest (1) the alleged violations, (2) the specific abatement periods, and/or (3) the penalties proposed by OSHA. The notice should state in detail those matters being contested. (See Appendixes 1A, 1B.)

For example, if there are two citations and the employer wishes to contest only one of them, the citation being contested should be identified. If there are six different items alleged as violations in a single citation and the employer wishes to contest items 3, 4, and 6, those items should be specified.

If the employer wishes to contest the entire penalty, or only the amount for one citation or specific items of one citation, or only the abatement period for some or all of the violations alleged, this should also be specified.

For any item (violation) not contested, the abatement requirements must be fully satisfied and any related penalty must be paid to the Department of Labor. If the employer contests whether a violation occurred, the abatement period and the proposed penalty for that item is suspended until the Commission issues a final decision.

NOTICE OF DOCKETING

The OSHA Area Director sends the notice of contest to the Commission. The Executive Secretary's Office then notifies the employer that the case has been received and assigns a docket number. This docket number must be printed on all documents sent to the Commission.

EMPLOYEE NOTIFICATION

At the time the employer receives the Notice of Docketing that the case has been filed and given a docket number, the Commission will furnish a copy of a notice to be used to inform affected employees of the case. A pre-printed post card is sent to the employer with this notice; the employer returns the post card to the Commission to inform it that affected employees have been notified.

EMPLOYEES MAY CONTEST ABATEMENT PERIOD

Unions or **affected employees** wishing to participate in a dispute may file a notice of contest (see Appendix 1C) challenging the reasonableness of the period of time given to the employer for abating (correcting) an alleged violation.

Even if the employer does not contest the citation, unions or affected employees can object to the abatement period. **This must be done within 15 working days of the employer's receipt of the citation**. The notice of contest should state that the signer is an affected employee or a union that represents affected employees and that the signer wishes to contest the reasonableness of the abatement period.

The employee or the union must mail the notice of contest to the Area Director of the OSHA office that mailed the citation, not the Commission. The Area Director's name and address will be listed on the citation. (See Section 10 of the Act and Rules 20, 22 and 33.)

EMPLOYEES MAY ELECT PARTY STATUS

Employees may also elect party status to a case by filing a written notice of election at least ten days before the hearing. A notice of election filed less than ten days prior to the hearing is ineffective unless good cause is shown for not timely filing the notice. It must be served on all other parties in accordance with Rule 7. (See Rule 20.)

PARTY REQUESTS FOR SIMPLIFIED PROCEEDINGS

Cases heard by Administrative Law Judges may proceed in one of two ways: conventional proceedings or simplified proceedings. Each method is described in detail in Sections 3 and 4 of this Guide. The Chief Administrative Law Judge may designate a case for simplified proceedings soon after the notice of contest is received at the Review Commission. Parties may also request simplified proceedings within 20 days of the date on the notice of docketing. If a case is not designated for simplified proceedings, conventional proceedings are in effect.

CHOOSING SIMPLIFIED PROCEEDINGS OR CONVENTIONAL PROCEEDINGS

Simplified proceedings are appropriate for cases that involve less complex issues and for which more formal procedures used in conventional proceedings are deemed unnecessary to assure the parties a fair and complete contest. Simplified proceedings are covered in Section 4 of this Guide and the Commission has developed a Guide to Simplified Proceedings that is published on the Commission's website at http://www.oshrc.gov or may be obtained by writing or calling:

Executive Secretary
U.S. Occupational Safety and Health
Review Commission
1120 20th Street, N.W., 9th Floor
Washington, D.C. 20036-3457
(202) 606-5400

SECTION 3 – AN OVERVIEW OF HEARINGS CONDUCTED UNDER CONVENTIONAL PROCEEDINGS

This section describes the major features of the Commission's hearings conducted under the Conventional Proceedings method as opposed to hearings conducted under Simplified Proceedings. Simplified Proceedings are explained briefly in Section 4 and in a separate guide that should be consulted by those persons interested in that method of hearing cases.

THE COMPLAINT

Within 20 calendar days of receipt of the employer's notice of contest, the Secretary of Labor must file a written complaint with the Commission. A copy must be sent to the employer and any other parties. The complaint sets forth the alleged violation(s), the abatement period and the amount of the proposed penalty. See Appendix 2A for an example of a complaint. (See Rule 34.)

THE ANSWER

The employer must file a written **answer** to the complaint **with the Commission within 20 calendar days** after receiving the complaint from the Secretary of Labor. The answer must contain a short, plain statement denying allegations of the complaint that the employer wishes to contest. **Any allegation not denied by the employer is considered to be admitted.** In addition, if the employer has a specific defense it wishes to raise, such as (1) the violation was due to employee error or failure to follow instructions, or (2) compliance with a standard was infeasible, or (3) compliance with a standard posed an even greater hazard, the answer must describe that defense. **If the employer fails to file an answer to the Complaint on time, its Notice of Contest may be dismissed, and the Citation and Penalties may become final.** The Answer must be filed with the Commission by mailing it to:

Executive Secretary
U.S. Occupational Safety and Health
Review Commission
1120 20th Street N.W., 9th Floor
Washington, D.C. 20036-3457

or to the judge, if the case has been assigned to one. A copy of the answer must also be sent to the Secretary of Labor. See Appendix 3. (See Rule 34.)

DISCOVERY

Discovery is a method used whereby one party obtains information from another party or person before a hearing. Discovery techniques in Commission cases include (1) written questions, called interrogatories; (2) oral statements taken under oath, which are depositions; (3) asking a party to admit the truth of certain facts, called requests for admissions; and (4) requests that another party produce certain documents or objects for inspection or copying. In conventional proceedings, any party can use these discovery techniques without the judge's permission, except for depositions, which require that that parties agree to take depositions or that the judge order the taking of depositions after a party files a motion requesting permission to do so. (See Rules 51–57.)

SCHEDULING ORDER OR CONFERENCE

In conventional cases, discovery takes place after the answer is filed and before the hearing date. After the answer to the complaint is filed, the judge will issue an order setting a schedule for the case and may also hold a conference with the parties to clarify the issues, consider settlement, or discuss other ways to expedite the hearing. (See Rule 51.)

WITHDRAWAL OF NOTICE OF CONTEST

A party wishing to withdraw its notice of contest to all or parts of a case may do so at any time. The **Notice of Withdrawal** must be served on all affected employees and all other parties. A copy must also be sent to the judge. See example at Appendix 8. The withdrawal terminates the proceedings before the Commission with respect to the citation or citation items covered by the notice of withdrawal. (See Rule 102.)

SETTLEMENT

The Commission Encourages the Settlement of Cases. Cases can be settled at any stage. The Secretary of Labor and the employer must agree to the settlement terms, and the affected employees or their union must be shown the settlement before it will be approved.

Any party can also request that a Settlement Judge be appointed to help facilitate a settlement. (See Rule 100.)

HEARINGS

Hearings are governed by Rules 60–74. The parties will be notified of the time and place of the hearing at least 30 days in advance. The employer must post the hearing notice if there are any employees who do not have a representative and served on all unions representing affected employees. The hearing is usually conducted as near the work place as possible.

At the hearing, a Commission Judge presides. The hearing enables the parties to present evidence on the issues raised in the complaint and answer. Each party to the proceedings may call witnesses, introduce documentary or physical evidence, and cross-examine opposing witnesses. In conventional proceedings, the Commission follows the Federal Rules of Evidence. Under these rules, evidence is only admitted into the record if it meets certain criteria that are designed to assure that the evidence is reliable and relevant.

HEARING TRANSCRIPTS

A transcript of the hearing will be made by a court reporter. A copy may be purchased from the reporter.

POST-HEARING BRIEFS

After the hearing is completed and before the judge reaches a decision, each party is given an opportunity to submit to the judge proposed findings of fact and conclusions of law with reasons why the judge should decide in its favor. Proposed findings of fact are what a party believes actually happened in the circumstances of a case based upon the evidence introduced at the hearing. Proposed conclusions of law are how a party believes the judge should apply the law to the facts of a case. The statement of reasons is known as a **brief**. (See Rule 74.)

Judge's Decision and Petition for Discretionary Review

After hearing the evidence and considering all arguments, the judge will prepare a decision based upon all of the evidence placed in the hearing record and mail copies of that decision to all parties. The parties then can object to the judge's decision by filing a **Petition for Discretionary Review** (See Appendix 6 for an example.) **Instructions for submitting such a petition will be stated in the judge's letter transmitting the decision and in a Notice of Docketing of Administrative Law Judge's Decision issued by the Executive Secretary's Office.** See Rule 91 for further information on filing **Petitions for Discretionary Review.**

Decisions Final in 30 Days

If a Commissioner does not order review of a judge's decision, it becomes a final order of the Commission 30 days after the decision has been filed. If a Commissioner does direct review, it will ultimately issue its own written decision and that decision becomes the final order of the Commission.

Any party who is adversely affected by a final order of the Commission can appeal to a United States Court of Appeals. However, the courts usually will not hear appeals from parties that have not taken advantage of all possible appeal rights earlier in the case. **Thus, a party who failed to file a petition for review of the judge's decision with the Commission likely will not be able to later appeal that decision to a court of appeals.**

SECTION 4 – SIMPLIFIED PROCEEDINGS – AN OVERVIEW FOR EMPLOYERS AND EMPLOYEES

What are Simplified Proceedings?

Simplified Proceedings are designed to resolve small and relatively simple cases in a less formal, less costly, and less time-consuming manner. **The Commission's Chief Administrative Law Judge ("Chief Judge") or the judge assigned to your case notifies you that your case will be heard under Simplified Proceedings.** Even though the legal process is streamlined, the proceedings are still a trial before an Administrative Law Judge with sworn testimony and witness cross-examination.

Major Features of Simplified Proceedings

Under Simplified Proceedings:

1. Early discussions among the parties and the Administrative Law Judge are required to narrow and define the disputes between the parties.

2. Motions, which are requests asking the judge to order some act to be done, such as having a party produce a document, are discouraged unless the parties try first to resolve the matter among themselves.
3. Disclosure. The Secretary is required to provide the employer with inspection details early in the process. In some cases, the employer will also be required to provide certain documents, such as evidence of their safety program, to the Secretary.
4. Discovery, which is the written exchange of information, documents and questionnaires between the parties before a hearing, is discouraged and permitted only when ordered by the judge.
5. Appeals of actions taken by the judge before the trial and decision, such as asking the Commission to rule on the judge's refusal to allow the introduction of a piece of evidence, called interlocutory appeals, are not permitted.
6. Hearings are less formal. The Federal Rules of Evidence, which govern other trials, do not apply. Each party may present oral argument at the close of the hearing. Post-hearing briefs (written arguments explaining your position in the case), will not be allowed except by order of the judge. (See Rule 209(e).) In some instances, the judge will render his or her decision "from the bench," which means the judge will state at the end of the hearing whether the evidence and testimony proved the alleged violations and will state the amount of the penalty the employer must pay, if a violation is found.

CASES ELIGIBLE FOR SIMPLIFIED PROCEEDINGS

It is possible that not all relatively small cases eligible for Simplified Proceedings will be selected. (See Rules 202 and 203(a).) **The Chief Judge will assign cases for Simplified Proceedings or, if your case is not selected, you may request that it be chosen.** Cases appropriate for Simplified Proceedings are those with one or more of the following characteristics:

- relatively simple issues of law or fact with relatively few citation items,
- total proposed penalty of not more than $30,000,
- a hearing that is expected to take less than two days, or
- a small employer whether appearing with or without an attorney.

Cases having willful or repeated violations or that involve a fatality are not deemed appropriate for Simplified Proceedings.

EMPLOYEE OR UNION PARTICIPATION

Affected employees or their unions who file a notice of contest may also request Simplified Proceedings. Unions or an affected employee (ones exposed to the alleged health or safety hazard) wishing to participate in a dispute may file a notice of contest (see Appendix 1C) challenging the reasonableness of the period of time given to the employer for abating (correcting) an alleged violation. Even if the employer does not contest the citation, unions or affected employees can object to the abatement period. **This must be done in writing within 15 working days of the employer's receipt of the citation.** You might consider Simplified Proceedings if you or your local union wish to avoid the time and expense of a full-blown hearing. You might also participate by electing party status after the employer files a notice of contest, but must do so promptly.

When affected employees or their unions contest the time allowed for abatement, and the employer does not contest the citation, the employer may in turn elect to participate. Once the abatement date has been contested, other employees or unions may likewise elect to participate.

An employee or a union must **mail a written notice of contest to the Area Director of the OSHA office that issued the citation, not the Commission.** First-class mail will be sufficient for this purpose. The Area Director's name and address will be listed on the citation. This process is governed by Section 10 of the Act and Commission Rules 20, 22 and 33.

SHOULD YOU ASK FOR SIMPLIFIED PROCEEDINGS?

If you are an employer, have received an OSHA citation, have filed a notice of contest, and the total proposed penalties in the citation are between $20,000 and $30,000, the Chief Judge may designate your case for Simplified Proceedings. If the penalties are $20,000 or less, you may file a request for Simplified Proceedings provided that there is no allegation of willfulness or a repeat violation, and the case does not involve a fatality.

You must file your request within 20 days of docketing of your case by the Executive Secretary's Office. The request must be in writing and it is sufficient if you state: "I request Simplified Proceedings." The Chief Judge or the assigned judge will then rule on your request.

Your case may be appropriate for Simplified Proceedings but that does not necessarily mean that your particular interests are best served by requesting Simplified Proceedings. In addition to considering time and expense, you should base your decision on the facts of your case, the nature of your objections to the citation, what you will try to show the judge at the hearing, the amount of paperwork involved if your case proceeds under conventional proceedings as compared to Simplified Proceedings, and whether you have legal representation.

You should also remember that, in most circumstances, your interests may be best served if you can reach a fair and equitable settlement of your case with OSHA before a hearing. Either way, Simplified Proceedings or conventional, the proceedings are legal and the Secretary of Labor will most likely be represented by an attorney. You have the right to represent yourself or to be represented by an attorney or by anyone of your choosing.

COMPLAINT AND ANSWER

Once your case is selected for Simplified Proceedings, the complaint and answer are not required. However, until an employer is notified that a case has been designated for Simplified Proceedings, conventional procedures should be followed and an answer must be filed. (See Rule 205(a).)

BEGINNING SIMPLIFIED PROCEEDINGS

You need not give any reasons for requesting Simplified Proceedings. A letter saying simply "I request Simplified Proceedings," and indicating the Docket Number assigned to your case, is sufficient. (See Appendix 4.) The letter must be sent to:

Executive Secretary
U.S. Occupational Safety and Health
Review Commission
1120 20th Street, N.W., 9th Floor
Washington, D.C. 20036-3457

NOTIFYING OTHER PARTIES

It is required that a copy of your request for Simplified Proceedings must be sent to the Regional Solicitor of the Department of Labor office for your region. The address is on your Notice of Docketing. All employee representatives, including an employee union, that have elected party status must also be sent a copy of your request for Simplified Proceedings. **A brief statement indicating to whom, when, and how your request was served on the parties** in the case **must be received with the request for Simplified Proceedings.** An example of such a "Certificate of Service" follows: (See Rule 203(b).)

Example: I certify that on October 1, 2004, a copy of my request for Simplified Proceedings was sent by first class mail to Jane Doe, Office of the Solicitor, U.S. Department of Labor, 123 Street, City, State Zip Code and to John Doe, President, Local 111, GHI International Union, 456 Street, City, State Zip Code. (See Appendix 2B.)

OBJECTIONS TO AND DISCONTINUING SIMPLIFIED PROCEEDINGS

Should you decide to object to the Chief Judge's assignment of your case to Simplified Proceedings or another party's request for Simplified Proceedings, all you need to do is file a brief written statement with the judge assigned to your case or, if the case has not been assigned to a judge, with the Chief Judge, explaining why your case is inappropriate for Simplified Proceedings. The judge is required to rule on a request for Simplified Proceedings within 15 days. Therefore, you must file your objections as soon as possible.

If you disagree with another party's request to discontinue Simplified Proceedings and you want your case to continue under Simplified Proceedings rules, you have seven days to file a letter explaining why you disagree. (See Rule 204(b).)

If it appears that a case is inappropriate for Simplified Proceedings, the use of this method may be discontinued by the judge at his or her discretion. A party may also request that the judge discontinue Simplified Proceedings. The request must explain why the requesting party believes that the case is inappropriate for Simplified

Proceedings. **If you agree with another party's request to discontinue Simplified Proceedings, you should submit a letter saying so.** When all parties agree that a case is inappropriate for Simplified Proceedings, the judge is required to grant the request. If the judge orders that a case be taken out of Simplified Proceedings, the case will proceed under the Commission's conventional procedures.

PRE-HEARING CONFERENCE

Soon after the parties exchange the required information, the judge will hold a pre-hearing conference to either reach settlement in the case or to find out which factual and legal issues the parties agree on. This discussion may be conducted in person but is usually conducted by a telephone conference call. The purpose of the pre-hearing conference is to settle the case or, if settlement is not possible, to determine what areas of dispute must be resolved at a hearing. Even if a settlement of the entire case cannot be reached, the parties are required to attempt agreement on as many facts and issues as possible. The discussion will include the following topics: (See Rule 207.)

1. **Narrowing of Issues.** The parties will be expected to discuss all areas in dispute and to resolve as many as possible. Where matters remain unresolved, the judge will list the issues to be resolved at the hearing.
2. **A Statement of Facts.** The parties are expected to agree on as many of the facts as possible. Examples of these facts may include: the size and nature of the business, its safety history, details of the inspection, and the physical nature of the worksite.
3. **A Statement of Defenses.** You will be required to list any specific defenses you might have to the citation. The burden is on the Secretary to establish that each violation occurred. However, you should be prepared to tell the judge all reasons why you believe that the Secretary's allegations are wrong.

   You might also have what is called an "affirmative defense." An affirmative defense is a recognized set of circumstances in which an employer will be found not in violation even though the employer did not comply with the cited standard. For example, you may believe that the alleged violation was the result of an employee acting contrary to a work rule that has been effectively communicated and enforced. Or, you may think that compliance with the standard was impossible or infeasible, or would have resulted in a danger to employees that was greater than the danger that the standard was designed to prevent.

   You should be aware that **the burden of proving an affirmative defense is on you, the employer.** Therefore, if you argue that the violation was the result of employee misconduct, at the hearing you will have to prove to the judge that you had an effectively communicated and enforced work rule. As will be discussed later, if you raise an affirmative defense, the judge may require you to provide the Secretary of Labor with certain documents before the hearing regarding the defense. For example, if you claim that an employee violated a written work rule, you will probably be required to provide the Secretary with a copy of your company's safety rules.

It is critical that you set forth your defenses at the pre-hearing conference. You may be prohibited from later asserting any defenses not raised at the pre-hearing conference. Remember, even if your defense does not excuse the violation, the judge may find it relevant in determining the penalty amount.

4. **Witnesses and Exhibits.** The parties are expected to list the witnesses they intend to call if there is a hearing, and to list any documents or physical evidence they intend to introduce to support their positions. For example, you should list any photographs that you believe show the existence of a safety device that the Secretary claims you failed to provide.

REVIEW OF THE JUDGE'S DECISION

Any party dissatisfied with the judge's decision may petition the Commission for review of that decision.

No particular form is required for the petition (see Appendix 6). However, it should clearly explain why you believe that the judge's decision is in error on either the facts or the law or both. **Review of a judge's decision is at the discretion of the Commission. It is not a right.** (See Rules 91 and 210.)

Your petition should be filed no later than 20 days after issuance of the judge's written decision. Under the law, the Commission cannot grant any petition for review more than 30 days after the judge's decision is filed. Therefore, **your petition must be filed as soon as possible to obtain maximum consideration.**

The Commission will notify you whether your petition has been granted (see Appendix 7). If it is granted, your case will then proceed under the Commission's conventional rules.

SECTION 5 – OTHER IMPORTANT THINGS TO KNOW

APPEARANCES IN COMMISSION PROCEDURES

Any employer, employee, or union that initially files a notice of contest is automatically a party to the proceedings. Affected employees or their union may also choose to participate as a party where the employer has filed a notice of contest. Any party may appear in a Commission proceeding either personally, through an attorney, or through another representative. (See Rule 22.) Such a person need not be an attorney. However, all representatives of parties must either enter an appearance by signing the first document filed on behalf of the party or intervenor, or thereafter by filing an entry of appearance. (See Rule 23.)

Every party to the case must serve every other party or representative with copies of every document it files with the Commission or judge. Service is made by first class mail, electronic transmission, or personal delivery. (See Rule 7(c).)

All notices the Commission sends to the parties will list the name and address of all parties or their representatives. (See Rule 22.) Parties must do the same.

PENALTIES

OSHA only **proposes** amounts which it believes are appropriate as penalties. These proposals automatically become penalties assessed against the cited employer when the enforcement action is not contested. Once a **citation or proposed penalty** is contested, the amount of the penalty for that citation, if any, will be decided by the Commission or a judge.

When a case goes to hearing before a Review Commission judge, the employer's evidence and argument on what penalty, if any, should be assessed, receives the same consideration as the evidence and argument of the Secretary of Labor.

The four factors that the law requires the Commission to consider in determining the appropriateness of civil penalties are:

- The size of the business of the employer being charged,
- The gravity of the violation,
- The good faith of the employer, and
- The employer's history of previous violations.

The amounts that may be assessed as civil penalties by the Commission under Section 17 of the Act are as follows:

- For a serious or non-serious violation: up to $7,000.00
- For violations committed willfully or repeatedly: up to $70,000.00
- For failure to correct a violation within the period permitted: up to $7,000.00 for each day it remains uncorrected.

PRIVATE (EX PARTE) DISCUSSIONS

Parties to cases before the Commission may not communicate ex parte (without the knowledge or consent of the other parties) with respect to the merits of a case with the judge (except a Settlement Judge), a Commissioner, or any employee of the Commission. In other words, no participant, directly or indirectly, may discuss the case or make any argument about a matter in a case to any of these people unless done in the presence of the other participants who are given an equal opportunity to present their side, or unless it is done in writing and copies are sent to all other parties. Violation of this rule may result in dismissal of the offending party's case before the Commission. This prohibition does not, however, preclude asking questions about the scheduling of a hearing, or other matters that deal only with procedures. (See Rule 105.)

PETITION FOR MODIFICATION OF ABATEMENT

An employer who does not contest a **citation** is required to correct all of the violations within the **abatement period** specified in the **citation**. If the Commission upholds a contested citation, the employer must then correct the violation, with the **abatement period** starting on the date of the Commission's final order. If the employer has made a good faith effort to correct a violation within the **abatement period**, but has not been able to do so because of reasons beyond his or her control, the employer may file a **Petition for Modification of Abatement (PMA)**. This petition is filed with the **OSHA** area director and should be filed no later than the end of the next working day following the day on which abatement was to have been completed. It must state why the abatement cannot be completed within the given time. The PMA must be posted in a conspicuous place where all affected employees can see it or near the location where the violation occurred. The PMA must remain posted for 10 days. The Secretary of Labor may not approve a PMA until the expiration of 15 working days from its receipt.

At the end of the 15-day period, if the Secretary of Labor, affected employees, or their union object to the petition, the Secretary of Labor is required to forward the PMA to the Commission. After notice by the Commission to the employer and the objecting parties of its receipt of the PMA, each objecting party has 10 calendar days in which to file a response to the PMA setting out the reasons for opposing it. Proceedings before the Commission are conducted in the same way as notice-of-contest cases, except that they are expedited. The employer must establish that abatement cannot be completed for reasons beyond the employer's control, and has the burden of proving the petition should be granted. In cases of this kind, the employer is called the Petitioner and the Secretary of Labor is called the Respondent. (See Rules 37 and 103.)

EXPEDITED PROCEEDINGS

In certain situations, time periods allowed for certain procedures are shortened. The Commission's Rules of Procedure provide that an **Expedited Proceeding** may be ordered by the Commission. If an order is made to speed up proceedings, all parties in the case will be specifically notified. All **Petitions for Modification of Abatement and all employee contests** are automatically expedited. (See Rule 103.) Expedited proceedings are different from Simplified Proceedings. (See Rule 103.)

MAINTAINING COPIES OF PAPERS FILED WITH THE JUDGE

In order that Affected Employees may have the opportunity to be kept informed of the status of the case, the employer must keep available at some convenient place copies of all pleadings and other documents filed in the case so they can be read at reasonable times by Affected Employees.

SECTION 6 – DESCRIPTIVE TABLE OF CONVENTIONAL PROCEEDINGS FOR CONTESTING AN OSHA CITATION

EVENTS COMMON TO ALL PROCEEDINGS

- Employer files notice of contest with OSHA office that mailed citation— within 15 working days of receiving the citation.
- Employer receives notification (notice of docketing) from Commission of case, docket number and forms to notify employees.
- Employer posts notification to employees of case in progress.
- Union and/or affected employees may contest reasonableness of abatement period; notice of contest is sent to citing OSHA office within 15 working days of the employer's receipt of the citation.

- If the Chief Administrative Law Judge has not assigned the case for Simplified Proceedings, and if a party has not requested Simplified Proceedings within 20 days of the notice of docketing and the request is not granted, conventional proceedings will be used. (See Rule 203.)

EVENTS PERTAINING TO CONVENTIONAL PROCEEDINGS

The Employer:

- Receives a complaint from OSHA's attorneys.
- Files an answer to the complaint within 20 calendar days of receiving the complaint.
- Discusses discovery techniques with the judge when applicable.
- Participates in a conference call to discuss issues and a possible settlement.
- Engages in discovery; exchanges interrogatories and depositions.
- Discusses settlement in another conference call with OSHA and judge. If not settled, then:
  - Prepares for and participates in the hearing.
  - May purchase a copy of the hearing transcript and may choose to submit a brief to the judge.
- Judge issues a decision. (If dissatisfied, any party may ask for Commission review of the decision).

SECTION 7 – DESCRIPTIVE TABLE OF CONTESTING AN OSHA CITATION AND CHOOSING SIMPLIFIED PROCEEDINGS

EVENTS COMMON TO ALL PROCEEDINGS

- Employer files notice of contest with OSHA office that mailed citation—within 15 working days of receiving the citation.
- Employer receives notification (notice of docketing) from Commission of case, docket number and forms to notify employees.
- Employer posts notification to employees of case in progress.
- Union and/or affected employees may contest reasonableness of abatement period; notice of contest is sent to citing OSHA office within 15 working days of citation's posting at work place.
- If the Chief Administrative Law Judge has assigned the case for Simplified Proceedings, or if a party has requested Simplified Proceedings and the request is granted, Simplified Proceedings will be in effect. (See Rule 203.)

EVENTS PERTAINING TO SIMPLIFIED PROCEEDINGS

If all disputed issues not resolved at the prehearing conference, then parties:

- List witnesses and exhibits.
- Prepare for and participate in the hearing.

- Present oral arguments at the close of the hearing.
- May purchase a copy of the hearing transcript.
- Decide whether to request permission to file a brief.

Judge issues decision.
- If dissatisfied, any party may ask for Commission review of the decision.

SECTION 8 – DESCRIPTIVE TABLE OF EVENTS

Pertaining to Review of an Administrative Law Judge's Decision

If an employer is dissatisfied with an administrative law judge's decision and wishes to seek review by the Commission members, the employer:

- Receives judge's decision; dissatisfied with the outcome.
- Files petition for discretionary review of the judge's decision.
- Receives notification from Commission that case is or is not directed for review.

If the case is not directed for review, the judge's decision is a final order of the Commission and the employer may file a petition for review in a Court of Appeals. If the case is directed for review, all parties:

- Receive a request from Commission for briefs on review.
- File briefs on review before Commission.
- Receive Commission decision that may supersede the judge's decision and affirm, modify, or reverse it. In some cases, the judge's decision may be remanded for further proceedings.
- Files petition for review in Court of Appeals if dissatisfied with Commission decision.

See also Rules 90–96

GLOSSARY

Abatement Period: Period of time specified in citation for correcting alleged workplace safety or health violation.

Affected Employee: An affected employee is one who has been exposed to or could be exposed to any hazard arising from the cited violations—that is the circumstances, conditions, practices, or operations creating the hazard.

Answer: Written document filed in response to a complaint, consisting of short plain statements denying the allegations in the complaint which the employer contests.

Authorized Employee Representative: A labor organization, such as a union, that has a collective bargaining relationship with the employer and represents affected employees or may be an affected employee(s) in cases where unions do not represent the employees.

Brief: A written document in which a party states what the party believes are the facts of the case and argues how the law should be applied.

Certificate of Service: A document stating the date and manner in which the parties were served (given) a document. See Appendix 2B for sample certificate. (Also see definition of 'service.')

Citation: Written notification from OSHA of alleged workplace violation(s), proposed penalty(ies), and abatement period.

Complaint: Written document filed by the Secretary of Labor detailing the alleged violations contained in a citation.

Conventional Proceedings: Typical Review Commission proceedings, which are similar to, but less formal than, court proceedings.

Discovery: The process by which one party obtains information from another party prior to a hearing.

Exculpatory Evidence: Information that may clear one of a charge or of fault or of guilt; in the context of OSHRC cases, information that might help the employer's case.

Exhibit: Something, e.g., a document, video, etc., that is formally introduced as evidence at a hearing.

File: To send papers to the Commission Executive Secretary, or to the judge assigned to a case, and to give copies of those papers to the other parties in the case.

Interlocutory Appeal: An appeal of a judge's ruling on a preliminary issue in a case that is made before the judge issues a final decision on the full case. These types of appeals are infrequently made and are infrequently allowed. One example of an issue often raised in an interlocutory appeal is whether certain material that a party wants kept confidential, such as an employer's trade secrets or employee medical records, should become part of the public record in a case.

Motion: A request asking that the judge direct some act to be done in favor of the party making the request or motion.

Notice of Appearance: A written letter informing the Review Commission of the name and address of the person or persons who will represent a party (that is, the employer or a union or OSHA) in a case.

Notice of Contest: Written document disagreeing with any part of an OSHA citation.

Notice of Docketing: Written document from the Review Commission's Executive Secretary telling an employer, the Secretary of Labor, and any other parties in a case that the case has been received by the Commission and given an OSHRC docket number.

Notice of Withdrawal: A written document from a party withdrawing its notice of contest or the citation and thus terminating the proceedings before the Commission (See Appendix 8).

Party: The Secretary of Labor, anyone who files a notice of contest, or a union or affected employee(s) that requests party status.

Petition for Discretionary Review: A written request from a party in a case asking the Commission in Washington, D.C. to review and change the judge's decision. The grounds on which a party may request discretionary review are: (1) it believes the judge made findings of material facts which are not supported by the evidence; (2) it believes that the judge's decision is contrary to law; (3) it believes that a substantial question of law, policy, or abuse of discretion is involved; or (4) it believes that a prejudicial error was committed.

Pro Se: Latin for without an attorney.

Secretary of Labor: The head of the U.S. Department of Labor. OSHA is part of that Department.

Service: Sending by first class mail or personal delivery a copy of documents filed in a case to all parties in the case. See Definition of "Certificate of Service." (See Rule 7.)

Settlement: An agreement reached by the parties resolving the disputed issues in a case.

Simplified Proceedings: Review Commission proceedings that are less formal than Conventional Proceedings and designed for smaller and relatively simple cases. A complaint and answer are not required and discovery occurs only if the judge permits it.

Solicitor of Labor: The U.S. Department of Labor's chief lawyer who has offices throughout the country. Lawyers from these offices represent the Secretary of Labor and OSHA in Review Commission cases

APPENDIXES/SAMPLE LEGAL DOCUMENTS

This section is not intended to be a manual of forms, and the sample legal documents here are limited in number. The sample legal documents are intended for illustration to familiarize the reader with the general nature of some of the documents received and issued. Many of the documents received by the Commission, such as those in Appendixes 2, 3, and 6 (Complaint, Answer, and Petition for Discretionary Review), vary significantly from case to case.

APPENDIX 1 – NOTICE OF CONTEST

Appendix 1A. Notice Of Contest To Citation And Proposed Penalties

XYZ Corp.
123 Street
City, State Zip Code

February 26, 2004

ABC, Area Director
Occupational Safety and Health Administration
U.S. Department of Labor, Federal Building
City, State Zip Code
Dear Mr. ABC:

This is to notify you that XYZ Corp. intends to contest all of the items and penalties alleged in the Citation and Proposed Penalty, received February 20, 2004, and dated February 19, 2004 (a copy is attached).

Very truly yours,

XYZ, President

APPENDIX 1B. NOTICE OF CONTEST TO PROPOSED PENALTIES ONLY

XYZ Corp.
123 Street
City, State Zip Code
September 14, 2004

ABC, Area Director
Occupational Safety and Health Administration
U.S. Department of Labor, Federal Building
City, State Zip Code

Dear Mr. ABC:

I wish to contest the amount of the Proposed Penalties of $1,200 issued September 9, 2004, based on the violations cited by you during your recent inspection.

Sincerely,

XYZ, President
General Manager

APPENDIX 1C. NOTICE OF CONTEST BY EMPLOYEE REPRESENTATIVE

GHI International Union
456 Street
City, State Zip Code
June 9, 2004

ABC, Area Director
Occupational Safety and Health Administration
U.S. Department of Labor, Federal Building
City, State Zip Code

Dear Mr. ABC:

We have been authorized by the employee representative, GHI International Union, to file this notice of contest to the OSHA citations issued on June 2, 2004, against the employer, XYZ Co. The abatement dates of June 27, 2004, for Items No. 1 and

No. 3 of the non-serious citation, and January 5, 2005, for Item No. 1 of the serious citation, are unreasonable and will continue to expose workers to safety hazards.

Sincerely,

JKL, Director
Safety Department
GHI International Union

APPENDIX 2 – COMPLAINT AND CERTIFICATE OF SERVICE

APPENDIX 2A. COMPLAINT

U. S. Occupational Safety and Health Review Commission

Secretary of Labor,
    Complainant,
    v.   OSHRC Docket No. 99-9999

XYZ Co.,
    Respondent,

COMPLAINT

This action is brought to affirm the Citations and Notifications of penalty issued under the Occupational Safety and Health Act of 1970, 29 U.S.C. § 651, et seq., hereinafter the Act, of violations of § 5(a) of the Act and the Safety and Health Regulations promulgated thereunder.

I

Jurisdiction of this action is conferred upon the Commission by § 10(a) of the Act.

II

Respondent, XYZ Co., is an employer engaged in a business affecting commerce within the meaning of § 3(5) of the Act.

III

The principal place of business of respondent is at 123 Street, City, State, Zip Code, where it was engaged in retail sales as of the date of the alleged violations.

IV

The violations occurred on or about June 9, 2004, at 123 Street, City, State, Zip Code (hereinafter "workplace").

V

As a result of an inspection at said workplace by an authorized representative of the complainant, respondent was issued three Citations and Notifications of Penalty pursuant to § 9(a) of the Act.

VI

The Citations and Notifications of Penalty, copies of which are attached hereto and made a part hereof as Exhibits "A", "B", and "C" (consisting of one page each) identify and describe the specific violations alleged, the corresponding abatement dates fixed, and the penalties proposed.

VII

On or about July 29, 2004, by a document dated July 26, 2004, the complainant received notification, pursuant to § 10(a) of the Act, of respondent's intention to contest the aforesaid Citations and Notifications of Penalty.

VIII

The penalties proposed, as set forth in Exhibits "A", "B", and "C" are appropriate within the meaning of § 17(j) of the Act. The abatement dates fixed were and are reasonable.

WHEREFORE, cause having been shown, complainant prays for an Order affirming the Citations and Notifications of Penalty, as aforesaid.

JKL, Attorney
Office of the Solicitor
U.S. Department of Labor, Federal Building
City, State Zip Code

Appendix 2B. Certificate of Service*

I certify that the foregoing Complaint was served this 19th day of August, 2004, by mailing true copies thereof, by first class mail to:

XYZ
XYZ Corp.
123 Street
City, State Zip Code

PQR
Attorney

* A similar document must accompany all other documents requiring a certificate of service.

APPENDIX 3 – ANSWER

U. S. Occupational Safety and Health Review Commission

Secretary of Labor,
        Complainant,
    v.   OSHRC Docket No. 99-9999

XYZ Corp.,
        Respondent,

ANSWER

I, II, III

Respondent admits Paragraphs I, II and III.

IV

Respondent denies Paragraph IV.

V

Respondent neither admits nor denies the allegations at Paragraph V.

VI

Respondent denies Paragraph VI.

VII

Respondent neither admits nor denies the allegations at Paragraph VII.

VIII

Respondent denies the allegations at Paragraph VIII. The penalties are excessive under § 17(j) of the Act based upon the small size of the employer, which has only twelve employees, and the low gravity of the alleged violations.

IX

Respondent pleads the affirmative defense of "greater hazard." Abatement of the alleged violations will increase the safety risk to employees. Respondent also pleads the affirmative defense of "unpreventable employee misconduct." The alleged conditions were the result of unauthorized actions by certain employees which resulted in the conditions referred to in the alleged violations.

RESPONDENT
By _____
Attorney
XYZ Corp.
123 Street
City, State Zip Code

APPENDIX 4 – REQUEST FOR SIMPLIFIED PROCEEDINGS

XYZ Corp.
123 Street
City, State Zip Code

March 26, 2004

Executive Secretary
U.S. Occupational Safety and Health
Review Commission
1120 20th Street, N.W., 9th Floor
Washington, D.C. 20036-3457

Dear Executive Secretary;

I request Simplified Proceedings. The Review Commission Docket Number assigned to my case is 99-9999.

Very truly yours,

XYZ, President

APPENDIX 5 – NOTICE OF DECISION

NOTICE OF DECISION

In Reference To:
Secretary of Labor v. XYZ Corp.
OSHRC Docket No. 99-9999

1. Enclosed is a copy of my decision. It will be submitted to the Commission's Executive Secretary on January 3, 2004. The decision will become the final order of the Commission at the expiration of thirty (30) days from the date of docketing by the Executive Secretary, unless within that time a member of the Commission directs that it be reviewed. All parties will be notified by the Executive Secretary of the date of docketing.

2. Any party that is adversely affected or aggrieved by the decision may file a petition for discretionary review by the Review Commission. A petition may be filed with the Judge within ten (10) days from the date of this notice. Thereafter, any petition must be filed with the Review Commission's Executive Secretary within twenty (20) days from the date of the Executive Secretary's notice of docketing. See Paragraph No. 1. The Executive Secretary's address is as follows:

Executive Secretary
Occupational Safety and Health
Review Commission
1120 20th Street, N.W. – 9th Floor
Washington, D.C. 20036-3457

3. The full text of the rule governing the filing of a petition for discretionary review is 29 C.F.R. 2200.91. It is appended hereto for easy reference, as are related rules prescribing post-hearing procedure.

MNO
Administrative Law Judge

December 1, 2004

APPENDIX 6 – PETITION FOR DISCRETIONARY REVIEW

U.S. Occupational Safety and Health Review Commission

Secretary of Labor,
　　　　Complainant,
　　　v.　　　　　　　　OSHRC Docket No. 99-9999
XYZ Corp.,
　　　　Respondent,

PETITION FOR DISCRETIONARY REVIEW

Comes now Respondent, XYZ Corp., being aggrieved by the Decision and Order of the Administration Law Judge in the above-styled matter, and hereby submits its Petition for Discretionary Review pursuant to 29 CFR 2200.91-Rule 91, Rules of Procedure of the Occupational Safety and Health Review Commission.

STATEMENT OF PORTIONS OF THE DECISION AND ORDER TO WHICH EXCEPTION IS TAKEN

1. XYZ Corp. takes exception to that portion of the Decision and Order wherein the Administrative Law Judge held XYZ Corp. in serious violation of the standard published at 29 CFR 1926.28(a) as alleged in Serious Citation 1, Item 1, in finding that

XYZ's employee John Jones was exposed to the alleged violation. (Judge's Decision at pp. 8–12.)

2. XYZ Corp. takes exception to that portion of the Decision and Order pertaining to Serious Citation 1, Item 1, wherein the Administrative Law Judge held that action of employee John Jones was not unpreventable employee misconduct. (Judge's Decision at pp. 13–17.)

STATEMENT OF REASONS FOR WHICH EXCEPTIONS ARE TAKEN

1. In his Decision, the Administrative Law Judge failed to follow the test set forth for the Fifth Circuit's Decision in Secretary of Labor v. RPQ Corp.for determining the existence of employee exposure. The testimony at transcript pages 25–45 clearly shows that John Jones was not in the zone of danger because he was on a work break and outside of the definition of the zone.

2. The evidence of record supports XYZ's position that the actions taken by employee John Jones were unpreventable. The Commission has set forth the test for determining unpreventable employee misconduct at Secretary of Labor v. ROM Corp. The testimony of XYZ's employees at transcript pp. 46–59 met all of the requirements of ROM Corp. to prove John Jones's actions were unpreventable.

For the reasons stated herein, XYZ Corp. hereby submits that the Occupational Safety and Health Review Commission should direct review of the Decision and Order of the Administrative Law Judge.

Respectfully submitted,

By _____
Attorney for
XYZ Corp.
123 Street
City, State Zip Code
Tel. No. (999) 999-9999

APPENDIX 7 – DIRECTION FOR REVIEW

U.S. Occupational Safety and Health Review Commission

Secretary of Labor,
        Complainant,

    v.   OSHRC Docket No. 99-9999

XYZ Corp.
        Respondent,

DIRECTION FOR REVIEW

Pursuant to 29 U.S.C. § 66(j) and 29 C.F.R. § 2200.92(a), the report of the Administration Law Judge is directed for review. A briefing order will follow.

COMMISSIONER

Dated:

Appendix 8 – Notice of Withdrawal

U.S. Occupational Safety and Health Review Commission

Secretary of Labor,
          Complainant,

          v.   OSHRC Docket No. 99-9999

XYZ Corp,
          Respondent,

RESPONDENT'S WITHDRAWAL OF NOTICE OF CONTEST

Respondent, XYZ Corp., by the undersigned representative, hereby withdraws its Notice of Contest in the case with the docket number above, pursuant to 29 CFR 2200.102 of the Rules of Procedure for the Commission.

XYZ
XYZ Corp.
123 Street
City, State Zip Code
March 30, 2004*

■■

---

* OSHRC website located at www.oshrc.gov.

8 OSHA Enforcement and Whistleblower Protections

******

Chapter Objectives:

1. Understanding the Occupational Safety and Health Administration (OSHA) enforcement process.
2. Appraise the OSHA penalty schedule.
3. Understand the use of criminal referrals under the OSH Act.
4. Understand the procedures, protocol, and processes in a whistleblower action.

******

Safety and health professionals should be aware that compliance officers are well-trained and educated in the OSHA standards, regulations, and directives and often possess more experience with the specific OSHA standard than the safety and health professional. However, safety and health professionals should be cognizant that the compliance officer is simply the "eyes and ears" for OSHA, and the decision-making authority as to whether or not to find a violation and issue proposed penalties lies at the area of regional director's level. Before, during, and after a compliance inspection, it is essential that the safety and health professional always acts in a professional, ethical, and business-like manner. The compliance officer is simply doing his/her job and following the prescribed procedures and protocols in the performance of his/her job function.

In the event that a violation is identified and a citation is issued, it is important for safety and health professionals to carefully and thoroughly review each alleged violation, the categorization of the proposed penalty, and the proposed monetary penalty. After the initial analysis, safety and health professionals may want to further analyze the categorization of the proposed penalty. As can be seen in detail below, the categories of potential monetary penalties include willful, repeat, serious, other-than-serious, failure to correct, and a posting requirement violation. Each of these categories carries a maximum penalty, and several also possess a minimum level for the monetary penalty.

Safety and health professionals should be aware that under the OSH Act in 1970, the maximum penalty was $10K per violation. In the late 1980s, the maximum monetary penalty was increased seven-fold to a maximum of $70K. Recent changes have increased the maximum monetary penalties to $129,336.00 per violation, and this monetary penalty amount will continue to increase annually as prescribed by the Department of Labor Federal Civil Penalties Inflation Adjustment Act of 2017.*

MONETARY PENALTIES

1903.15(A)

After, or concurrent with, the issuance of a citation, and within a reasonable time after the termination of the inspection, the Area Director shall notify the employer by certified mail or by personal service by the Compliance Safety and Health Officer of the proposed penalty in accordance with paragraph (d) of this section, or that no penalty is being proposed. Any notice of proposed penalty shall state that the proposed penalty shall be deemed to be the final order of the Review Commission and not subject to review by any court or agency unless, within 15 working days from the date of receipt of such notice, the employer notifies the Area Director in writing that he intends to contest the citation or the notification of proposed penalty before the Review Commission.

1903.15(B)

The Area Director shall determine the amount of any proposed penalty, giving due consideration to the appropriateness of the penalty with respect to the size of the business of the employer being charged, the gravity of the violation, the good faith of the employer, and the history of previous violations, in accordance with the provisions of section 17 of the Act and paragraph (d) of this section.

1903.15(C)

Appropriate penalties may be proposed with respect to an alleged violation even though after being informed of such alleged violation by the Compliance Safety and Health Officer, the employer immediately abates, or initiates steps to abate, such alleged violation. Penalties shall not be proposed for de minimis violations which have no direct or immediate relationship to safety or health.

1903.15(D)

Adjusted civil monetary penalties. The adjusted civil penalties for penalties proposed after January 2, 2018 are as follows:

1903.15(D)(1)

Willful violation. The penalty per willful violation under section 17(a) of the Act, 29 U.S.C. 666(a), shall not be less than $9,239 and shall not exceed $129,336.

---

* 1902. 1902.4(c)(2)(xi); 1902.37(b)(12); 1903.15(d)(1); 1903.15(d)(2); 1903.15(d)(3); 1903.15(d)(4); 1903.15(d)(5); 1903.15(d)(6); 1956.11(c)(2)(x).

1903.15(D)(2)

Repeated violation. The penalty per repeated violation under section 17(a) of the Act, 29 U.S.C. 666(a), shall not exceed $129,336.

1903.15(D)(3)

Serious violation. The penalty for a serious violation under section 17(b) of the Act, 29 U.S.C. 666(b), shall not exceed $12,934.

1903.15(D)(4)

Other-than-serious violation. The penalty for an other-than-serious violation under section 17(c) of the Act, 29 U.S.C. 666(c), shall not exceed $12,934.

1903.15(D)(5)

Failure to correct violation. The penalty for a failure to correct a violation under section 17(d) of the Act, 29 U.S.C. 666(d), shall not exceed $12,934 per day.

1903.15(D)(6)

Posting requirement violation. The penalty for a posting requirement violation under section 17(i) of the Act, 29 U.S.C. 666(i), shall not exceed $12,934.

[81 FR 43453, July 1, 2016; 82 FR 5382, Jan. 18, 2017; 83 FR 14, Jan. 2, 2018]*

■■■

In determining the categorization of the alleged violation as well as the proposed monetary penalty, the area or regional director should consider: the size of the employer, the gravity of the alleged violation, and the employer's safety and inspection history, as well as the good faith of the employer. When assessing the proposed monetary penalty and in preparation for the informal conference, safety and health professionals may want to prepare arguments supported by documentation to show the history of an effective safety and health program and/or "good faith" through the immediate correction of the identified violation. Although these arguments will not vacate the violation, it can go a long way in reducing the proposed monetary penalty at informal conference.

On appeal, OSHA has the burden of proof for any alleged violation. Safety and health professionals should carefully analyze the categorization of each violation to ensure that OSHA possesses the appropriate level of proof for the identified category. For example, an alleged violation which is categorized as willful requires OSHA to prove the violation was intentional (i.e., to prove intent). If OSHA cannot prove the required element, i.e., intent, this may be grounds to reduce or vacate the alleged violation and proposed penalties. Safety and health professionals can find the specific

---

* OSHA website located at www.osha.gov.

requirements for each category of monetary penalties in the OSHA *Field Operations Manual* located on the OSHA website.

When entering into a settlement at informal conference, safety and health professionals should pay careful attention to the potential for future repeat violations which can carry up to a $12,934.00 per day monetary penalty. Additionally, safety and health professionals should exercise caution when settling alleged violations which may impact a key production process or operation which can impact production and quality of the company's product. Prudent safety and health professionals in situations which may not comply with all aspects of an applicable standard may want to consider pursuit of a variance prior to any compliance inspection to potentially reduce the risk of violation while maintaining a safe and healthful work environment.

In the area of whistleblower protection, the Occupational Safety and Health Administration is responsible for enforcement of the anti-retaliation or whistleblower provisions of a number of environmental, consumer, investor, and other statutes, as well as safety and health-related laws. In general, anti-retaliation provisions within these regulations and laws are designed to protect individual workers who report illegalities or wrongdoing within their company or agency from retaliation by the company or agency. Protection from workplace retaliation means that an employer cannot take an "adverse action" against workers, such as: firing or laying off, blacklisting, demoting, denying overtime or promotion, disciplinary actions, denial of benefits, failure to hire or rehire, intimidation/harassment, making threats, reassignment affecting prospects for promotion, and reducing pay or hours.* OSHA has established a protocol with various committees, investigators, and others for review and evaluation of whistleblower cases.

■■■

STATUTES

The statutes enforced by OSHA are listed below. They contain whistleblower or anti-retaliation provisions that generally provide that employers may not discharge or retaliate against an employee because the employee has filed a complaint or has otherwise exercised any rights provided to employees. Each law requires that complaints be filed within a certain number of days after the alleged retaliation. Complaints may be filed orally or in writing, and OSHA will accept the complaint in any language.

Statute	Title
29 U.S.C. 218C	Affordable Care Act (ACA), Section 1558
15 U.S.C. §2651	Asbestos Hazard Emergency Response Act (AHERA)
42 U.S.C. §7622	Clean Air Act (CAA)
42 U.S.C. §9610	Comprehensive Environmental Response, Compensation and Liability Act (CERCLA)

* www.whistleblower.gov.

12 U.S.C.A. §5567	Consumer Financial Protection Act of 2010 (CFPA), Section 1057 of the Dodd–Frank Wall Street Reform and Consumer Protection Act of 2010
15 U.S.C. §2087	Consumer Product Safety Improvement Act (CPSIA)
42 U.S.C. §5851	Energy Reorganization Act (ERA)
21 U.S.C. 399d	FDA Food Safety Modernization Act (FSMA), Section 402
49 U.S.C. §20109	Federal Railroad Safety Act (FRSA)
33 U.S.C. §1367	Federal Water Pollution Control Act (FWPCA)
46 U.S.C. §80507	International Safe Container Act (ISCA)
49 U.S.C. §30171	Moving Ahead for Progress in the 21st Century Act (MAP-21)
6 U.S.C. §1142	National Transit Systems Security Act (NTSSA)
29 U.S.C. §660	Occupational Safety and Health Act (OSH Act), Section 11(c)
49 U.S.C. §60129	Pipeline Safety Improvement Act (PSIA)
42 U.S.C. §300j-9(i)	Safe Drinking Water Act (SDWA)
18 U.S.C.A. §1514A	Sarbanes–Oxley Act (SOX)
46 U.S.C. §2114	Seaman's Protection Act (SPA), as amended by Section 611 of the Coast Guard Authorization Act of 2010, P.L. 111-281
42 U.S.C. §6971	Solid Waste Disposal Act (SWDA)
49 U.S.C. §31105	Surface Transportation Assistance Act (STAA)
15 U.S.C. §2622	Toxic Substances Control Act (TSCA)
49 U.S.C. §42121	Wendell H. Ford Aviation Investment and Reform Act for the 21st Century (AIR21)

In the event that an employee is disciplined, terminated or otherwise believes he/she has been retaliated against by the company for reporting a safety and health violation to OSHA or reporting a violation under one of the above noted laws, OSHA has developed an online complaint filing method as well as an investigation process which can accessed through the OSHA website.

Safety and health professionals should be aware that whistleblowers' actions give them

> protection from workplace retaliation [which] means that an employer cannot take an "adverse action" against workers, such as: firing or laying off, blacklisting, demoting, denying overtime or promotion, disciplining, denial of benefits, failure to hire or rehire, intimidation/harassment, making threats, reassignment affecting prospects for promotion, reducing pay or hours

or other retaliatory actions resulting from an employee reporting an unsafe or illegal activity.*

For most safety and health professionals, whistleblower complaints result from actions or inactions by the employer after an employee reports an alleged violation to OSHA. The safety and health professional may not be aware of the employee reporting the alleged violation to OSHA until notified by the employer. When an action or inaction taken by the employer which the employee believes is retaliatory and a result of their reporting an alleged violation to OSHA, the initial step is for the

---

* www.whistleblower.gov.

employee or his/her representative to file a whistleblower compliant with OSHA. As can be seen below, this written complaint can be filed via mail, telephone, or online at www.whistleblower.gov. The employee or representative is required to provide current contact information as well as other information regarding the alleged retaliation by the employer.

Under the proscribed procedure,

> OSHA will interview the Complainant [the Employee who filed the Complaint] to obtain information about the alleged retaliation, and will determine whether the allegation is sufficient to initiate an investigation under one or more of the twenty-two whistleblower protection statutes administered by OSHA. Regardless of the statute under which the complaint is filed, the conduct of the investigation is generally the same.* ... If the allegation is sufficient to proceed with an investigation, the complaint will be assigned to an OSHA whistleblower investigator who is a neutral fact-finder who does not represent either party. The investigator will notify the Complainant (employee), Respondent (employer) and appropriate federal partner agency that OSHA is opening an investigation.†

At this point in the process, the safety and health professional is usually provided with notice of the whistleblower's complaint by OSHA or through the contact department within the company. As usually identified in the communications from OSHA, it is important that the safety and health professional acquire and maintain all potential evidence regarding the matter and circumstances identified in the complaint. Although every situation is different, this can include emails, phone logs, text messages, voicemails, individual personnel files, meeting minutes, training presentations, contracts or agreements, and other documentation. Safety and health professionals should be aware that the OSHA whistleblower investigator will usually request that copies of all documents submitted to OSHA also be provided to the opposing party, as well as exchanging lists of potential witnesses and contact information.

The OSHA whistleblower investigator usually asks the Respondent (employer) to provide a "position paper" or written defenses to each of the employee's allegations. Additionally, safety and health professionals can expect further inquiries and requests from the investigator throughout the investigation. It should be noted that the safety and health professional normally is provided with an opportunity to see the responses provided by the employee and is usually given an opportunity to provide evidence or otherwise rebut the employee's allegations.

At the conclusion of the OSHA whistleblower investigator's investigation, he/she will analyze the evidence and make a recommendation to his/her supervisor at OSHA. The primary question usually involves the issue as to whether or not, from the evidence and information provided by the parties, the investigator believes there is reasonable cause that the Respondent (employer) violated the statute at issue. The OSHA supervisor will concur with the investigator's findings that there is merit to the complaint or move to dismiss the action. At this point, OSHA issues a finding

---

* Id.
† Id.

letter to the parties which includes information about the remedies and the right to appeal the decision to an administrative law judge (ALJ).*

Safety and health professionals should be aware that a retaliation complaint can be settled at any time during the investigation with the approval of OSHA. Additionally, safety and health professionals should be aware that an Alternate Dispute Resolution program is available from OSHA and the Complainant (employee) does possess the right to "kick out" of the investigation, and "file the retaliation complaint in federal court if OSHA does not complete its investigation within a specified time period – 180 days or 210 days depending on the statute."†

When notified of a whistleblower complaint, prudent safety and health professionals should immediately notify their company and involve their legal department or outside counsel.

Additionally, it is vital to document and preserve all information and assist counsel in responding to all allegations. Safety and health professionals should follow the sound advice of their legal counsel and have counsel present at any time they are communicating with or being interviewed by the OSHA whistleblower investigator.

****

OSHA ONLINE WHISTLEBLOWER COMPLAINT FORM

EMERGENCY NOTICE: Do Not Report an Emergency Using this Form or Email!

To report an emergency, fatality, or imminent life threatening situation please contact our toll free number immediately:

1-800-321-OSHA (6742)

TTY 1-877-889-5627

INTRODUCTION AND INSTRUCTIONS

OSHA administers more than twenty whistleblower protection laws, including Section 11(c) of the Occupational Safety and Health (OSH) Act, which prohibits retaliation against employees who complain about unsafe or unhealthful conditions or exercise other rights under the Act. Each law has a filing deadline, varying from 30 days to 180 days, which starts when the retaliatory action occurs.

A whistleblower complaint must allege four key elements:

- The employee engaged in activity protected by the whistleblower protection law(s) (such as reporting a violation of law);
- The employer knew about, or suspected, that the employee engaged in the protected activity;

---

* Hearings before an ALJ are not permitted in cases under OSHA 11©, AHERA, or ISCA.
† www.whistleblower.gov.

- The employer took an adverse action against the employee;
- The employee's protected activity motivated or contributed to the adverse action.

Filing with this form is not required, as OSHA accepts whistleblower complaints made orally (telephone or walk-in at any OSHA office) or in writing, and in any language. If you choose to use this form, you must complete the screens and fields that are marked as "required"; all other screens and fields are optional.

If you file a complaint, OSHA will contact you to determine whether to conduct an investigation. You **must** respond to OSHA's follow-up contact or your complaint will be dismissed.

A whistleblower complaint filed with OSHA cannot be filed anonymously. If OSHA proceeds with an investigation, OSHA will notify your employer of your complaint and provide the employer with an opportunity to respond. Because your complaint may be shared with the employer, **do not include witness names or their contact information on this form**; you will have the opportunity to offer evidence in support of your complaint during the investigation.

If you have any questions about the complaint filing or investigative processes, please do not hesitate to call 1-800-321-OSHA (6742).

If you think your job is unsafe and you want to ask for an inspection, you can call 1-800-321-OSHA (6742), or file a "Notice of Alleged Safety or Health Hazards" by clicking here.

Do you want to file an online whistleblower complaint now?

Yes, Launch the Online Whistleblower Complaint Form

No, Return to www.whistleblowers.gov

PRIVACY ACT STATEMENT

This form requests personal information that is relevant and necessary to determine whether and how to conduct an investigation. OSHA collects this information in order to process complaints under its statutory and regulatory authority. Once a complaint is filed, the individual's name and information about the allegations of retaliation will be disclosed to the employer. During the course of an OSHA investigation, information contained in an investigative case file may be disclosed to the parties in order to resolve the complaint. During an investigation, information about the complaining party and the employer will not be released to the public except to the extent allowed under the Freedom of Information Act (FOIA). However, once a case is closed, it is possible that information contained in the complaint or a case file may be released to the public as required by the FOIA. Any such documents will be redacted as appropriate under the FOIA and the Privacy Act.

PAPERWORK REDUCTION ACT STATEMENT

According to the Paperwork Reduction Act, an Agency may not conduct or sponsor, and no persons are required to respond to a collection of information unless such collection displays a valid OMB control number. Public reporting burden for this voluntary collection of information is estimated to be one hour per response, including the time for reviewing instructions, searching existing data sources, gathering and maintaining the data needed, and completing and reviewing the collection of information. Please send comments regarding this burden estimate or any other aspect of this collection of information, including suggestions for reducing this burden to OSHA.DWPP@dol.gov or to the Directorate of Whistleblower Protection Programs, Department of Labor, Room N4624, 200 Constitution Ave., NW, Washington, DC; 20210; Attn: Paperwork Reduction Act Comment. (This address is for comments only; do not send completed complaint forms to this office.)

OMB Approval # 1218-0236; Expires: 03-31-2020

OSHA 8-60.1. (Rev.06/17)

9 Anti-Discrimination Laws Impacting Safety and Health

******

Chapter Objectives:

1. Acquire an understanding of the historical underpinnings of U.S. anti-discrimination laws.
2. Develop an understanding of the fabric of U.S. laws prohibiting discrimination in the workplace.
3. Acquire an understanding of the safety and health professional's role and responsibilities in creating and maintaining diversity in the workplace.
4. Increase safety and health professional's knowledge and recognition of discrimination in the workplace and the applicable anti-discrimination laws.

******

Safety and health professionals should be aware that throughout history, there has been harassment and discrimination in societies throughout the world, and often specifically related to the workplace, based on a variety of factors including, but not limited to, ethnicity, religion, skin color, nationality, gender, pregnancy, disability, and numerous other characteristics or beliefs. One of the unique strengths of the United States since the beginning has been its consistent ability to welcome diversity in society, and specifically in the workplace, and build upon the strengths of individuals, rather than selectively eliminating individuals and groups from the workforce based upon their beliefs, physical make-up, or heritage. However, despite the best efforts of the American society, government, and most employers, discrimination still exists. Over time, a patchwork of laws and regulations have been enacted to address and redress current and past discriminations, primarily in the workplace, and they provide a broad spectrum of protections against discrimination to the American worker.

Safety and health professionals should be aware that, as a member of the management team, your company or organizational entity possesses a duty and responsibility to protect employees against discrimination in their workplace. Safety professionals are often on the front lines in identifying potential issues and situations where potential discrimination may exist in the workplace. Although most safety and health professionals do not possess direct responsibilities within the area of anti-discrimination policy enforcement and most safety professionals are not "experts" as to the laws and regulations addressing discrimination in the workplace, it is important that safety and health professionals acquire a knowledge of the various laws and regulations, as well as a working knowledge of the processes and procedures to follow when an issue involving potential discrimination in the workplace arises.

As with the Occupational Safety and Health Act (hereinafter referred to as the "OSH Act") enacted on April 28, 1971,* many of the federal laws prohibiting discrimination in the workplace had their origins in the turbulent decade of the 1960s. However, since the beginning of this century, there has been a slow progression in the development in laws, executive orders, and regulations through which the federal government attempted to regulate and thus prohibit discrimination in the American workplace which culminated with the relatively recent enactment of a patchwork of various laws specifically enacted to address the relevant issues involving a specific discriminatory act, issue, or action.

Safety and health professionals should be aware that each of the protected classes, and thus each of the forms of discrimination, have been regulated and the laws enacted at different times in our history, incorporating different philosophies, and often utilizing different methods to regulate and prohibit discrimination in the workplace. This patchwork of laws, executive orders, and regulations, at the federal, state, and local levels, as well as court decisions, have created the quilt of protections against discrimination for workers in today's workplace. Additionally, safety and health professionals should be aware that this patchwork of regulatory protection is a "work in progress" and is constantly moving and adjusting to address current issues, as well as changes in our society and workplace. Although we will focus on the laws and regulations enacted at the federal level in this chapter, safety and health professionals should be aware that most states also have enacted anti-discrimination laws and regulations which can provide far greater protection or prohibitions against discrimination than the laws and regulations at the federal level, and local governments have also provided protection against discrimination within their limited jurisdiction. Additionally, safety professionals should be aware that the Constitution of the United States also addresses discrimination in the Fifth, Thirteenth and Fourteenth Amendments. The federal level patchwork of anti-discriminations laws and regulations which have developed over time include the following:

- Title VII of the Civil Rights Act of 1964 (commonly referred to as "Title VII or 7").[†]
- The Americans with Disabilities Act (commonly referred to as the "ADA").[‡]
- The Civil Rights Act of 1866.[§]
- The Civil Rights Act of 1871.[¶]

---

* 29 U.S.C.A. § 654(a)(2) and (b) (1971).
† 42 U.S.C.A. §§ 200e–2000e17 (1964).
‡ 42 U.S.C. §§ 12101–12102, 12111–12117, 12201–12213 (1990).
§ 42 U.S.C. §§ 1981 and 1981a (1866).
¶ 42 U.S.C. § 1983 (1871).

- The Civil Rights Act of 1991.*
- The Age Discrimination in Employment Act (commonly referred to as "ADEA").†
- The Equal Pay Act (commonly referred to as the "EPA").‡
- The Genetic Information Nondiscrimination Act (commonly referred to as "GINA").§
- Lilly Ledbetter Fair Pay Act of 2009.¶
- Vocational Rehabilitation Act of 1973.**
- Family and Medical Leave Act of 1993.††
- Immigration Reform and Control Act of 1986.‡‡
- Glass Ceiling Act of 1991.§§

As can be seen, this patchwork of laws, regulations, and executive orders prohibit discrimination, primarily focused on the workplace, based upon race, color, national origin, religion, sex, age, physical or mental disability, and ethnicity, while also providing protections for genetic information, wages, and the ability to be hired and promoted in the workplace, can impact the safety function.

Although safety and health professionals seldom possess direct responsibility for compliance with these laws and regulations, the safety and health function often interacts, either directly or indirectly, with each and every one of these laws and regulations at some point throughout the course of the safety working activities. The safety and health professional does not function in a vacuum and thus, through the interaction with management, employees, and the daily safety functions, the safety professionals will often be confronted with issues sounding in safety, but that primarily are focused in one or more of the above laws or regulations.

Safety and health professionals should also be aware that these antidiscrimination laws were often specifically focused on the workplace and are designed to keep the employer from making employment-related decisions that could disadvantage employees based on their race, sex, age, color, disability, or any of the other protected classes. Safety and health professionals should be aware that most issues of discrimination involve hiring, promotion, training, termination, demotion, layoffs, or other terms and conditions involved in the workplace, including safety issues.

The primary federal antidiscrimination law encompassing a broad spectrum of protections is Title VII of the Civil Rights Act of 1964¶¶ (hereinafter referred to as "Title VII"). Title VII is general in its application and extends protections to all races,

---

* Pub.L.No. 102–166, 105 Stat. 1071(1991).
† 29 U.S.C. §§ 621–634. Taken from the EEOC website located at www.eeoc.gov (1967).
‡ 29 U.S.C. § 206(d) (1963).
§ Pub.L.No. 110–233, 122 Stat. 881(2008) (Codified at 42 U.S.C. § 000ff).
¶ Pub.L.No. 110–535, 122 Stat. 3553 (2008).
** 29 U.S.C. §§ 705, 791, 793–794a.
†† 29 U.S.C. §§ 2601–2619, 2651–26454 (1993).
‡‡ 8 U.S.C. § 1324b (1986).
§§ Pub. L.No. 102–166, 105 Stat. 1081 (1991) (Codified at 42 U.S.C. § 2000e)
¶¶ 42 U.S.C.A. §§ 2000E through 2000E-17.

including Caucasian, as well as protections in the areas of religion, pregnancy, sex, color, and national origin. At the time of its creation by Congress in 1964, Title VII was the most sweeping civil rights legislation ever enacted and contains 11 different titles addressing discrimination in public accommodations, voting, education, and, most importantly to safety professionals, in the employment setting. Safety and health professionals should be aware that the Civil Rights Act has been modified and amended on several occasions since 1964, with the most recent being the Civil Rights Act of 1991.* In the Civil Rights Act of 1991,† Congress reacted to a series of court decisions by the U.S. Supreme Court changing the landscape of discrimination law and precedent. With the Civil Rights Act of 1991, Congress amended several of the statutes enforced by the Equal Employment Opportunity Commission (EEOC) and added jury trials, and compensatory and punitive damages in Title VII (and ADA) lawsuits involving intentional discrimination, and added statutory caps on damages awarded for future losses, pain, and suffering, and punitive damages depending on the size of the employer. Additionally, Congress codified the disparate impact theory of discrimination and expanded the coverage of Title VII to cover employees of American-controlled companies or organizations with operations outside of the United States.

As identified above, safety and health professionals should be aware that most states also possess individual state antidiscrimination laws primarily focused on the employment setting which can encompass more than the federal law. Section 708 of Title VII permits each state to have parallel state legislation and regulation in the employment setting, as long as the law does not conflict with Title VII. Safety and health professionals should be aware that employers with 15 or fewer employees, employees employed for each workday in 20 or more calendar weeks in a calendar year and engaged in interstate commerce can be considered outside of Title VII's jurisdiction.‡ For most companies or organizations which possess a full-time safety and health professional, this small employer requirement is not applicable.

More specifically, safety and health professionals should be aware that Title VII identified it as unlawful employment practice to:

- *"fail or refuse to hire or to discharge any individual, or otherwise to discriminate against any individual with respect to his or her compensation, terms, conditions, or privileges of employment, because of such individual's race, color, religion, sex or national origin; or*
- *Limit, segregate, or classify his or her employment or applicants for employment in any way which would deprive or tend to deprive any individual of employment opportunities or otherwise adversely affect his or her status as an employee, because of such individual's race, color, religion, sex or national origin."*§

---

* Public Law 102–166 (1991).
† *Id.*
‡ 42 U.S.C.A. § 2000e(b).
§ 42 U.S.C.A. § 2000-2(a).

Safety and health professionals should be aware that in addition to employers, labor unions, employment agencies, and joint labor-management training committees are also required to comply with Title VII. For employers, labor organizations, and others covered under Title VII, retaliation is prohibited in any manner. Safety and health professionals should be aware that it is considered an unlawful employment practice "to discriminate against the respective employees, applicants, members, or other related individuals because of their opposition to an unlawful employment practice or because of their filing a charge, testifying, assisting, or participating in any manner in an investigation, proceeding, or hearing under the Act."* Additionally, safety and health professionals should be cognizant of the scope and application of the protections extended to individuals under Title VII and be able to identify situations and issues where the potential of discrimination may be present. Being aware of the protections provided to individuals under Title VII is the first step in preventing discrimination in your workplace.

Within the categories of race and color, safety and health professionals should be aware that Title VII includes protections to all races, including Caucasians.† In the category of national origin, safety and health professionals should be aware that language requirements are often found to be improper unless job-relatedness can be shown,‡ and prohibiting employees of foreign descent from using their native language at work has also been found to be an unlawful employment practice, unless the business necessity it can be proven by the employer.§ Title VII prohibits discrimination against male or female individuals based on their sex.¶ Safety and health professionals should be aware that the prohibition against discrimination based on sex extends beyond the hiring process to include such other areas as life insurance programs, profit-sharing plans, bonus programs, and all other plans or programs.**

With the emergence of the "Me Too" movement, safety and health professionals must be cognizant of sexual harassment in their workplace, especially with activities involving managerial or exempt personnel and hourly or non-exempt personnel. The EEOC has established special obligations on companies or organizations to ensure that their supervisory personnel, as well as other employees, do not engage in the unlawful employment practice of sexually based harassment. Safety and health professionals should be aware that sexual harassment can include "any unwelcome sexual advances, requests for sexual favor, or other verbal or physical conduct of a sexual nature."†† Additionally, safety and health professionals should be aware that sexual advances or requests for sexual favors in the workplace substantially and detrimentally affects an employee's work performance or can create an "intimidating, hostile or offensive work environment."‡‡

---

* 42 U.S.C.A. § 2000e-2(a).

† Note: This is often referred to as "reverse discrimination" actions.

‡ EEOC decision No. 73-0377 (1972).

§ *Saucedo v. Brothers Well Service, Inc.*, 464 F.Supp. 919 (DC Tex., 1979).

¶ 42 U.S.C.A. § 2000e-2.

** 29 C.F.R. § 1604.9.

†† EEOC, *Sex Discrimination Guidelines*, 29 C.F.R. § 1604.11.

‡‡ 29 C.F.R. §§ 1604.2 and 1604.11.

Safety and health professionals should also be aware that petty slights, annoyances, and isolated incidents (unless extremely serious) usually will not rise to the level of illegality. To be unlawful, the conduct must create a work environment that would be intimidating, hostile, or offensive to reasonable people. Offensive conduct may include, but is not limited to, offensive jokes, slurs, epithets or name-calling, physical assaults or threats, intimidation, ridicule or mockery, insults or put-downs, offensive objects or pictures, and interference with work performance. Harassment can occur in a variety of circumstances, including, but not limited to, the following:

"The harasser can be the victim's supervisor, a supervisor in another area, an agent of the employer, a co-worker, or a non-employee.
The victim does not have to be the person harassed, but can be anyone affected by the offensive conduct.
Unlawful harassment may occur without economic injury to, or discharge of, the victim."*

Safety professionals should be aware that Title VII defines "religion" as including all aspects of religious observances, practices, or beliefs.† The EEOC expanded this definition in its regulations to include moral and ethical beliefs not confined to theistic concepts or to traditional precepts which are sincerely held by individuals with the strength of traditional religious views and beliefs.‡ Safety professionals should be aware that Title VII goes beyond simple neutrality in the workplace providing for companies or organizations to provide reasonable accommodation of an employee's religious observance or practice.§ However, safety professionals should be aware that Title VII also requires the employee seeking to observe his or her religious beliefs to provide proper notification to the company or organization as to their religious needs and an obligation to resolve any conflicts between job requirements and religious observances or practices.¶ Of particular importance for safety professionals is that requiring an applicant or employee to wear clothing or other apparel other than the clothing required by the employee's religion may constitute an unfair employment practice.**

Also of particular importance for safety and health professionals is the Pregnancy Discrimination Act, which was added as an amendment to Title VII. Safety and health professionals should be aware that unlawful employment practices in this category can include exclusion from medical or insurance programs, denial of a leave of absence, or discrimination based on the time or duration of a leave of absence. Of particular note is that protections under the Pregnancy

---

* EEOC website located at www.eeoc.gov.
† 42 U.S.C.A. § 2000e(j).
‡ 29 C.F.R. § 1605.1.
§ 42 U.S.C.A. § 2000e(j); 29 C.F.R. § 1605.2.
¶ *Chrysler v. Mann*, 561 F.2d 1282 (CA-8, 1977).
** EEOC decision No. 71-2620 (1971).

Discrimination Act extend not only to female employees, but also to the spouses of male employees.*

As can be seen, Title VII provides a wide spectrum of protections for individuals working within an employment setting. Safety and health professionals should be cognizant of these protections and be able to identify situations and issues which may possess the potential of discrimination and take the appropriate actions to avoid such discrimination. As identified above, Title VII offers protection against discrimination based upon race, sex, color, religion, and pregnancy within the employment setting. Most companies and organizations have addressed the requirements of Title VII and strictly adhere to the letter of this law. Safety and health professionals should not only be aware of the requirements of Title VII, but also their internal company policies and procedures when addressing issues and circumstances of potential discrimination in the workplace.

Within the scope of the anti-discrimination laws, safety and health professionals should be aware that there is a difference between discrimination and harassment. The laws and regulations prohibiting discrimination in the workplace are primarily focused on policies and decisions. Harassment, on the other hand, is unwelcome conduct that creates a hostile work environment. Harassment is only illegal if it is against one or more of the employees who are in one or more of the protected classes created by one of the federal or state laws. A hostile work environment is a form of harassment where the employee may not have suffered a "tangible employment action," but the employee is subject to unwelcome conduct which: (1) is based on the employee's protected classification (such as age or sex); (2) is severe or pervasive that it alters the terms and conditions of the employee's employment situation; and (3) creates a hostile or abusive environment for the employee. Safety and health professionals should be aware that conduct by other employees or management may be considered "unwelcome" if the employee doesn't invite the comments, actions, etc. and/or if the employee doesn't want the actions, comments, etc. to happen. Some examples of unwelcome conduct include unwelcome touching, racial slurs, cartoons of an offensive nature, jokes, horseplay, insults, and gestures.

In general, safety and health professionals should be aware that there are three (3) predominant theories under which discrimination can take form. The first theory is that of Disparate Treatment. Under the Disparate Treatment theory, the discrimination takes the form of intentionally treating individuals differently due to their race, sex, or other protected classification or characteristic. For example, the disparate treatment theory would apply where an employer denies a job to an individual because the individual is of a specific race, or the safety professional only permits the male members of the safety team to travel to the safety conference.

Under the Disparate Impact theory, safety and health professionals should be aware that conduct that appears fair on its face, but that can detrimentally affect a number of individuals in a protected class can also constitute discrimination. However, safety and health professionals should be aware that if the employer has

---

* *Newport News Shipping & Dry Dock Co. v. EEOC*, 32 FEP 1 (S.Ct. 1983).

a job-related or business reason for this different treatment, this conduct may be permissible. For example, the safety and health professional's company is incurring a number of back injuries in the dock area where employees are lifting constantly. The company, in an effort to reduce back injuries, requires all employees to be able to lift X lbs. This conduct may disproportionally affect female employees. If the company possesses a valid business reason, such as reduction of back injuries, this may be appropriate; however, even with a valid business reasons, this practice may constitute discrimination if an alternative method, such as lifting assist equipment, exists which would not disproportionally affect the female employees.

The third theory of discrimination which safety and health professionals should be aware is that of Failure to Accommodate. This theory is especially very pertinent for safety and health professionals under the Americans with Disabilities Act (and correlating state laws). The Failure to Accommodate theory comes into play for safety and health professionals when their company is required to accommodate a qualified individual with a disability, a religious practice, or other protected activity, and the company fails to accommodate the individual's disability or practice.

Of particular importance for safety and health professionals is the fact that discrimination is not always blatantly obvious, such as failing to promote an employee due to their race or sex. In fact, the type of discrimination where there is "smoking gun" direct evidence of discrimination is very rare. In most cases, evidence of discrimination is circumstantial or "piecemeal," suggesting that the logical conclusion is that the action or inaction by the employer must be the result of discrimination. Safety and health professionals should be aware that an employee does not automatically win a discrimination case simply by filing a claim of discrimination. The employee has the burden of proving several elements in order to be successful with their claim. In general, an employee filing a claim of discrimination must prove: (1) he or she is qualified for the job; (2) he or she is a member of a protected class; (3) the employee suffered an adverse action; and (4) there is a reason to suspect the company or management team member possessed an improper motive.

The primary federal government agency charged with enforcement of most of the anti-discrimination laws is the Equal Employment Opportunity Commission (known as the "EEOC").

Title VII, ADA and related federal laws are enforced by the EEOC and correlating state antidiscrimination laws usually have a correlating state agency. Safety and health professionals should be aware that the laws and rules may be significantly different between the federal and state laws. Additionally, safety and health professionals should be aware that other federal government agencies may have jurisdiction for particular antidiscrimination laws.

Safety and health professionals should be aware that there are a number of different laws and regulations prohibiting discrimination in the workplace, and each and every one of these laws and regulations impact, whether directly or indirectly, the safety and health function. Although safety and health professionals do not need to be discrimination experts, it is imperative that safety and health professionals possess a working knowledge of these laws and regulations in order to avoid any type of potential discrimination within their work activities, actions, or decisions. Through knowledge and diligence, safety and health professionals can avoid even a hint of any type of discrimination in their safety actions or inactions which may impact on the important goals and objective of creating and maintaining a safety and healthful workplace for all employees.

With safety and health professionals often serving as the "eyes and ears" of the company within the workplace, prudent safety and health professionals should identify the reporting structure, often to the human resource or legal counsel, for reporting incidents or issues which potentially may be discriminatory within the workplace. A proactive approach to identifying and eliminating possible discrimination in the workplace is essential. Situations or circumstances involving possible discrimination of any type should be reported to the appropriate function with the organization. Safety and health professionals are not antidiscrimination experts, but the ability to recognize and react to possible discriminatory language, actions, or circumstances is important in order that the appropriate human resource or legal function can address the circumstances.

Additionally, prudent safety and health professionals may wish to examine each and every aspect of their safety, environmental, or related programs to ensure that any potential for discrimination has been eliminated. From the hiring process through the termination process and all safety and related practices, programs, and procedures in between, a careful analysis and evaluation of issues and situations where discrimination potential could be present should be undertaken and those identified issues or situations addressed to avoid any potential for discrimination within the safety and correlating function.

Case Study: This case may be modified for the purposes of this text.

809 F.3d 780
United States Court of Appeals,
Third Circuit.

Sandra CONNELLY, Appellant
v.
LANE CONSTRUCTION CORPORATION, Defendant.

No. 14–3792.
|
Argued Sept. 15, 2015.
|
Filed: Jan. 11, 2016.

SYNOPSIS

Background: Female former employee, who worked as truck driver, brought action against former employer, alleging claims of gender-based disparate treatment, sexual harassment, hostile work environment, and retaliation under Title VII and the Pennsylvania Human Relations Act. The United States District Court for the Western District of Pennsylvania, Terrence F. McVerry, J., 2014 WL 3735634, dismissed the action. Employee appealed.

Holdings: The Court of Appeals, Jordan, Circuit Judge, held that:

[1] allegations paraphrasing the statutory language or elements of the claims were legal conclusions that would be excluded in deciding motion to dismiss for failure to state a claim;

[2] employee stated plausible Title VII gender discrimination claim against employer; and

[3] employee stated plausible Title VII retaliation claim.

Vacated and remanded.

OPINION

JORDAN, Circuit Judge.

Sandra Connelly appeals the dismissal of the employment discrimination claims she brought against her former employer, Lane Construction Corporation ("Lane"). We disagree with the District Court's assessment that Connelly failed to plead plausible claims and, accordingly, will vacate the order of dismissal and remand for further proceedings.

I. BACKGROUND

A. Factual History*

Lane is a construction company operating in 20 states. In May 2006, it hired ***784** Sandra Connelly as a union truck driver at its Pittsburgh, Pennsylvania facility, and she worked during construction seasons—normally from March or April until October or November of each year—until near the close of the season in October 2010. During Connelly's tenure with the company, Lane employed seven union truck drivers at that location. Connelly ranked fifth in seniority and was the only woman. Since October 2010, Lane has employed no female truck drivers at its Pittsburgh facility.

Sometime after May 2007, and allegedly because Connelly had ended a romance with a man who also worked at Lane, her male co-workers began "curs[ing] at Connelly and belittl[ing] her on a daily basis." (App. 29.) Some male drivers refused to speak directly to her. In the summer of 2007, another Lane employee told Connelly that Connelly's former boyfriend, truck driver Mark Nogy, was making "increasingly frequent and disparaging" comments about her. (App. 29.) The employee went on to say that he had complained about Nogy's behavior to Charlie Ames, a Lane executive. Connelly herself told several supervisors at Lane about the hostile treatment

---

* Because the District Court addressed Connelly's Amended Complaint upon a motion to dismiss, we recount the facts as alleged in that pleading and draw all reasonable inferences in favor of Connelly. *Phillips v. Cty. of Allegheny*, 515 F.3d 224, 231 (3d Cir.2008).

she was experiencing. She called the company's Connecticut headquarters and, a day later, Ames and another Lane executive met with her to discuss the harassment problem. Following the meeting, Lane suspended Nogy for three days but did not discipline or warn any other Lane employees, who continued to harass and disparage Connelly.

In early 2009, Connelly learned that Lane employees could make job-related complaints through the company's "Ethics Line," which she called multiple times to report further harassment from Nogy, to make complaints about her male co-workers drinking on the job, and to report "discriminatory treatment due to her gender and her previous complaints about the hostile work environment." (App. 31.)

In or around May 2010, Lane foreman George Manning made an unwanted physical advance to Connelly, coming close to her and saying, "[O]ne day I'm going to kiss you." (App. 31.) Connelly backed away and said "No," and she reported the incident to the Ethics Line a few days later. (App. 31.) She also reported the incident to supervisor Jeremy Hostetler, requesting that he transfer her to another work site because she was now uncomfortable working with Manning. Hostetler expressed disbelief that Manning would "do something like that." (App. 32.) Although Hostetler told Connelly that he wanted to meet with her and Manning together, no such meeting occurred. After Connelly again called the Ethics Line about the situation, Hostetler agreed to transfer her to another job site, although it appears that Connelly continued to work from Lane's Pittsburgh facility. Connelly's relationship with both her supervisors and her male co-workers became "increasingly strained" throughout 2010, during which time she made numerous complaints to the Ethics Line and to local management at the Pittsburgh facility. (App. 32.)

In October 2010, Lane supervisor Jerry Schmittein became "incensed" at Connelly when she refused to drive a truck that had a flat tire and steering problems. (App. 32.) Schmittein "persisted in berating Connelly," despite her explanation that she could not safely operate the truck. (App. 32.) Connelly contacted Ames, who instructed her to leave the job site. A short time later, and despite her seniority, Connelly was laid off before the end of the ***785** construction season and before any of the other union truck drivers. Lane has never recalled her to work.

Lane did, however, recall Connelly's male truck driver co-workers in 2011, and it continues to employ them. In April or May of 2011, after Connelly saw several of her co-workers working at a job site, she repeatedly telephoned Ames to ask why she had not been recalled. Ames cited the bad economy and told her that no work was available. In one conversation, Ames told her that he would recall her if Lane "got more work." (App. 33.)

Connelly had observed that all six of her male truck driver co-workers were working for Lane, so she called Ames and asked why union drivers with less seniority than her had been recalled before she was. In Connelly's experience, between 2006 and 2010, Lane had always recalled truck drivers in order of seniority. Ames told Connelly that the truck driver with the least seniority had been permitted to return to work as a general laborer because "he needed to work." (App. 33.) Lane had not offered any such accommodation to Connelly. Ames also explained that the other driver with less seniority than Connelly had been recalled to operate what was known as the "tack" truck because Connelly did not have the requisite training to operate that type of vehicle. (App. 33.) Connelly asked why the most senior driver, who was the primary tack truck operator, was no longer driving that truck. Ames answered that that driver was the "senior man—he can choose what he drives." (App. 33.) However, Lane had not previously permitted truck drivers to choose their work assignment based on seniority, and the union's collective bargaining agreement provided that "[d]rivers in accordance with their qualifications and seniority shall be offered the highest rate classification of work but cannot choose their equipment or work assignments." (App. 33.) Connelly was qualified to operate—and routinely had operated—all of the trucks used by Lane other than the tack truck.

Connelly also observed non-union truck drivers working at Lane sites in the spring and summer of 2011. In addition, she saw Lane employing rental trucks from other companies and using Lane laborers to drive trucks. Prior to 2011, Lane had only resorted to that when no Lane drivers were available, and never when a Lane driver was waiting to be recalled.

B. Procedural History

On September 26, 2013, Connelly filed her original complaint in the United States District Court for the Western District of Pennsylvania, alleging claims of gender-based disparate treatment, sexual harassment, hostile work environment, and retaliation under Title VII of the Civil Rights Act of 1964, 42 U.S.C. § 2000e *et seq.*, as amended ("Title VII"), and the Pennsylvania Human Relations Act, 43 P.S. § 951 *et seq.* ("PHRA"). Lane responded by filing an answer along with a motion to partially dismiss the complaint. The Court dismissed as time-barred all but the retaliation claim, which related to Lane's failure to rehire Connelly in April 2011, but granted Connelly's request to file an amended complaint.

Connelly then filed her Amended Complaint, alleging separate counts of disparate treatment and retaliation under both Title VII and the PHRA. Lane promptly moved to dismiss the Amended Complaint under Federal Rule of Civil Procedure 12(b)(6) and, after briefing, the District Court granted that motion. The Court held that, with respect to her disparate treatment claims, Connelly had "failed to plead a sufficiently plausible inference that she was not rehired due to her gender." ***786** (App. 12.) Similarly, the Court held that the Amended Complaint failed to allege sufficient facts to establish a plausible claim of retaliation. It also denied Connelly's request to file a second amended complaint. The District Court thus dismissed all of Connelly's claims with prejudice. She timely appealed.

II. DISCUSSION*

[1] Connelly asserts two claims of error. First, she says that the District Court erred in holding that her Amended Complaint failed to meet the plausibility standard set

---

* The District Court had jurisdiction over the federal claims under 28 U.S.C. §§ 1331 and 1343, and supplemental jurisdiction over the related state law claims under 28 U.S.C. § 1367. We have appellate jurisdiction over the final decision of the District Court pursuant to 28 U.S.C. § 1291. We review

forth in *Bell Atlantic Corp. v. Twombly*, 550 U.S. 544, 127 S.Ct. 1955, 167 L.Ed.2d 929 (2007), and *Ashcroft v. Iqbal*, 556 U.S. 662, 129 S.Ct. 1937, 173 L.Ed.2d 868 (2009). Second, she argues that the District Court should have granted her leave to further amend the Amended Complaint. Because we agree with her on the first point, we need not reach the second.*

A. Standards for Pleading Sufficiency

[2] A complaint may be dismissed under Rule 12(b)(6) for "failure to state a claim upon which relief can be granted." But detailed pleading is not generally required. The Rules demand "only 'a short and plain statement of the claim showing that the pleader is entitled to relief,' in order to 'give the defendant fair notice of what the ... claim is and the grounds upon which it rests.'" *Twombly*, 550 U.S. at 555, 127 S.Ct. 1955 (quoting *Conley v. Gibson*, 355 U.S. 41, 47, 78 S.Ct. 99, 2 L.Ed.2d 80 (1957)). "To survive a motion to dismiss, a complaint must contain sufficient factual matter, accepted as true, to state a claim to relief that is plausible on its face." *Iqbal*, 556 U.S. at 678, 129 S.Ct. 1937 (citation and internal quotation marks omitted). "A claim has facial plausibility when the plaintiff pleads factual content that allows the court to draw the reasonable inference that the defendant is liable for the misconduct alleged." *Id.; see also Sheridan v. NGK Metals Corp.*, 609 F.3d 239, 262 n. 27 (3d Cir.2010). Although the plausibility standard "does not impose a probability requirement," *Twombly*, 550 U.S. at 556, 127 S.Ct. 1955, it does require a pleading to show "more than a sheer possibility that a defendant has acted unlawfully," *Iqbal*, 556 U.S. at 678, 129 S.Ct. 1937. A complaint that pleads facts "merely consistent with a defendant's liability ... stops short of the line between possibility and plausibility of entitlement to relief." *Id.* (citation and internal quotation marks omitted). The plausibility determination is "a context-specific task that requires the reviewing court to draw on its *787 judicial experience and common sense." *Id.* at 679, 129 S.Ct. 1937.

Under the pleading regime established by *Twombly* and *Iqbal*, a court reviewing the sufficiency of a complaint must take three steps.† First, it must "tak[e] note of the elements [the] plaintiff must plead to state a claim." *Iqbal*, 556 U.S. at 675, 129 S.Ct.

---

the District Court's decision to grant a motion to dismiss under a plenary standard. *Fowler v. UPMC Shadyside*, 578 F.3d 203, 206 (3d Cir.2009). We are "required to accept as true all allegations in the complaint and all reasonable inferences that can be drawn from them after construing them in the light most favorable to the nonmovant." *Foglia v. Renal Ventures Mgmt., LLC*, 754 F.3d 153, 154 n. 1 (3d Cir.2014) (quotation marks and citations omitted). However, as more fully described herein, we disregard legal conclusions and recitals of the elements of a cause of action supported by mere conclusory statements. *Santiago v. Warminster Twp.*, 629 F.3d 121, 128 (3d Cir.2010).

* Connelly only sought a curative amendment if the District Court decided to dismiss the Amended Complaint under Rule 12(b)(6). In that event, she asked for leave to "bolster the factual allegations related to her retaliation and disparate treatment claims." (App. 14.) Because we conclude that Connelly's pleadings were sufficient to survive the motion to dismiss, no curative amendment is necessary.

† Although *Ashcroft v. Iqbal* described the process as a "two-pronged approach," 556 U.S. 662, 679, 129 S.Ct. 1937, 173 L.Ed.2d 868 (2009), the Supreme Court noted the elements of the pertinent claim before proceeding with that approach, *id.* at 675–79, 129 S.Ct. 1937. Thus, we have described the process as a three-step approach. *Burtch v. Milberg Factors, Inc.*, 662 F.3d 212, 221 n. 4 (3d Cir.2011) (citing *Santiago*, 629 F.3d at 130).

1937. Second, it should identify allegations that, "because they are no more than conclusions, are not entitled to the assumption of truth." *Id.* at 679, 129 S.Ct. 1937. *See also Burtch v. Milberg Factors, Inc.*, 662 F.3d 212, 224 (3d Cir.2011) ("Mere restatements of the elements of a claim are not entitled to the assumption of truth." (citation and editorial marks omitted)). Finally, "[w]hen there are well-pleaded factual allegations, [the] court should assume their veracity and then determine whether they plausibly give rise to an entitlement to relief." *Iqbal*, 556 U.S. at 679, 129 S.Ct. 1937.

B. The Elements Necessary to State a Claim

We thus begin by taking note of the elements Connelly must plead to state her claims. With respect to her disparate treatment claim, Title VII makes it an "unlawful employment practice for an employer ... to discriminate against any individual ..., because of such individual's race, color, religion, sex, or national origin." 42 U.S.C. § 2000e–2(a) (1). *See also Desert Palace, Inc. v. Costa*, 539 U.S. 90, 92–93, 123 S.Ct. 2148, 156 L.Ed.2d 84 (2003). In 1991, Congress amended Title VII to further specify that, "[e] xcept as otherwise provided in this subchapter, an unlawful employment practice is established when the complaining party demonstrates that race, color, religion, sex, or national origin was a motivating factor for any employment practice, even though other factors also motivated the practice." 42 U.S.C. § 2000e–2(m). In *Watson v. Southeastern Pennsylvania Transportation Authority*, we interpreted that amendment to apply only to the category of discrimination cases that involve a "mixed-motive." 207 F.3d 207, 214–20 (3d Cir.2000). Generally speaking, in a "mixed-motive" case a plaintiff claims that an employment decision was based on both legitimate and illegitimate reasons. Such cases are in contrast to so-called "pretext" cases, in which a plaintiff claims that an employer's stated justification for an employment decision is false.

[3] [4] [5] A Title VII plaintiff may make a claim for discrimination "under either the pretext theory set forth in *McDonnell Douglas Corp. v. Green*[, 411 U.S. 792, 93 S.Ct. 1817, 36 L.Ed.2d 668 (1973)], or the mixed-motive theory set forth in *Price Waterhouse v. Hopkins*[, 490 U.S. 228, 109 S.Ct. 1775, 104 L.Ed.2d 268 (1989)], under which a plaintiff may show that an employment decision was made based on both legitimate and illegitimate reasons."* ***788** *Makky v. Chertoff*, 541 F.3d 205, 213 (3d Cir.2008). As we recognized in *Watson*, the "pretext" and "mixed-motive" labels can be misleading

---

* An employee proceeding under the *McDonnell Douglas* pretext framework bears the initial burden of establishing a *prima facie* case by showing: (1) that she was a member of a protected class, (2) that she was qualified for the job, and (3) another person, not in the protected class, was treated more favorably. *See Scheidemantle v. Slippery Rock Univ. State Sys. of Higher Educ.*, 470 F.3d 535, 539 (3d Cir.2006). If the employee establishes a *prima facie* case, the burden shifts to the employer to establish a legitimate nondiscriminatory reason for its employment action. *Id.* If the employer provides such a reason, the burden shifts back to the employee to show that the proffered reason was mere pretext for actual discrimination. *Id.* Notwithstanding this burden-shifting framework, a plaintiff who produces "direct evidence" of discrimination may proceed under the mixed-motive framework of *Price Waterhouse v. Hopkins*. 490 U.S. 228, 276, 109 S.Ct. 1775, 104 L.Ed.2d 268 (1989) (O'Connor, J., concurring). As we explained in *Armbruster v. Unisys Corp.*:

> [I]n the *Price Waterhouse* framework ... the evidence the plaintiff produces is so revealing of discriminatory animus that it is not necessary to rely on any presumption from the *prima facie* case to shift the burden of production. Both the burden of production and the risk of non-persuasion are shifted to the defendant who ... must persuade the factfinder that even

because, even in a case that does not qualify for a burden-shifting instruction under *Price Waterhouse*, the employer's challenged conduct may nevertheless result from two or more motives, and the plaintiff "need not necessarily show 'pretext' but may prevail simply by showing, through direct or circumstantial evidence, that the challenged action resulted from discrimination." 207 F.3d at 214 n. 5 (3d Cir.2000) (citations omitted). Under either theory of discrimination, the plaintiff must establish that her protected status was a factor in the employer's challenged action. The difference is in the degree of causation that must be shown: in a "mixed-motive" case, the plaintiff must ultimately prove that her protected status was a "motivating" factor, whereas in a non-mixed-motive or "pretext" case, the plaintiff must ultimately prove that her status was a "determinative" factor. *See id.* at 214–20 (summarizing the distinction in standards of causation that apply to "pretext" and "mixed-motive" cases and concluding that the 1991 amendment to Title VII did not alter that distinction).

Connelly's Amended Complaint does not specify whether she intends to proceed under a "mixed-motive" or a "pretext" theory, and understandably so. The distinction between those two types of cases "lies in the kind of proof the employee produces on the issue of [the employer's] bias," *Starceski v. Westinghouse Elec. Corp.*, 54 F.3d 1089, 1097 (3d Cir.1995), and identifying the proof before there has been discovery would seem to put the cart before the horse. Indeed, we have said that, even at trial, an employee "may present his case under both theories," provided that, prior to instructing the jury, the judge decides whether one or both theories applies. *Id.* at 1098 (internal citation omitted); *see also Radabaugh v. Zip Feed Mills, Inc.*, 997 F.2d 444, 448 (8th Cir.1993) (stating that "[w]hether a case is a pretext case or a mixed-motives case is a question for the court once all the evidence has been received"). Thus, for purposes of noting the elements Connelly must plead to state a disparate treatment claim, we take it as given that she may advance either a mixed-motive or a pretext theory.

[6] [7] The District Court, however, incorrectly evaluated the Amended Complaint as if Connelly were confined to showing pretext. Moreover, the Court's analysis proceeded with a point-by-point consideration of the elements of a prima facie case required under a pretext theory. It is thus worth reiterating that, at least for purposes of pleading sufficiency, a complaint need not establish a prima facie case in order to survive a motion to dismiss.* *789 A prima facie case is "an evidentiary standard, not a pleading requirement," *Swierkiewicz v. Sorema, N.A.*, 534 U.S. 506, 510, 122 S.Ct. 992, 152 L.Ed.2d 1 (2002), and hence is "not a proper measure of

---

if discrimination was a motivating factor in the adverse employment decision, it would have made the same employment decision regardless of its discriminatory animus.

32 F.3d 768, 778 (3d Cir.1994).

* In *Makky v. Chertoff*, we held that the plaintiff could not avoid dismissal of his mixed-motive discrimination claim if there was "unchallenged objective evidence" that he did not possess the "baseline qualifications" to do his job, because such a plaintiff would inevitably fail to establish a prima facie case of employment discrimination after the pleading stage. 541 F.3d 205, 215 (3d Cir.2008). However, our analysis explicitly assumed the sufficiency of the plaintiff's pleadings, *id.* at 214, and we limited our "necessarily narrow" holding to those rare mixed-motive cases in which the plaintiff's lack of baseline qualifications is "capable of objective determination before discovery," as when the job requires consideration of a license or similar prerequisite, *id.* at 215. Thus, that opinion expressly recognized that the prima facie case is a separate inquiry that generally cannot occur until after discovery.

whether a complaint fails to state a claim." *Fowler v. UPMC Shadyside*, 578 F.3d 203, 213 (3d Cir.2009). As we have previously noted about pleading in a context such as this,

> [a] determination whether a *prima facie* case has been made ... is an evidentiary inquiry—it defines the quantum of proof [a] plaintiff must present to create a rebuttable presumption of discrimination. Even post-*Twombly*, it has been noted that a plaintiff is not required to establish the elements of a *prima facie* case ....
> *Id.* at 213 (citation omitted). Instead of requiring a *prima facie* case, the post-*Twombly* pleading standard "'simply calls for enough facts to raise a reasonable expectation that discovery will reveal evidence of' the necessary element[s]." *Phillips v. Cty. of Allegheny*, 515 F.3d 224, 234 (3d Cir.2008) (quoting *Twombly*, 550 U.S. at 556, 127 S.Ct. 1955).

Should her case progress beyond discovery, Connelly could ultimately prevail on her disparate treatment claim by proving that her status as a woman was either a "motivating" or "determinative" factor in Lane's adverse employment action against her. Therefore, at this early stage of the proceedings, it is enough for Connelly to allege sufficient facts to raise a reasonable expectation that discovery will uncover proof of her claims.

[8] For the same reasons, Connelly's retaliation claim may survive Lane's motion to dismiss if she pleads sufficient factual allegations to raise a reasonable expectation that discovery will reveal evidence of the following elements: (1) she engaged in conduct protected by Title VII; (2) the employer took adverse action against her; and (3) a causal link exists between her protected conduct and the employer's adverse action. *Charlton v. Paramus Bd. of Educ.*, 25 F.3d 194, 201 (3d Cir.1994).

C. Excluding Conclusory Allegations

[9] [10] At the second step in our pleading analysis, we identify those allegations that, being merely conclusory, are not entitled to the presumption of truth. *Twombly* and *Iqbal* distinguish between legal conclusions, which are discounted in the analysis, and allegations of historical fact, which are assumed to be true even if "unrealistic or nonsensical," "chimerical," or "extravagantly fanciful." *Iqbal*, 556 U.S. at 681, 129 S.Ct. 1937. Put another way, *Twombly* and *Iqbal* expressly declined to exclude even outlandish allegations from a presumption of truth except to the extent they resembled a "formulaic recitation of the elements of a ... claim" or other legal conclusion.*
Id. (internal quotation marks *790 omitted); *see also Firestone Fin. Corp. v. Meyer*, 796 F.3d 822, 827 (7th Cir.2015) (concluding that allegations that were "neither legal assertions nor conclusory statements reciting the elements of a cause of action" were "entitled to a presumption of truth" under *Iqbal*). Perhaps "some allegations, while

---

* The Court in *Iqbal* clarified that it was only the conclusory nature of certain allegations—that is, their mere recitation of formulaic legal elements—that rendered them excludable: "[W]e do not reject these bald allegations on the ground that they are unrealistic or nonsensical .... It is the conclusory nature of [the] allegations, rather than their extravagantly fanciful nature, that disentitles them to the presumption of truth." *Iqbal*, 556 U.S. at 681, 129 S.Ct. 1937.

not stating ultimate legal conclusions, are nevertheless so threadbare or speculative that they fail to cross the line between the conclusory and the factual," but the clearest indication that an allegation is conclusory and unworthy of weight in analyzing the sufficiency of a complaint is that it embodies a legal point. *Peñalbert-Rosa v. Fortuño-Burset*, 631 F.3d 592, 595 (1st Cir.2011) (citation and internal quotation marks omitted).

[11] Although the District Court considered the Amended Complaint to be "extremely vague and conclusory," it did not specifically identify any allegations that, being mere legal conclusions, should have been discounted. (App. 10.) In our plenary review of the motion to dismiss, we consider the following allegations in the Amended Complaint to be disentitled to any presumption of truth: (1) that Connelly's supervisors at Lane "subjected her to disparate treatment based on her gender and retaliation for making complaints about discrimination and sexual harassment" (App. 26); (2) that Lane, "[b]y subjecting Connelly to discrimination based on her gender and retaliation," violated Title VII and the PHRA (App. 26–27); (3) that Connelly was an "employee" of Lane "within the meaning of Title VII and the PHRA" (App. 27); (4) that "[a]t all times relevant to this case, [Lane] was an 'employer' within the meaning of Title VII and the PHRA" (App. 27); and (5) that "Connelly has exhausted her federal and state administrative remedies." (App. 36). All of these allegations paraphrase in one way or another the pertinent statutory language or elements of the claims in question. To the extent that Connelly's allegation that she "was sexually harassed" by Manning states a legal conclusion, that is also excluded, although her factual allegations describing Manning's behavior and her reaction to him, along with her allegation that his threatened physical contact was "unwanted," are accepted as true. (App. 31.)

D. Construing the Historical Facts in the Plaintiff's Favor

Even after *Twombly* and *Iqbal*, a complaint's allegations of historical fact continue to enjoy a highly favorable standard of review at the motion-to-dismiss stage of proceedings. *See Phillips*, 515 F.3d at 231 (noting that *Twombly* "leaves intact" the pleading standard under which "detailed factual allegations" are not required). Although a reviewing court now affirmatively disregards a pleading's legal conclusions, it must still—as we have already emphasized—assume all remaining factual allegations to be true, construe those truths in the light most favorable to the plaintiff, and then draw all reasonable inferences from them. *Foglia v. Renal Ventures Mgmt., LLC*, 754 F.3d 153, 154 n. 1 (3d Cir.2014); *see also Phillips*, 515 F.3d at 231 (holding that *Twombly* did not "undermine [the] principle" that all reasonable inferences are to be drawn in favor of the plaintiff, and reaffirming that "the facts alleged must be taken as true and a complaint may not be dismissed merely because it appears unlikely that the plaintiff *791 can prove those facts or will ultimately prevail on the merits").

1. The Disparate Treatment Claim

[12] [13] With respect to Connelly's disparate treatment claim,* the Amended Complaint set forth sufficient factual allegations to raise a reasonable expectation

---

* While Connelly advances a disparate treatment claim under both Title VII and the PHRA, we refer to

that discovery would reveal evidence that Connelly was a member of a protected class and that she suffered an adverse employment action when Lane did not rehire her in 2011. More specifically, Connelly has alleged that (i) during her tenure at Lane, she was the only female truck driver at the Pittsburgh facility; (ii) she was qualified to drive all but one of Lane's trucks; (iii) Lane failed to rehire her at the start of the 2011 construction season, despite recalling the six other union truck-drivers—all male, and two with less union seniority than Connelly; and (iv) since failing to rehire Connelly, Lane has employed no other female truck drivers. Once accepted as true and construed in the light most favorable to the plaintiff, those allegations raise a reasonable expectation that discovery will reveal evidence that Connelly's protected status as a woman played either a motivating or determinative factor in Lane's decision not to rehire her. That is enough for Connelly's disparate treatment claim to survive a motion to dismiss. *Cf. Fowler*, 578 F.3d at 211–12 ("Although [the] complaint is not as rich with detail as some might prefer, it need only set forth sufficient facts to support plausible claims.").

Connelly has also alleged that Lane apparently deviated from its own past hiring norms and work assignments during the 2011 construction season by employing rental trucks and allowing a less senior driver to operate the tack truck. Once accepted as true and construed in the light most favorable to Connelly, those factual allegations would also permit the reasonable inference that Lane's proffered explanation that it failed to rehire Connelly for lack of work was pretextual. But, to be clear, at this stage Connelly is not obliged to choose whether she is proceeding under a mixed-motive or pretext theory, nor is she required to establish a prima facie case, much less to engage in the sort of burden-shifting rebuttal that *McDonnell Douglas* requires at a later stage in the proceedings. It suffices for her to plead facts that, construed in her favor, state a claim of discrimination that is "plausible on its face." *Iqbal*, 556 U.S. at 678, 129 S.Ct. 1937 (quoting *Twombly*, 550 U.S. at 570, 127 S.Ct. 1955). She has done that.

2. The Retaliation Claim

[14] [15] Turning to the elements of Connelly's retaliation claim, the facts alleged in the Amended Complaint, taken as true, also raise a reasonable expectation that discovery will reveal evidence both that Connelly engaged in activity protected by Title VII and that Lane took an adverse employment action against her.* To the latter point, Lane took an *792 adverse employment action against Connelly when it failed to rehire her at the start of the 2011 construction season. To the former, Connelly engaged in protected activity when she filed multiple complaints of sexual

---

those claims in the singular, as they are governed by essentially the same legal standards. *See Goosby v. Johnson & Johnson Med., Inc.*, 228 F.3d 313, 317 n. 3 (3d Cir.2000) ("The analysis required for adjudicating [plaintiff's discrimination] claim under PHRA is identical to a Title VII inquiry, and we therefore do not need to separately address her claim under the PHRA.") (internal citation omitted).

* Again, although Connelly's retaliation claims are advanced under both Title VII and the PHRA, we refer to those claims in the singular because the same framework for analyzing retaliation claims applies to both. *Cf. Krouse v. Am. Sterilizer Co.*, 126 F.3d 494, 500 (3d Cir.1997) ("[W]e analyze ADA retaliation claims under the same framework we employ for retaliation claims arising under Title VII.").

harassment—including and most obviously her May 2010 complaint that Manning, a company foreman, had made unwanted physical advances toward her.*

The District Court held that Connelly's retaliation claim came short of plausibility by "fail[ing] to plead a causal connection between the failure to rehire Connelly in April 2011 and her alleged protected activity." (App. 13.) In pertinent part, the District Court concluded that there was "no temporal proximity (as pled, her last report of sexual harassment was in May 2010, almost a year prior to the failure to rehire her), and no pattern of antagonism by Lane management." (App. 13.)

[16] Given the seasonal character of Connelly's work, we question the District Court's conclusion about temporal proximity. Because Lane only hired Connelly during construction seasons, traditionally laying workers off in October or November and then rehiring them in March or April of the following year, it may be that a retaliatory decision to not rehire her would not become apparent until after the off-season that ran from October 2010 to March 2011.[†]

In any case, the question of temporal proximity does not render Connelly's retaliation claim facially implausible. Connelly alleged that, after she complained of Manning's unwanted advances, and after overcoming another supervisor's resistance to her grievance by complaining directly to the Ethics Line, her relationship with both her supervisors and male co-workers became "increasingly strained" throughout *793 the year. (App. 32.) Thus, Connelly has alleged facts that could support a reasonable inference of a causal connection between her protected activity in May 2010 and the gradual deterioration of her relationship with her employer until she was laid off in October 2010.

In finding no causal connection between Connelly's protected acts and Lane's failure to rehire her in 2011, the District Court noted that Lane continued to rehire

---

* To be protected from retaliation under Title VII, the protected activity must relate to employment discrimination charges brought under that statute, implicating "discrimination on the basis of race, color, religion, sex, or national origin." *Slagle v. Cty. of Clarion*, 435 F.3d 262, 268 (3d Cir.2006). For that reason, we agree with the District Court that Connelly's other complaints, to the extent they implicated only safety issues, were not protected activity for purposes of her retaliation claim.

† As we have already stated, no showing of proof is necessary at this stage of the proceedings, but even if the record ultimately produced no evidence of temporal proximity suggestive of retaliation, that would not necessarily be fatal to Connelly's claim. *See Robinson v. Se. Pa. Transp. Auth.*, 982 F.2d 892, 894 (3d Cir.1993) ("The mere passage of time is not legally conclusive proof against retaliation."); *Kachmar v. SunGard Data Sys., Inc.*, 109 F.3d 173, 178 (3d Cir.1997) ("It is important to emphasize that it is causation, not temporal proximity itself, that is an element of plaintiff's prima facie case, and temporal proximity merely provides an evidentiary basis from which an inference can be drawn.").

> Where the time between the protected activity and adverse action is not so close as to be unusually suggestive of a causal connection standing alone, courts may look to the intervening period for demonstrative proof, such as actual antagonistic conduct or animus against the employee, or other types of circumstantial evidence, such as *inconsistent reasons given by the employer* for terminating the employee or the employer's treatment of other employees, that give rise to an inference of causation when considered as a whole.

Marra v. Phila. Hous. Auth., 497 F.3d 286, 302 (3d Cir.2007) (citations omitted and emphasis added). Even at this stage, if one accepts as true all of Connelly's factual allegations about her union seniority, Lane's past hiring practices, the company's traditional distribution of labor, and her personal observations of Lane's 2011 workforce, one could reasonably draw the inference that Lane gave Connelly inconsistent and false reasons for declining to rehire her.

Connelly for four consecutive years despite her many complaints, and even encouraged her to continue calling the Ethics Line. While we agree that those facts could be viewed as cutting against Connelly, that is not what the applicable standard of review allows at this point in the case. We must adhere to the requirement that all alleged facts be construed in the light most favorable to the plaintiff, which, if done, permits the view that gender discrimination was a motivating factor or determinative factor in the decision not to recall Connelly in 2011. Likewise, the fact that Lane continued to rehire Connelly for four years despite her complaints about co-workers, but declined to rehire her at the first such opportunity after she complained of harassment by a supervisor, can be construed to support a reasonable inference of a causal connection between the protected act and the adverse employment action.

Therefore, even if one believed it "unlikely that the plaintiff can prove those facts or will ultimately prevail on the merits," *Phillips*, 515 F.3d at 231 (citing *Twombly*, 550 U.S. at 563 n. 8, 127 S.Ct. 1955), it must still be said that Connelly—under a favorable standard of review—has raised a reasonable inference that discovery will reveal evidence of the elements necessary to establish her claims.

III. CONCLUSION

Because Connelly has alleged facially plausible claims sufficient to survive a motion to dismiss, we will vacate the District Court's Order dismissing the Amended Complaint and remand for further proceedings consistent with this opinion.

ALL CITATIONS

809 F.3d 780, 128 Fair Empl.Prac.Cas. (BNA) 970, 99 Empl. Prac. Dec. P 45,475

10 Family and Medical Leave Act

******

Chapter Objectives:

1. Acquire a general understanding of the requirements of the Family and Medical Leave Act (FMLA).
2. Acquire an understanding of the National Defense Authorization Act.
3. Acquire an understanding of the Uniformed Services Employment and Reemployment Rights Act.
4. Acquire an understanding of the interaction of the FMLA with the safety and health function.

******

Safety and health professionals with responsibilities for the management of staff personnel and/or the management and administration of the workers' compensation function should become well versed in the requirements of the Family and Medical Leave Act (hereinafter "FMLA"),* as well as the National Defense Authorization Act (hereinafter "NDAA").† In general, the FMLA

"provides a means for employees to balance their work and family responsibilities by taking unpaid leave for certain reasons ... and is intended to promote the stability and economic security of families as well as the nation's interest in preserving the integrity of families."‡

The NDAA amends the military leave entitlements of the FMLA, expanding coverage to family members, military caregivers, and spouses, sons, daughters, parents, or next-of-kin for certain veterans with serious injuries or illnesses.§

For safety and health professionals working in the private sector, the basic requirements for an employee to be eligible for a leave of absence of up to 12 workweeks, which is unpaid but job-protected, starts with employment. Under the FMLA, the employee must meet the eligibility requirements which include having to:

---

* 29 U.S.C. § 2601; Also see, Dept. of Labor website located at www.dol.gov.
† Public Law 111-84 (2010)
‡ Department of Labor website located at www.dol.gov.
§ Public Law 110-84.

- *Be employed by a covered employer and work at a worksite within 75 miles of which that employer employs at least 50 people;*
- *Have worked at least 12 months (which does not have to be consecutive) for the employer; and*
- *Have worked at least 1,250 hours during the 12 months immediately before the date FMLA leave begins.**

For most safety and health professionals, the majority of the private sector companies or organizations meet this requirement. However, safety professionals should be aware that some companies and organizations have established policies in which the requirements to qualify for FMLA are substantially lower than the federal minimum requirements.

For an employee to be eligible to take a job-protected, unpaid FMLA leave of absence, the employee must identify a qualifying event which includes:

- *Birth and care of the employee's child, within one year of birth;*
- *Placement with the employee of a child from adoption or foster care, within one year of the placement; Care of an immediate family member (spouse, child parent) who has a serious health condition;*
- *The employee's own health condition that makes the employee unable to perform the essential functions of his or her job;*
- *Any qualifying exigency arising out of the fact that the employee's spouse, son, daughter, or parent is on active duty or has been notified of an impending call or order to active duty in the U.S. National Guard or Reserves in support of a contingency operation.*
- *Twenty-six workweeks of leave during a single 12-month period to care for a covered service member with a serious injury or illness if the employee is a spouse, son, daughter, parent, or next-of-kin of the service member (Military Caregiver Leave).†*

Safety and health professionals should be aware that most private sector companies or organizations have policies and procedures established, usually through the human resource department, to address all requests for FMLA leave of absences. Safety and health professionals often see requests for FMLA leave for non-work-related injuries or illnesses by employees or family members; however, FMLA leave can also be requested in cases involving work-related injuries or illnesses. Safety professionals should be aware that although the employee is eligible for up to 12 workweeks of unpaid leave, employees may take the leave of absence on an intermittent basis or on a reduced leave schedule if medically necessary, or due to a qualifying exigency. Additionally, safety and health professionals should be aware that intermittent leave for the placement for adoption, foster care of a child, or birth of a child is subject to approval by the company or organization. However, safety and health professionals should be aware that approval by the company or organization is

---

* Department of Labor website at www.dol.gov – Elaws.
† *Id.*

not required for intermittent schedules or reduced work schedules that are medically necessary due to pregnancy, serious health conditions, or serious illness or injury of a covered service member, or in exigent circumstances.

Although many instances in which an FMLA leave of absence is requested are unforeseen, safety and health professionals should be aware that if the requested leave of absence is foreseeable, the employee is required to provide the company or organization at least 30 days' notice, or as soon as practicable. The employee is required to comply with the company or organization's policies and procedural requirements for requesting an FMLA leave of absence and provide sufficient information to permit the company or organization to make a reasonable determination whether or not FMLA applies to the leave of absence request. Safety and health professionals should be aware that in situations where the company or organization has previously granted an FMLA leave of absence and the employee is requesting a second FMLA leave, the employee must specifically reference the qualifying reason or need for a second FMLA leave of absence.

Although it is often difficult for an employee to financially support twelve weeks without a continuous weekly/biweekly paycheck in these difficult financial times, safety and health professionals should be aware that the company or organization can request that the employee requesting an FMLA leave of absence for a serious health condition provide verification and certification of the employee's health condition or the health condition of the family member or service member. Additionally, safety and health professionals should be aware that companies or organizations granting an FMLA leave of absence can require a periodic report on the employee's health status or the family member or service member's health status. Of importance to safety and health professionals, the company or organization may require that the employee returning from an FMLA leave of absence due to a serious health condition provide a certification from the employee's physician or healthcare provider of the employee's ability to return to full or unrestricted work duty.

Although usually addressed by the company or organization's human resource department, safety and health professionals should be aware that FMLA provides job protections to employees while they are on a qualified leave of absence. However, when an employee returns from an FMLA leave of absence, the employee is entitled to be restored to the same or an equivalent job with equivalent pay, benefits, and other terms and conditions of employment. The company or organization is required to return the employee to employment with the same benefits and at the same level as before the leave of absence; however, the employee is not entitled to any additional benefits accrued during the unpaid leave of absence. Safety and health professionals should also note that your employees may not be employed by other companies or organizations while on FMLA leave from your company or organization.

Safety and health professionals should be aware that the company or organization is required to maintain any group health insurance benefits during the FMLA leave of absence. The company or organization is also required to maintain the group health insurance benefits at the same level and in the same manner if the employee returns to work. Safety and health professionals should also note that employees can elect, or the company or organization can require, that the employee use accrued paid vacation, sick pay, personal days, and other accrued leave for the period of the unpaid FMLA leave of absence. For many companies or organizations, the accrued leave is required to be utilized concurrently with the FMLA leave of absence period of time. Prudent safety and health professionals should check with their human resource department to identify the company or organization's policy and procedure regarding concurrent FMLA and accrued leave.

The governing federal agency for FMLA is the Wage and Hour Division of the U.S. Department of Labor. Safety and health professionals should be aware that FMLA requires notices and posters to be displayed to inform employees of their FMLA rights, as well as recordkeeping requirements. FMLA does not currently possess any reporting requirements. Safety and health professionals should note these requirements, and may want to incorporate the FMLA poster and notice requirements in their annual audits, wherein the FMLA poster and notice requirements can be verified at the same time as the OSHA posters and other poster requirements.

The National Defense Authorization Act (NDAA)* amended the FMLA to allow eligible employees to take up to 12 workweeks of job-protected leave for "qualifying exigency" arising from active duty or call to active duty status of a spouse, son, daughter, or parent. The NDAA also amended the FMLA to permit up to 26 workweeks within a 12-month period to care for a service member with a serious injury or illness. Safety and health professionals should be aware that these amendments to the FMLA are often referred to as "military family leave."[†]

Safety and health professionals should be aware that the eligibility requirements and employee coverage requirements for NDAA are the same as FMLA. Under NDAA,

"a covered service member is a current member of the Armed Forces, including a member of the National Guard or Reserves, who is undergoing medical treatment, recuperation, or therapy, is otherwise in outpatient status, or is otherwise on the temporary disability retired list, for a serious injury or illness."[‡]

Safety professionals should be aware that the company or organization is required to provide up to 12 unpaid workweeks of leave during the normal 12-month period for qualifying exigencies arising from the employee's spouse, son, daughter, or parent being called to active military duty. However, safety professionals should be aware that under the terms of NDAA, qualifying exigency leave is only available to family members in the National Guard or Reserves and does not extend to family members of military service personnel in the regular Armed Forces.

A qualifying exigency under the NDAA includes:

- *Short notice deployment of a military service member for the seven days from the date of notification;*
- *Military events, ceremonies, programs, and events sponsored by the military or support organizations;*

* Public Law 110-181 (2008)
† Department of Labor website located at www.dol.gov.
‡ *Id.*

- *Certain daycare and related activities;*
- *Making financial or legal arrangements;*
- *Attending counseling;*
- *Up to five days rest and recuperation leave during deployment;*
- *Attending post-deployment activities;*
- *And other events the employee and company agree to be considered a qualifying exigency event.*

Safety and health professionals should be aware that many of the posting and recordkeeping requirements mirror the FMLA, and companies or organizations can require verification from the employee requesting an NDAA leave of absence. Safety and health professionals should be aware that where the military service member and his or her spouse both work for your company or organization, the spouse is limited to a combined total of 26 workweeks in any 12-month period if the leave is to care for a covered service member with a serious injury or illness and for the birth and care of a newborn child, adoption, or foster care of a child, or care of a parent who has a serious health condition.*

Safety and health professionals should be aware that in the vast majority of the FMLA and NDAA situations, companies and employers are more than accommodating to the situation of the employee and the FMLA or NDAA leave is immediately granted. However, it is important for safety professionals to ensure that all forms and verifications required under your company or organization's policies are completed and all necessary certifications required upon the return of the employee.

Similar to FMLA and NDAA is the Uniformed Service Employment and Reemployment Rights Act[†] (hereinafter "USERRA") which requires that the returning veterans receive all benefits of employment that they would have obtained if they had been continuously employed with the company or organization. Safety and health professionals should be aware that one of these benefits is the ability to receive an FMLA leave of absence. Thus, if a veteran returned from a tour of duty and is reemployed by your company or organization in the same 40 hour per week job as he or she had prior to serving, the employee should receive credit toward the 1,250-hour requirements for the time on active duty and thus would be eligible for FMLA leave.

Safety and health professionals should be aware that there has been a substantial amount of litigation since the enactment of the FMLA addressing many of the major elements within this statute. Safety and health professionals are encouraged to review the information on the U.S. Department of Labor website, as well as the current cases addressing many of the issues involving the FMLA and military leave provisions. Safety and health professionals should possess a level of understanding of their company or organization's policies and procedures with regards to FMLA and military leave, especially in the areas which impact the safety function such as medical certifications, overlap of FLMA with workers' compensation or short-term disability leave, training requirements, and poster requirements. As always, prudent safety and health

---

* *Id.*
[†] 38 U.S.C. §§ 4301–4333 (1994).

professionals should review any requests related to FMLA, NDAA, or USERRA with their human resource department or legal counsel before proceeding.

Case Study: This case may be modified for the purposes of this text.

259 F.3d 1112
United States Court of Appeals,
Ninth Circuit.

Penny BACHELDER; Mark Bachelder, Plaintiffs–Appellants,
v.
AMERICA WEST AIRLINES, INC., Defendant–Appellee.

No. 99–17458.
|
Argued and Submitted April 11, 2001
|
Filed Aug. 8, 2001

SYNOPSIS

Employee sued former employer, alleging that it violated the Family and Medical Leave Act (FMLA) when it terminated her for poor attendance. The United States District Court for the District of Arizona, Roslyn O. Silver, J., granted partial summary judgment to former employer, finding that employee was not entitled to the Act's protection for the 16 absences that occurred in the year of her firing, and, following bench trial, entered judgment for former employer, finding that, in deciding to fire employee, it did not impermissibly consider FMLA-protected leave that she took in prior years. Employee appealed. The Court of Appeals, Berzon, Circuit Judge, held that: (1) *McDonnell Douglas* burden-shifting approach was inapplicable to employee's claim that her taking of FMLA-protected leave was a negative factor in the decision to terminate her; (2) former employer did not adequately inform its employees that it had chosen the retroactive rolling "leave year" method for calculating their eligibility for FMLA leave; (3) because former employer did not fulfill its notification obligation, it "fail[ed] to select" a calculating method, thus requiring application of the option most beneficial to employee, the calendar year method, pursuant to which employee's 16 absences were covered by the Act; and (4) employee was not required to show that the other reasons that former employer proffered for firing her were pretextual.
Reversed and remanded with directions.

OPINION

BERZON, Circuit Judge:

[1] Penny Bachelder* claims that her employer, America West Airlines, violated the Family and Medical Leave Act of 1993 ("FMLA" or "the Act") when it terminated

* Penny's husband, Mark Bachelder, is also a plaintiff and appellant in this case. The district court found

her in 1996 for poor attendance. The district court granted partial summary judgment to America West, holding that Bachelder was not entitled to the Act's protection for her 1996 absences. Bachelder also appeals from the district court's subsequent finding, after a bench *1119 trial, that, in deciding to fire her, America West did not impermissibly consider FMLA-protected leave that she took in 1994 and 1995. This appeal requires us to interpret both the Act and the regulations issued pursuant to it by the Department of Labor.

I. BACKGROUND

A. THE FAMILY AND MEDICAL LEAVE ACT OF 1993

[2] [3] [4] The FMLA provides job security to employees who must be absent from work because of their own illnesses, to care for family members who are ill, or to care for new babies. 29 U.S.C. § 2612. Congress recognized that, in an age when all the adults in many families are in the work force, employers' leave policies often do not permit employees reasonably to balance their family obligations and their work life. The result, Congress determined, is "a heavy burden on families, employees, employers and the broader society." S.Rep. No. 103–3 at 4, 103d Cong., 2d Sess. (1993). As for employees' own serious health conditions, Congress found that employees' lack of job security during serious illnesses that required them to miss work is particularly devastating to single-parent families and to families which need two incomes to make ends meet. *Id.* at 11–12. As Congress concluded, "it is unfair for an employee to be terminated when he or she is struck with a serious illness and is not capable of working." *Id.* at 11. In response to these problems, the Act entitles covered employees* to up to twelve weeks of leave each year for their own serious illnesses or to care for family members, and guarantees them reinstatement after exercising their leave rights. 29 U.S.C. §§ 2612(a)(1), 2614(a)(1).†

---

that Mark Bachelder has standing to sue because, under Arizona law, he has a community property interest in Penny's earnings. America West has not contested the district court's standing decision. Although we normally must satisfy ourselves that a party has standing before proceeding to the merits of the case, even if the parties have not disputed standing, *see Friends of the Earth, Inc. v. Laidlaw Envt'l Servs. Inc.,* 528 U.S. 167, 180, 120 S.Ct. 693, 145 L.Ed.2d 610 (2000), the standing question is irrelevant in this case because Penny Bachelder unquestionably has standing to sue, and Mark's presence as a plaintiff has no effect on the relief available. For convenience, we refer in this opinion only to Penny Bachelder.

* The FMLA covers employees who have worked for a covered employer for at least 12 months and for at least 1,250 hours during the previous 12-month period. 29 U.S.C. § 2611(2). A covered employer is "any person engaged in commerce or in any industry or activity affecting commerce who employs 50 or more employees for each working day during each of 20 or more calendar workweeks in the current or preceding calendar year." 29 U.S.C. § 2611(4)(A)(i).

† 29 U.S.C. § 2612(a)(1) provides:
Subject to section 2613 of this title [allowing employers to require medical certification], an eligible employee shall be entitled to a total of 12 workweeks of leave during any 12-month period for one or more of the following:
(A) Because of the birth of a son or daughter of the employee and in order to care for such son or daughter.
(B) Because of the placement of a son or daughter with the employee for adoption or foster care.
(C) In order to care for the spouse, or a son, daughter, or parent, of the employee, if such spouse, son, daughter, or parent has a serious health condition.

[5] The FMLA was the culmination of several years of negotiations in Congress to achieve a balance that reflected the needs of both employees and their employers. While recognizing employees' need for job security at the times when they most needed time off from work, Congress, *1120 in enacting the FMLA, also took employers' legitimate prerogatives into account:

It is the purpose of this Act—

1. to balance the demands of the workplace with the needs of families, to promote the stability and economic security of families, and to promote national interests in preserving family integrity;
2. to entitle employees to take reasonable leave for medical reasons, for the birth or adoption of a child, and for the care of a child, spouse, or parent who has a serious health condition;
3. to accomplish the purposes described in paragraphs (1) and (2) in a manner that accommodates the legitimate interests of employers.

29 U.S.C. § 2601(b). The twelve-week limitation on employees' protected leave time—protected in the sense that the employee is entitled to reinstatement upon the end of the leave—as well as other provisions in the final Act, demonstrates that Congress wanted to ensure that employees' entitlement to leave and reinstatement did not unduly infringe on employers' needs to operate their businesses efficiently and profitably.*

The regulations implementing the twelve-week leave provision reflect this concern for employers' administrative efficiency and convenience needs. *See* Family and Medical Leave Act of 1993, 60 Fed.Reg. 2180, 2199 (Jan. 6, 1995) ("The choice of options was intended to give maximum flexibility for ease in administering FMLA in conjunction with other ongoing employer leave plans, given that some

---

(D) Because of a serious health condition that makes the employee unable to perform the functions of the position of such employee.
29 U.S.C. § 2614(a)(1) provides:
Except as provided in subsection (b) of this section [allowing employers to deny reinstatement to certain highly compensated employees], any eligible employee who takes leave under section 2612 of this title for the intended purpose of the leave shall be entitled, on return from such leave—
(A) to be restored by the employer to the position of employment held by the employee when the leave commenced; or
(B) to be restored to an equivalent position with equivalent employment benefits, pay, and other terms and conditions of employment.
* Versions of the legislation defeated in earlier Congresses would have entitled employees to longer periods of protected leave. *See* H.R. 4300, 99th Cong. (1986) (entitling employees to up to 26 weeks of leave in a 12-month period for the employee's own serious health condition and up to 18 weeks in a two-year period for the birth or adoption of a child or to care for an ill family member); H.R. 925, 100th Cong. (1988) (entitling employees to up to 10 weeks of leave in a two-year period for the birth or adoption of a child or to care for an ill family member and up to 15 weeks of leave in a 12-month period for the employee's own serious health condition).

Similarly, the earlier, unsuccessful family leave bills covered more employers than the law enacted in 1993. *Compare* H.R. 4300 § 101 (covering employers with 15 or more employees); H.R. 925 § 101(5)(A) (covering employers with 50 or more employees for the first three years the legislation would have been in effect, and thereafter, employers with 35 or more employees); *with* 29 U.S.C. § 2611(4)(A) (covering employers with 50 or more employees).

employers establish a 'leave year' and because of state laws that may require a particular result."). Consistent with that concern, the regulations provide employers with a menu of choices for how to determine the "twelve-month period" during which an employee is entitled to twelve weeks of FMLA-protected leave:

An employer is permitted to choose any one of the following methods for determining the "12-month period" in which the 12 weeks of leave entitlement occurs:

1. The calendar year;
2. Any fixed 12-month "leave year," such as a fiscal year, a year required by State law, or a year starting on an employee's "anniversary" date;
3. The 12-month period measured forward from the date an employee's first FMLA leave begins; or,
4. A "rolling" 12-month period measured backward from the date an employee uses any FMLA leave.

29 C.F.R. § 825.200(b). This "leave year" regulation is at the heart of Bachelder's appeal.

B. FACTS

Bachelder began working for America West as a customer service representative ***1121** in 1988. From 1993 until her termination in 1996, she was a passenger service supervisor, responsible for several gates at the Phoenix Sky Harbor Airport.

From 1994 to 1996, Bachelder was often absent from work for various health and family-related reasons. In 1994, she took five weeks of medical leave to recover from a broken toe, and in mid-1995, she took maternity leave for approximately three months. It is undisputed that these two leaves were covered by, and protected by, the FMLA. In addition to these extended absences, Bachelder also called in sick several times in 1994 and 1995.

On January 14, 1996, one of America West's managers had a "corrective action discussion" with Bachelder regarding her attendance record. Among the absences that concerned the company were several occasions on which Bachelder had called in sick and the 1994 and 1995 FMLA leaves. Bachelder was advised to improve her attendance at work and required to attend pre-scheduled meetings at which her progress would be evaluated.

In February 1996, Bachelder was absent from work again for a total of three weeks. During that time, she submitted two doctor's notes to America West indicating her diagnosis and when she could return to work. Bachelder's attendance was flawless in March 1996, but in early April, she called in sick for one day to care for her baby, who was ill. Right after that, on April 9, Bachelder was fired. The termination letter her supervisor prepared gave three reasons for the company's decision: (1) Bachelder had been absent from work 16 times since being counseled about her attendance in mid-January; (2) she had failed adequately to carry out her responsibilities for administering her department's Employee of the Month program; and (3) her personal on-time performance and the on-time performance in the section of the airport for which she was responsible were below par.

In due course, Bachelder filed this action, alleging that America West impermissibly considered her use of leave protected by the FMLA in its decision to terminate her.* In response, America West maintained that it had not relied on FMLA-protected leave in firing Bachelder, because none of her February 1996 absences were protected by the Act, and because her 1994 and 1995 FMLA leaves did not factor into its decision. None of Bachelder's February 1996 absences were covered by the Act, argued America West, because the company used the retroactive "rolling" year method—the fourth of the four methods permitted by the leave year regulation—to calculate its employees' eligibility for FMLA leave. If that method was used, Bachelder had exhausted her full annual allotment of FMLA leave as of June 1995,† and was entitled, according to the company, to no more such leave until twelve months had elapsed from the commencement of her 1995 maternity leave. Therefore, America West maintained, Bachelder's February 1996 absences could not have been protected by the Act.

Bachelder countered that according to the regulations implementing the FMLA, she was entitled to have her leave eligibility calculated by the method most favorable *1122 to her. Under a calendar year method of calculating leave eligibility, she contended, her February 1996 absences were protected by the FMLA, and America West had violated the Act by relying on those absences in deciding to fire her.

The district court granted America West's motion for summary judgment in part, deciding that none of Bachelder's 1996 absences were protected by the FMLA. The court nonetheless determined that a factual dispute remained as to whether America West had impermissibly considered Bachelder's 1994 medical leave and her 1995 maternity leave, which all agreed were covered by the FMLA, in its decision to fire her. Because it found that Bachelder had failed timely to request a jury trial, the court submitted this issue to a bench trial. Following the trial, the district court found that America West had not considered Bachelder's 1994 and 1995 FMLA-protected leaves in making the firing decision, and entered judgment for America West. Bachelder appeals from both the summary judgment and the judgment following the bench trial.

II. DISCUSSION

A. Prohibition on Considering Use of FMLA Leave in Making Employment Decisions

[6] [7] [8] The FMLA creates two interrelated, substantive employee rights: first, the employee has a right to use a certain amount of leave for protected reasons, and second, the employee has a right to return to his or her job or an equivalent job after using protected leave. 29 U.S.C. §§ 2612(a), 2614(a).‡ Congress intended that these

---

* Bachelder also asserted various claims under the Americans with Disabilities Act, Title VII, and the Arizona Civil Rights Act. The district court granted America West's motion for summary judgment as to all of these claims, and Bachelder has not appealed those rulings.

† Whether Bachelder in fact exhausted her full allotment of FMLA leave in 1995 is disputed. We have no need to resolve this dispute, as, under the applicable legal standards, the length of the 1995 leave does not matter.

‡ The FMLA also entitles employees to retain any employer-paid health benefits while using FMLA-protected leave, subject to the proviso that if the employee fails to return to work at the end of his or

new entitlements would set "a minimum labor standard for leave," in the tradition of statutes such as "the child labor laws, the minimum wage, Social Security, the safety and health laws, the pension and welfare benefit laws, and other labor laws that establish minimum standards for employment." S.Rep. No. 103–3 at 4.

[9] [10] Implementing this objective, Congress made it unlawful for an employer to "interfere with, restrain, or deny the exercise of or the attempt to exercise, any right provided" by the Act. 29 U.S.C. § 2615(a)(1).* The regulations explain that this prohibition encompasses an employer's consideration of an employee's use of FMLA-covered leave in making adverse employment decisions:

> [E]mployers *cannot use the taking of FMLA leave as a negative factor in employment actions*, such as hiring, promotions or disciplinary actions; nor can FMLA leave be counted under "no fault" attendance policies.
> 29 C.F.R. § 825.220(c) (emphasis added). We find, for the following reasons, that this rule is a reasonable interpretation of *1123 the statute's prohibition on "interference with" and "restraint of" employee's rights under the FMLA.†

[11] Section 2615's language of "interference with" and "restraint of" the exercise of the rights it guarantees to employees largely mimics that of § 8(a)(1) of the National Labor Relations Act. *See* 29 U.S.C. § 158(a)(1) (providing that it is an unfair labor practice for an employer "to interfere with, restrain, or coerce employees in the exercise of the rights guaranteed" by § 7 of the NLRA). Like the NLRA, the FMLA entitles employees to engage in particular activities—under the FMLA, taking leave from work for FMLA-qualifying reasons—that will be shielded from employer interference and restraint. *Compare* 29 U.S.C. § 157 (endowing employees with the rights "to self-organization, to form, join, or assist labor organizations, to bargain collectively ..., and to engage in other concerted activities for the purpose of collective bargaining or other mutual aid or protection, and ... the right to refrain from any or all of such activities") *with* 29 U.S.C. § 2612 (providing that eligible employees "shall be entitled to a total of 12 workweeks of leave during any 12-month period" for qualifying reasons).

---

her leave, the employer may recover from the employee the premiums paid for maintaining coverage during the employee's absence. 29 U.S.C. § 2614(c).

* It is also unlawful:
   to discharge or in any other manner discriminate against any individual for opposing any practice made unlawful by the Act, 29 U.S.C. § 2615(a)(2), or:
   to discharge or in any other manner discriminate against any individual because such individual—
   (1) has filed any charge, or has instituted or caused to be instituted any proceeding, under or related to this subchapter;
   (2) has given, or is about to give, any information in connection with any inquiry or proceeding relating to any right provided under this subchapter; or
   (3) has testified, or is about to testify, in any inquiry or proceeding relating to any right provided under this subchapter.
   29 U.S.C. § 2615(b).

† Congress authorized the Department of Labor to promulgate regulations implementing the FMLA. 29 U.S.C. § 2654. The department's reasonable interpretations of the statute are therefore entitled to deference under *Chevron USA Inc. v. Natural Resources Defense Council*, 467 U.S. 837, 843–44, 104 S.Ct. 2778, 81 L.Ed.2d 694 (1984). *See Ninilchik Traditional Council v. United States*, 227 F.3d 1186, 1191 (9th Cir.2000) (agency's interpretation of statute pursuant to statutory delegation of authority is entitled to *Chevron* deference); *Duckworth v. Pratt & Whitney, Inc.*, 152 F.3d 1, 5 (1st Cir.1998) (Labor Department's FMLA regulations were promulgated pursuant to statutory delegation and are entitled to *Chevron* deference).

[12] Because the FMLA's language so closely follows that of the NLRA, the courts' interpretation of § 8(a)(1) of the NLRA helps to clarify the meaning of the statutory terms "interference" and "restraint." *Northcross v. Bd. of Educ. of Memphis City Schs.*, 412 U.S. 427, 428, 93 S.Ct. 2201, 37 L.Ed.2d 48 (1973) (per curiam) (similarity of statutory language is strong indication that statutes should be interpreted in the same manner). The Supreme Court has held that, for example, an employer's award of preferential seniority rights to striker replacements interferes with employees' rights under the NLRA, *NLRB v. Erie Resistor Corp.*, 373 U.S. 221, 231, 83 S.Ct. 1139, 10 L.Ed.2d 308 (1963) (observing that the practice's "destructive impact upon the strike and union activity cannot be doubted"), as does an employer's threat to shut down its plant in retaliation if its employees should elect to form a union. *NLRB v. Gissel Packing Co.*, 395 U.S. 575, 616–20, 89 S.Ct. 1918, 23 L.Ed.2d 547 (1969). Similarly, this circuit has held—giving just a few examples—that literature distributed by an employer indicating that job losses will be inevitable if employees vote to form a union "interferes" with employees' rights, *NLRB v. Four Winds Indus. Inc.*, 530 F.2d 75, 78–79 (9th Cir.1976), as does an employer's surveillance of its employees meeting with a union organizer outside the workplace. *California Acrylic Indus. Inc. v. NLRB*, 150 F.3d 1095, 1099 (9th Cir.1998).

[13] The basis for these holdings, as *California Acrylic* stated, is that "the courts have long recognized that employers violate section 8(a)(1) ['s prohibition on interfering with or restraining employee rights] by engaging in activity that tends to chill an employee's freedom to exercise his [ ] rights." *Id.* For, "[a] protected activity acquires a precarious status if innocent employees can be discharged [for] ***1124** engaging in it [.] ... It is the tendency of those discharges to weaken or destroy [ ] right that is controlling." *NLRB v. Burnup & Sims, Inc.*, 379 U.S. 21, 23–24, 85 S.Ct. 171, 13 L.Ed.2d 1 (1964).

[14] As a general matter, then, the established understanding at the time the FMLA was enacted was that employer actions that deter employees' participation in protected activities constitute "interference" or "restraint" with the employees' exercise of their rights. Under the FMLA, as under the NLRA, attaching negative consequences to the exercise of protected rights surely "tends to chill" an employee's willingness to exercise those rights: Employees are, understandably, less likely to exercise their FMLA leave rights if they can expect to be fired or otherwise disciplined for doing so. The Labor Department's conclusion that employer use of "the taking of FMLA leave as a negative factor in employment actions," 29 C.F.R. § 825.220(c), violates is the Act is therefore a reasonable one.

[15] The pertinent regulation uses the term "discrimination" rather than "interfere" or "restrain" in introducing the "negative factor" prohibition. *See* 29 U.S.C. § 2615(a) (1); 29 C.F.R. § 825.220(c).* In the case before us and in similar cases, the issue is

---

* Some of the case law applying § 2615 erroneously uses the term "discriminate" to refer to interference with exercise of rights claims. This semantic confusion has led many courts to apply anti-discrimination law to interference cases, instead of restricting the application of such principles—assuming they are applicable to FMLA at all—to "anti-retaliation" or "anti-discrimination" cases under §§ 2615(a)(2) and (b). *See Morgan v. Hilti, Inc.*, 108 F.3d 1319, 1323 (10th Cir.1997); *Hodgens v. General Dynamics Corp.*, 144 F.3d 151, 160–61 (1st Cir.1998); *King v. Preferred Tech. Group*, 166 F.3d 887, 891 (7th Cir.1999); *Chaffin v. John H. Carter Co.*, 179 F.3d 316, 319 (5th Cir.1999); *Gleklen*

one of *interference* with the exercise of FMLA rights under § 2615(a)(1), not retaliation *or* discrimination: Bachelder's claim does not fall under the "anti-retaliation" or "anti-discrimination" provision of § 2615(a)(2), which prohibits *"discriminat[ion]* against any individual for opposing any practice made unlawful by the subchapter" (emphasis added); nor does it fall under the anti-retaliation or anti-discrimination provision of § 2615(b), which prohibits discrimination against any individual for instituting or participating in FMLA proceedings or inquiries. By their plain meaning, the anti-retaliation or anti-discrimination provisions do not cover visiting negative consequences on an employee simply because he has used FMLA leave. Such action is, instead, covered under § 2615(a)(1), the provision governing "Interference [with the] Exercise of rights." *See Diaz v. Ft. Wayne Foundry Corp.*, 131 F.3d 711, 712 (7th Cir.1997) (holding that a claim by a former employee that he was denied the use of FMLA leave is a claim of a substantive right, covered under (a)(1), and not (a) (2); *Rankin v. Seagate Techs., Inc.*, 246 F.3d 1145, 1148 (8th Cir.2001) (same)).

[16] The regulation we apply in this case, 29 C.F.R. 825.220, implements all the parts of 29 U.S.C. § 2615. As noted, the particular provision of the regulations prohibiting the use of FMLA-protected leave as a negative factor in employment decisions, 29 C.F.R. 825.220(c), refers to "discrimination," but actually pertains to the "interference with the exercise of rights" section of the statute, § 2615(a)(1), not the anti-retaliation or anti-discrimination sections, §§ 2615(a)(2) and (b). While the unfortunate intermixing of the two different statutory concepts is confusing, there is no doubt that 29 C.F.R. 825.220(c) **1125** serves, at least in part, to implement the interference with the exercise of rights section of the statute. *See* 29 C.F.R. 825.220(b) ("Any violations of the Act or of these regulations constitute interfering with, restraining, or denying the exercise of rights provided by the Act.").

[17] Consequently, our analysis is fairly uncomplicated. Much as it should be obvious that the "FMLA is not implicated and does not protect an employee against disciplinary action based upon [ ] absences" if those absences are not taken for one of the reasons enumerated in the Act, *Rankin*, 246 F.3d, at 1147 (8th Cir.2001); *see also Marchisheck v. San Mateo County*, 199 F.3d 1068 (9th Cir.1999) (determining that a terminated employee had no cause of action under the FMLA because the absences for which she was fired were not protected by the Act); *Diaz* 131 F.3d, at 713–14 (7th Cir.1997) (same), the FMLA *is* implicated and does protect an employee against disciplinary action based on her absences if those absences are taken for one of the Act's enumerated reasons. *See, e.g., Victorelli v. Shadyside Hosp.*, 128 F.3d 184, 190–91 (3d Cir.1997) (reversing grant of summary judgment for the employer where there was a triable issue whether the absence that triggered the plaintiff's termination was covered by the FMLA); *Rankin*, 246 F.3d at 1148–49; *Price v. City of Ft. Wayne*, 117 F.3d 1022, 1023–27 (7th Cir.1997).

[18] America West contends for quite a different approach, arguing that we should apply a *McDonnell Douglas*-style shifting burden-of-production analysis, familiar from anti-discrimination law, to determine whether the company illegally "retaliated" against Bachelder for using leave that was protected by the FMLA. *See*

---

v. Democratic Congressional Campaign Comm., 199 F.3d 1365, 1368 (D.C.Cir.2000); *Brungart v. BellSouth Telecommunications, Inc.*, 231 F.3d 791, 798 (11th Cir.2000).

McDonnell Douglas Corp. v. Green, 411 U.S. 792, 93 S.Ct. 1817, 36 L.Ed.2d 668 (1973); *see Reeves v. Sanderson Plumbing Products*, 530 U.S. 133, 142–43, 120 S.Ct. 2097, 147 L.Ed.2d 105 (2000) (the *McDonnell Douglas* framework only affects the burden of production, not the burden of persuasion). The *McDonnell Douglas* approach is inapplicable here, however.

[19] [20] The regulation promulgated by the Department of Labor, 29 C.F.R. 825.220(c) plainly prohibits the use of FMLA-protected leave as a negative factor in an employment decision. In order to prevail on her claim, therefore, Bachelder need only prove by a preponderance of the evidence that her taking of FMLA-protected leave constituted a negative factor in the decision to terminate her. She can prove this claim, as one might any ordinary statutory claim, by using either direct or circumstantial evidence, or both. *See e.g., Lambert v. Ackerley*, 180 F.3d 997 (9th Cir.1999) (en banc) (using both direct and circumstantial evidence to prove prohibited act under the Fair Labor Standards Act); *Davis Supermarkets, Inc. v. NLRB*, 2 F.3d 1162 (D.C.Cir.1993) (using both direct and circumstantial evidence to prove unfair labor practice under NLRA); *Reeves*, 530 U.S. at 142–43, 120 S.Ct. 2097 (2000) (circumstantial evidence, including evidence that the employer's explanation of its decision was false, can meet an employee's burden of persuasion in a Title VII case). No scheme shifting the burden of production back and forth is required.*

In the case before us, there is direct, undisputed evidence of the employer's motives: *1126 America West told Bachelder when it fired her that it based its decision on her sixteen absences since the January 1996 corrective action discussion. If those absences were, in fact, covered by the Act, America West's consideration of those absences as a "negative factor" in the firing decision violated the Act. The pivotal question in this case, then, is only "whether the plaintiff has established, by a preponderance of the evidence, that [s]he is entitled to the benefit [s]he claims." *Diaz*, 131 F.3d at 713.

B. FMLA Coverage of Bachelder's 1996 Leave

1. Calculating FMLA Leave Eligibility

Construing the statutory language and the Department of Labor's regulations, the district court held that Bachelder's February 1996 absences were not protected by the FMLA. We conclude that the district court's understanding of the statutory and regulatory scheme was erroneous.

[21] [22] [23] [24] The "leave year" regulation, 29 C.F.R. § 825.200, allows employers, at their option, to calculate the twelve-month period in which an employee is limited to twelve weeks of protected leave by one of four methods. Under the two fixed-year methods, the employee could use up to twelve weeks of leave at any time during the

---

* In contrast, the "anti-retaliation" provisions of FMLA prohibit "[discrimination] against any individual for opposing any practice made unlawful by this subchapter," (a)(2), and discrimination against any individual for instituting or participating in FMLA proceedings, (b), prohibitions which are not at issue in this case. 29 U.S.C. § 2615(a)(2). Whether or not the *McDonnell Douglas* anti-discrimination approach is applicable in cases involving the "anti-retaliation" provisions of FMLA, is a matter we need not consider here.

twelve-month period selected by the employer. 29 C.F.R. § 825.200(c).* For example, an employee whose employer had adopted the calendar year method could, consistently with the Act, "take 12 weeks of leave at the end of the year and 12 weeks at the beginning of the following year." *Id.* On January 1, this employee would be entitled to a full bank of FMLA-protected leave, no matter how recently, or how much, she had exercised her entitlement to protected leave the previous year.

[25] Under the rolling method, "each time an employee takes FMLA leave the remaining leave entitlement would be any balance of the 12 weeks which has not been used during the immediately preceding 12 months." *Id.* Thus, if an employee used her full allotment of twelve weeks of FMLA leave starting on February 1, she would be entitled to no additional days of FMLA leave until February 1 of the following year.

[26] [27] [28] The FMLA "leave year" regulation, while allowing employers flexibility in deciding how to comply with the Act, also includes various safeguards for employees. First, the employer must apply its chosen calculating method consistently to all employees. 29 C.F.R. § 825.200(d)(1). Second, if the employer has failed to select a calculating method, the regulations state that the method "that provides the most beneficial outcome for the employee will be used." 29 C.F.R. § 825.200(e). By preventing employers from calculating FMLA leave eligibility in their own favor on an ad hoc, employee-by-employee basis, the "leave year" regulation encourages the employer to choose its calculating method prospectively. By doing so, the regulation not only prevents unfairness to employees through retroactive manipulation of the "leave year," but also encourages a system under which both employees and employers can plan for future leaves in an orderly fashion.†

*1127 2. Notice Requirement

[29] [30] [31] The regulations allow employers to choose among four methods for calculating their employees' eligibility for FMLA leave, but they do not specifically state how an employer indicates its choice. America West contends, correctly, that the FMLA's implementing regulations do not expressly embody a requirement that employers inform their employees of their chosen method for calculating leave eligibility. The regulations nonetheless plainly contemplate that the employer's selection of one of the four calculation methods will be an open one, not a secret kept from the employees, the affected individuals.

First, the regulations require covered employers who provide "any written guidance to employees concerning employee benefits or leave rights, such as in an employee handbook," to "incorporate information on FMLA rights and responsibilities *and*

---

* The calculating method based on the employee's first leave request is a hybrid method, unique to each employee. *See* 29 C.F.R. § 825.200(c) ("Under the method in paragraph (b)(3) of this section, an employee would be entitled to 12 weeks of leave during the year beginning on the first date FMLA leave is taken; the next 12-month period would begin the first time FMLA leave is taken after completion of any previous 12-month period.").

† For example, parents may want to plan the time of an adoption, or of elective surgery, to coincide with the availability of FMLA leave under their employers' chosen calculation method. Similarly, an employer who chooses in advance a calendar year method is assured that no employee can take more than twelve weeks of FMLA leave in the calendar year, and can make staffing plans accordingly.

the employer's policies regarding the FMLA" therein. 29 C.F.R. § 825.301(a)(1) (emphasis added).* Because America West has an employee handbook, it is bound by § 825.301(a)(1).

Scattered throughout the Act and the regulations are choices for employers in how to comply with the statute. *See, e.g.*, 29 U.S.C. § 2612(d)(2) (permitting employers to require employees to use their accrued paid leave time for FMLA-qualifying purposes); 29 C.F.R. § 825.207(b) (same); 29 U.S.C. § 2613(a)(4) (permitting employers to require employees to provide medical certification that the employee can return to work after FMLA-qualifying leave); 29 C.F.R. § 825.310 (same). Section 825.301(a)(1), by its terms, requires employers to notify employees of the choices they have made. As the Department of Labor explained in announcing § 825.301(a)(1):

> The purpose of this provision is to provide employees the opportunity to learn from their employers of the manner in which that employer intends to implement FMLA and what company policies and procedures are applicable so that employees may make FMLA plans fully aware of their rights and obligations. It was anticipated that to some large degree these policies would be peculiar to that employer.

60 Fed.Reg. at 2219.

The rule allowing employers a choice of calculating methods is one example of the flexibility afforded to employers in complying with the FMLA. Section 825.301(a)(1) requires employers to notify their employees of this choice, just as it requires employers to notify their employees of other policies adopted to comply with the Act.[†]

*1128 Moreover, the "leave year" rule expressly requires notice in particular situations. Although these notice requirements do not explicitly require that employees be informed of the initial selection, they would be meaningless if the regulations as a whole allowed employers to conceal the initial selection from their employees.

For example, the "leave year" regulation provides that "[a]n employer *wishing to change to another alternative* [for calculating employees' FMLA leave eligibility] is required to give at least 60 days' notice to all employees, and the transition must take place in such a way that the employees retain the full benefit of 12 weeks of leave under whichever method affords the greatest benefit to the employee." 29 C.F.R. § 825.200(d)(1) (emphasis added). The 60-day rule demonstrates that employees are

---

* Employers who do not have employee handbooks must "provide written guidance to an employee concerning all the employee's rights and obligations under the FMLA." 29 C.F.R. § 825.301(a)(2).

† America West's argument that it satisfied any notice requirements by complying with the FMLA's general posting rule is unavailing. Covered employers are required conspicuously to post a notice "explaining the Act's provisions and providing information concerning the procedures for filing complaints of violations of the Act with the Wage and Hour Division" of the Labor Department. 29 C.F.R. § 825.300(a). The sample poster that satisfies this requirement does not mention the methods by which employers shall calculate leave eligibility. 29 C.F.R. § 825 App. C. But to conclude that complying with the posting rule satisfied *all* of an employer's notice obligations under the Act, as America West argues, would render a nullity the subsequent rule, 29 C.F.R. § 825.301, describing the "other notices to employees ... required of employers under the FMLA." Moreover, the Labor Department explicitly indicated that compliance with the posting rule would not suffice to meet all of the employer's notice requirements. *See* 60 Fed.Reg. at 2219 ("The posting of the notice [required by § 825.300] is but one of the notice requirements applicable to employers.").

entitled to act in reliance on their employer's choice of a calculating method in, for example, scheduling elective surgery or deciding which spouse will stay home to care for a seriously ill family member. Employees cannot reasonably act in reliance on an employer's initial policy choice if that choice was kept secret from them. Moreover, notifying employees of a *change* of methods is only meaningful if they are aware that another method was previously in use. For both these reasons, the regulations clearly contemplate that the employees not be kept in the dark concerning their employer's initial selection.

By the same token, "[i]f an employer fails to select one of the options, … [t]he employer may subsequently select an option only by providing the 60-day notice to all employees of the option the employer intends to implement." 29 C.F.R. § 825.200(e). Employees would not realize that their employer had "fail[ed] to select" a calculating method, such that they would be entitled to notice of a belated selection, unless the employer had a duty to provide timely information initially regarding its selection. Rather, the employer's "failure to select" a method is best understood to include the failure to inform employees of its selection.

The only sensible reading of the regulations taken as a whole, therefore, is that an employer's "selection" of a calculating method must be an open rather than a secret act, necessarily carrying with it an obligation to inform its employees thereof.* That the Labor Department so understood its own regulations is confirmed by the Department's statement, when announcing the regulations, that "[e]mployers must inform employees of the applicable method for determining FMLA leave entitlement when informing employees of their FMLA rights." 60 Fed.Reg. at 2200.

Further, as to any leave request made before the employer has thus selected a calculating method, the employer may properly be held to the rule that "the option that provides the most beneficial outcome for the employee" shall be used. 29 C.F.R. § 825.200(e). To hold otherwise would force employees to bear the risk of their employer's failure properly to inform them of the calculating method that will be used.

*1129 We therefore conclude that an initial selection of a method for calculating the leave year must be an open—not a secret—one before it can be applied to an employee's disadvantage.

3. Adequacy of Notice

[32] [33] [34] The question remains whether America West adequately notified its employees that it had chosen the retroactive rolling "leave year" calculation method. America West contends, and the district court agreed, that, because its employee handbook states that "employees are entitled to up to twelve calendar weeks of unpaid [FMLA] leave within any twelve-month period," it provided sufficient notice to its employees that it uses the "rolling method" for calculating leave eligibility. We disagree.

This statement from the America West handbook does nothing more than parrot the language of the Act. *See* 29 U.S.C. § 2612(a)(1) (providing that "an

---

* We note that no negative implication arises from the fact that the regulations are explicit in requiring 60-day notices in the event of a change in or failure to implement a leave year policy. In both of these instances, the nub of the regulation is the requirement that there be a 60-day period before a newly selected policy can take effect. A 60-day *advance* notice is not implicit in the "selection" requirement as read against the regulations' more general notice provisions and therefore had to be spelled out.

eligible employee shall be entitled to a total of 12 workweeks of leave during any 12-month period"). Pursuant to the authority granted to it by Congress, however, the Labor Department determined that the "rolling method" is *not* the only system permitted by the statute; the Department interpreted the statutory language to allow for three other calculating methods as well. So, the Department construed the statute's reference to "any 12-month period" to include a variety of differently-calculated 12-month periods, as chosen by the employer, thereby promoting employer flexibility. The Department then proceeded to enumerate four methods of determining FMLA leave eligibility, each of which, it necessarily determined, was consistent with the statute's "any 12-month period" language.*

True, in the preamble to its final rule, the Labor Department noted that the rolling method "most literally tracks" the Act's language. *See* 60 Fed.Reg. at 2200 ("While many comments were received opposing [the rolling] method, it has been retained as one of the available options because it is the one method that most literally tracks the statutory language."). But the very fact that the regulation permits employers to use any of four calculating methods is fatal to America West's argument: Because the statute can reasonably be read to allow the four different methods spelled out, merely parroting the statutory language cannot possibly inform employees of the method the employer has chosen. By paraphrasing the statutory language, in other words, America West has done no more than announce its intention to comply with the Act.

Because choosing a calculating method carries with it an obligation to inform employees of that choice and America West has failed to fulfill this obligation, it has "fail[ed] to select" a calculating method. 29 C.F.R. § 825.200(e). Thus, "the option that provides the most beneficial outcome for the employee" must be used to determine whether Bachelder's 1996 absences were covered by the FMLA. *Id.*

The calendar year method provides the most favorable outcome to Bachelder. 29 C.F.R. § 825.200(b)(1). Under this approach, it is immaterial that Bachelder had utilized her full allotment of FMLA-protected leave between April and June 1995 (and it is unnecessary for us to resolve the dispute whether she used every single day of FMLA leave to which she was entitled ***1130** in 1995). Because she began 1996 with a fresh bank of FMLA-protected leave, Bachelder's February 1996 absences were covered by the Act.†

C. America West's Additional Arguments

[35] America West nonetheless contends that "Bachelder's termination could not have been for her exercise of FMLA rights in 1996 because … both she and [America West] believed she had exhausted all of her FMLA leave." Whether either America

---

* Neither party suggests that there is any question concerning the validity of the regulation's interpretation of the statute to include various different 12-month periods. In light of the statute's use of the term "any," we also can perceive no basis for limiting employers to the single 12-month rolling period, rather than one of the other options enumerated by the regulation.

† To establish that her February 1996 absences qualify as FMLA leave, Bachelder also had to have suffered from a "serious health condition" and have been employed by America West for at least 1,250 hours in the preceding twelve months. 29 U.S.C. § 2611(2) (defining "eligible employee"). America West has not disputed that these conditions were met.

West or Bachelder believed at the time that her February 1996 absences were protected by the FMLA is immaterial, however, because the company's liability does not depend on its subjective belief concerning whether the leave was protected.

[36] [37] [38] [39] [40] First, the employer's good faith or lack of knowledge that its conduct violated the Act is, as a general matter, pertinent only to the question of damages under the FMLA, not to liability. An employer who violates the Act is liable for damages equal to the amount of any lost wages and other employment-related compensation, as well as any actual damages sustained as a result of the violation, such as the cost of providing care, and interest thereon. 29 U.S.C. § 2617(a)(1)(A). The employer is also liable for liquidated damages equal to the amount of actual damages and interest, unless it can prove that it undertook in good faith the conduct that violated the Act *and* that it had "reasonable grounds for believing that [its action] was not a violation" of the Act. 29 U.S.C. § 2617(a)(1)(A)(iii). Under such circumstances, it is within the district court's discretion to limit damages to only the amount of actual damages and interest thereon. *Id.* An employer who acts in good faith and without knowledge that its conduct violated the Act, therefore, is still liable for actual damages regardless of its intent.*

[41] [42] [43] [44] Second, it is the employer's responsibility, not the employee's, to determine whether a leave request is likely to be covered by the Act. Employees must notify their employers in advance when they plan to take foreseeable leave for reasons covered by the Act, *see* 29 U.S.C. § 2612(e), and as soon as practicable when absences are not foreseeable. *See* 29 C.F.R. § 825.303(a). Employees need only notify their employers that they will be absent under circumstances which indicate that the FMLA might apply:

> The employee need not expressly assert rights under the FMLA or even mention the FMLA, but may only state that leave is needed [for a qualifying reason]. The employer should inquire further of the employee if it is necessary to have more information about whether FMLA leave is being sought by the employee, and obtain the necessary details of the leave to be taken. In the case of medical conditions, the employer may find it ***1131** necessary to inquire further to determine if the leave is because of a serious health condition and may request medical certification to support the need for such leave.

29 C.F.R. § 825.302(c); *see also Price*, 117 F.3d at 1026 ("The FMLA does not require that an employee give notice of a desire to invoke the FMLA. Rather, it requires that the employee give notice of *need* for FMLA leave.") (emphasis in original); *Manuel v. Westlake Polymers Corp.*, 66 F.3d 758, 761–62 (5th Cir.1995). In

---

* That an employer's good-faith mistake as to whether its action violates the law is not a defense to liability is, similarly, commonplace in other areas of employment law. *See, e.g., Trans World Airlines, Inc. v. Thurston*, 469 U.S. 111, 105 S.Ct. 613, 83 L.Ed.2d 523 (1985) (holding that an employer's actions violating the Age Discrimination in Employment Act were taken in good faith and without knowledge of the violation barred an award of liquidated damages, but had no effect on liability); *Burnup & Sims*, 379 U.S. at 22–23, 85 S.Ct. 171 (employer's good faith is not a defense to liability for interfering with employees' rights under § 8(a)(1) of the NLRA).

short, the employer is responsible, having been notified of the reason for an employee's absence, for being aware that the absence may qualify for FMLA protection.*

Bachelder provided two doctor's notes to America West regarding her absences in February 1996.† The company was therefore placed on notice that the leave might be covered by the FMLA, and could have inquired further to determine whether the absences were likely to qualify for FMLA protection.

[45] Finally, America West argues that Bachelder failed to show that the other two reasons it initially put forward for firing her-her failure adequately to administer the Employee of the Month program and her unsatisfactory on-time performance-were pretextual. As we have already explained, however, there is no room for a *McDonnell Douglas* type of pretext analysis when evaluating an "interference" claim under this statute. The question here is not whether America West had additional reasons for the discharge, but whether Bachelder's taking of the 1996 FMLA-protected leave was used as a negative factor in her discharge. We know that the taking of the leave for the period in question was indeed used as a negative factor because America West so announced at the time of the discharge and does not deny that fact now. Moreover, America West does not seriously contend that, even though it considered an impermissible reason in firing Bachelder, it would have fired her anyway for the other two reasons alone. Even had it made such an argument, of course, the regulations clearly prohibit the use of FMLA-protected leave as a negative factor *at all.* Therefore no further inquiry on the question whether America West violated the statute in discharging Bachelder is necessary.‡

*1132 III. CONCLUSION

Because we hold that Bachelder's February 1996 absences were protected by the FMLA, and because America West used these absences as a negative factor in its decision to fire her, we reverse the district court's grant of summary judgment for

---

* In contrast, where an employee completely fails to give notice that she is absent for a potentially FMLA-qualifying reason, several circuits have held that the absence is not protected by the Act. *See, e.g., Strickland v. Water Works & Sewer Bd.,* 239 F.3d 1199, 1207 (11th Cir.2001) (affirming summary judgment for employer where there was no evidence that the employee told his supervisor he was leaving work for a medical reason); *Brungart,* 231 F.3d at 800 (holding that the employer was not liable because the decisionmaker who fired the plaintiff did not know that she was about to take leave at all, protected or otherwise); *Bailey v. Amsted Indus. Inc.,* 172 F.3d 1041 (8th Cir.1999) (affirming judgment for the employer because it lacked notice that plaintiff's frequent absences were for medical reasons).

† Although there is some dispute whether Bachelder's supervisors received both of these doctor's notes, the record shows that her February 1996 absences were recorded in her personnel file as "Medical Leave of Absence," a designation that, according to her supervisor's testimony, applied to leaves taken for medical reasons.

‡ We note that it appears fairly clear in any event that Bachelder would *not* have been fired had she not taken the protected leave. The supervisor who recommended that Bachelder be fired admitted in his deposition that "the basis for her termination, for the most part, was availability," and characterized her on-time performance and Employee of the Month deficiencies as "minor performance issues." Moreover, America West's witnesses testified at the trial that Bachelder's attendance was the primary reason for firing her, and the district court ultimately found that Bachelder failed to contradict their testimony that "the likely reason for her termination ... was because of her continued unavailability in 1996."

America West, direct the court to grant Bachelder's cross-motion for summary judgment as to liability, and remand for further proceedings.*

REVERSED and REMANDED.

ALL CITATIONS

259 F.3d 1112, 81 Empl. Prac. Dec. P 40,689, 143 Lab.Cas. P 34,320, 7 Wage & Hour Cas.2d (BNA) 286, 21 NDLR P 127, 00 Cal. Daily Op. Serv. 6809, 2001 Daily Journal D.A.R. 8373

---

* The district court's finding after the bench trial that America West did not impermissibly consider Bachelder's 1994 and 1995 FMLA-protected leaves in deciding to fire her appears to be supported by the record, and the court's refusal to grant Bachelder's inadvertently untimely jury demand was correct. *See Pacific Fisheries Corp. v. HIH Cas. & Gen. Ins. Ltd.*, 239 F.3d 1000, 1002 (9th Cir.2001). In light of our holding on the 1996 absences, however, these two rulings are beside the point.

11 The Americans with Disabilities Act and Safety

******

Chapter Objectives:

1. Acquiring an understanding of the requirements and protections afforded under the Americans with Disabilities Act (ADA).
2. Acquire an understanding of the overall purpose and scope of the ADA.
3. Acquire an understanding of the ADA Amendment Act and Regulations.
4. Gain an understanding of the issues and liabilities under the ADA within the safety and health function.

******

Outside of the Occupational Safety and Health Act, the Americans with Disabilities Act* (hereinafter "ADA") impacts and intersects with the safety and health function more than any other law. It is important that safety and health professionals acquire a working knowledge of the ADA, as well as key areas in which the ADA and safety and health function may intersect, in order to be able to recognize when the ADA may be applicable to the situation. Safety and health professionals should be able to recognize when a potentially qualified individual is requesting an accommodation and address the situation appropriately in order to comply with the ADA.

Within the safety and health function, the ADA can impact and intersect in the areas of workers' compensation, restricted duty programs, facility modifications, safety training, equipment design, personal protective equipment (PPE), and other safety functions, creating duties and responsibilities for the safety and health professional as well as potential liabilities for the company or organization. Additionally, safety and health professionals should continuously educate themselves in the changing requirements of the ADA, as well as the ADA Amendment Act and Equal Employment Opportunity Commission (EEOC) regulations and decisions, in order to ensure that the safety function as well as the company or organization is in compliance with the antidiscrimination protections provided to qualified individuals with disabilities.

---

* 42 U.S.C. Section 12101-12102, 12111-12117,12201-12213.

Safety and health professionals should be aware that the ADA is one of the more extensive laws which can provide protections to a very large number of applicants, employees, and others. In a nutshell, the ADA prohibits discriminating against qualified individuals with physical or mental disabilities in all employment settings. In theory, this sounds relatively simple; however, the ADA can be a minefield for safety professionals who are uninformed of the requirements of this law.

Safety and health professionals should be aware that although the ADA became law in 1990, the ADA is still transitioning through court decisions, agency regulations, and interpretations, and was substantially amended by the ADA Amendments Act of 2008* (hereinafter "ADAAA) and the Civil Rights Act of 1991.† Safety and health professionals should be aware that, effectively, the ADAAA makes significant changes to the term "disability" as defined in the ADA by rejecting several Supreme Court decisions and portions of previous EEOC's ADA regulations. The ADAAA retains the basic definition of "disability" as defined in the ADA as being an impairment that substantially limits one or more major life activities, a record of such an impairment, or being regarded as possessing an impairment. More specifically, the ADAAA provided the following:

The Act requires the EEOC to revise the section of their regulations which define the term "substantially limits";

The Act expands the definition of "major life activities" by including two non-exhaustive lists of which list 1 includes many activities that the EEOC has previously recognized (such as walking) as well as activities that the EEOC previously did not specifically recognize (such as reading, bending and communicating) and the second list includes major bodily functions (such as functions of the immune system, normal cell growth, digestive system, bowel, bladder, neurological, brain, respiratory, circulatory, endocrine and reproductive functions.

The Act states that mitigating measures such as "ordinary eyeglasses or contact lenses" shall not be considered in assessing whether an individual possesses a disability;

The Act clarified that an impairment that is episodic or in remission is a disability if it would substantially limit a major life activity when active; the Act provides that an individual subjected to an action prohibited by the ADA (such as failure to hire) because of an actual or perceived impairment will meet the "regarded as" definition of disability, unless the impairment is transitory or minor;

The Act provides that individuals covered only under the "regarded as" prong of the ADA test are not entitled to reasonable accommodation; And the Act emphasizes that the definition of "disability" should be interpreted broadly.‡

---

* Public Law 110-325.
† 42 U.S.C. Section 1981(a) & 42 U.S.C. Section 2000e(2)(k)-(n).
‡ Notice Concerning The Americans With Disabilities (ADA) Amendment Act of 2008, EEOC website located at www.eeoc.gov.

Safety and health professionals should acquire a firm grasp of the requirements of the ADA, as well as the interpretations and changes resulting from the ADAAA, the Civil Rights Act of 1991, and the EEOC interpretations. The initial place to start for safety and health professionals is with the federal or state agency responsible for administration and enforcement of the law. For the ADA, safety and health professionals should be aware that the federal agency responsible for administration and enforcement of the ADA is the Equal Employment Opportunity Commission (hereinafter "EEOC") and information can be found on their website located at www.eeoc.gov. Safety professionals are encouraged to monitor the EEOC website as well as recent case decisions, regulations, and interpretations to ensure the most current information regarding the ADA. Additionally, safety and health professional should also be aware that individual states may also possess laws which parallel or are more stringent than the federal level ADA and information can usually be found on the individual state agency's website.

Safety and health professionals should be aware that the ADA is divided into five titles, and all titles have the potential to substantially impact the safety function in covered public or private sector organizations. Title I contains the employment provisions that protect all individuals with disabilities who are in the United States, regardless of their national origin or immigration status. Title II prohibits discriminating against qualified individuals with disabilities or excluding them from the services, programs, or activities provided by public entities. Title II contains the transportation provisions of the Act. Title III, entitled "Public Accommodations," requires that goods, services, privileges, advantages, and facilities of any public place be offered "in the most integrated setting appropriate to the needs of the individual."* Title IV also covers transportation offered by private entities and addresses telecommunications. Title IV requires that telephone companies provide telecommunication relay services and that public service television announcements that are produced or funded with federal money include closed captions. Title V includes the miscellaneous provisions. This Title notes that the ADA does not limit or invalidate other federal and state laws providing equal or greater protection for the rights of individuals with disabilities, and addresses related insurance, alternate dispute resolution, and congressional coverage issues.

Safety and health professionals in the private sector should pay careful attention to the scope and potential impact of Title I of the ADA on safety functions. Title I prohibits covered employers from discriminating against a "qualified individual with a disability" with regard to job applications, hiring, advancement, discharge, compensation, training, and other terms, conditions, and privileges of employment.† Of particular importance for safety and health professionals is Section 101(8). This section defines a "qualified individual with a disability as any person who, with or without reasonable accommodation, can perform the essential functions of the employment position that such individual holds or desires ... consideration shall be given to the employer's judgment as to what functions of a job are essential, and if an employer has prepared a written description before advertising or interviewing applicants for the job, this description shall be considered evidence of the essential function of the job."‡ The Equal Employment Opportunity Commission (EEOC)

---

* ADA Section 305.
† ADA Section 102(a); 42 U.S.C. Section 12122.
‡ ADA Section 101(8).

provides additional clarification of this definition by stating, "an individual with a disability who satisfies the requisite skill, experience and educational requirements of the employment position such individual holds or desires, and who, with or without reasonable accommodation, can perform the essential functions of such position."* Congress did not provide a specific list of disabilities covered under the ADA because "of the difficulty of ensuring the comprehensiveness of such a list."[1] Under the ADA, an individual has a disability if he or she possesses:

- *a physical or mental impairment that substantially limits one or more of the major life activities of such individual,*
- *a record of such an impairment, or*
- *is regarded as having such an impairment.*[†]

Safety and health professionals should be aware that the ADA utilizes the broader language of "disability" rather than the term "handicapped" adopted under the Rehabilitation Act. For an individual to be considered "disabled" under the ADA, the physical or mental impairment must limit one or more "major life activities." Under the U.S. Justice Department's regulation issued for Section 504 of the Rehabilitation Act, "major life activities" are defined as, "functions such as caring for one's self, performing manual tasks, walking, seeing, hearing, speaking, breathing, learning and working."[2] Congress clearly intended to have the term "disability" broadly construed. However, this definition does not include simple physical characteristics, nor limitations based on environmental, cultural, or economic disadvantages.[3]

The second prong of this definition is "a record of such an impairment disability." The Senate Report and the House Judiciary Committee Report each stated:

This provision is included in the definition in part to protect individuals who have recovered from a physical or mental impairment which previously limited them in a major life activity. Discrimination on the basis of such a past impairment would be prohibited under this legislation. Frequently occurring examples of the first group (i.e., those who have a history of an impairment) are people with histories of mental or emotional illness, heart disease or cancer; examples of the second group (i.e., those who have been misclassified as having an impairment) are people who have been misclassified as mentally retarded.[4]

The third prong of the statutory definition of a disability extends coverage to individuals who are "being regarded as having a disability." The ADA has adopted the same "regarded as" test that is used in Section 504 of the Rehabilitation Act:

"Is regarded as having an impairment" means (A) has a physical or mental impairment that does not substantially limit major life activities but is treated ... as constituting such a limitation; (B) has a physical or mental impairment that substantially limits

---

* EEOC Interpretive Rules, 56 Fed. Reg. 35 (July 26, 1991).
† ADA Section 101(8).

major life activities only as a result of the attitudes of others toward such impairment; (C) has none of the impairments defined (in the impairment paragraph of the Department of Justice regulations) but is treated ... as having such an impairment.*

Safety and health professionals should be aware that a "qualified individual with a disability" under the ADA is any individual who can perform the essential or vital functions of a particular job with or without the employer accommodating the particular disability. Safety and health professionals should be aware that companies or organizations are provided with the opportunity to determine the "essential functions" of the particular job before offering the position through the development of a written job description. This written job description will be considered evidence of which functions of the particular job are essential and which are peripheral. In deciding the "essential functions" of a particular position, the EEOC will consider the company or organization's judgment, whether the written job description was developed prior to advertising or beginning the interview process, the amount of time spent performing the job, the past and current experience of the individual to be hired, relevant collective bargaining agreements, and other factors.[5]

The EEOC defines the term "essential function" of a job as meaning "primary job duties that are intrinsic to the employment position the individual holds or desires" and precludes any marginal or peripheral functions which may be incidental to the primary job function. The factors provided by the EEOC in evaluating the "essential functions" of a particular job include the reason that the position exists, the number of employees available, and the degree of specialization required to perform the job. This determination is especially important to safety professionals who may be required to develop the written job descriptions or to determine the "essential functions" of a given position.†

Safety and health professionals should recognize that they can be placed in a difficult position when the issue involved is whether or not the qualified individual creates a direct threat to the safety and health of themselves or others in the workplace. This issue may require the safety professional to evaluate and render a decision which will not only impact the individual with a disability, but also the company or organization. Safety and health professionals should be aware that the ADA does identify that any individual who poses a direct threat to the health and safety of others that cannot be eliminated by reasonable accommodation may be disqualified from the particular job.‡ The term "direct threat" to others is defined by the EEOC as creating "a significant risk of substantial harm to the health and safety of the individual or others that cannot be eliminated by reasonable accommodation."[6] The determining factors that safety and health professionals should consider in making this determination include the duration of the risk, the nature and severity of the potential harm, and the likelihood that the potential harm will occur.[7] Safety and

---

* EEOC Interpretive Rules, 56 Fed. Reg. 35 (July 26, 1991).
† Id.
‡ ADA Section 103(b).

health professionals when addressing this issue should also consider the EEOC's Interpretive Guidelines, which state:

> *[If] an individual poses a direct threat as a result of a disability, the employer must determine whether a reasonable accommodation would either eliminate the risk or reduce it to an acceptable level. If no accommodation exists that would either elimi-nate the risk or reduce the risk, the employer may refuse to hire an applicant or may discharge an employee who poses a direct threat.*[8]

Safety and health professionals should note that Title I provides that if a company or organization does not make the reasonable accommodations for the known limitations of a qualified individual with disabilities, this action or inaction is to be considered discrimination. However, if the company or organization can prove that providing the accommodation would place an undue hardship on the operation of the business, the claim of discrimination be disproved. Section 101(9) defines a "reasonable accommodation" as:

1. *making existing facilities used by employees readily accessible to and usable by the qualified individual with a disability and includes:*
2. *job restriction, part-time or modified work schedules, reassignment to a vacant position, acquisition or modification of equipment or devices, appropriate adjustments or modification of examinations, training mate-rials, or policies, the provisions of qualified readers or interpreters and other similar accommodations for ... the QID (qualified individual with a disability).**

The EEOC further defines "reasonable accommodation" as:

1. *Any modification or adjustment to a job application process that enables a qualified individual with a disability to be considered for the position such qualified individual with a disability desires, and which will not impose an undue hardship on the ... business; or*
2. *Any modification or adjustment to the work environment, or to the manner or circumstances which the position held or desired is customarily per-formed, that enables the qualified individual with a disability to perform the essential functions of that position and which will not impose an undue hardship on the ... business; or*
3. *Any modification or adjustment that enables the qualified individual with a disability to enjoy the same benefits and privileges of employment that other employees enjoy and does not impose an undue hardship on the ... business.*†

---

* ADA Section 101(9).
† EEOC Interpretive Rules, supra. note 11.

Of particular importance for safety and health professionals is the area of "reasonable accommodation" for qualified individuals with disabilities. Safety and health professionals should be aware that the company or organization would be required to make "reasonable accommodations" for any/all known physical or mental limitations of the qualified individual with a disability, unless the employer can demonstrate that the accommodations would impose an "undue hardship" on the business, or that the particular disability directly affects the safety and health of that individual or others. Safety and health professionals should also be aware that included under this section is the prohibition against the use of qualification standards, employment tests, and other selection criteria that can be used to screen out individuals with disabilities, unless the employer can demonstrate that the procedure is directly related to the job function. In addition to the modifications to facilities, work schedules, equipment, and training programs, the company or organization is required to initiate an "informal interactive (communication) process" with the qualified individual to promote voluntary disclosure of his or her specific limitations and restrictions to enable the employer to make appropriate accommodations that will compensate for the limitation.*

Additionally, safety and health professionals should pay careful attention to Title I, Section 102(c)(1). This section prohibits discrimination through medical screening, employment inquiries, and similar scrutiny. Safety professionals should be aware that underlying this section was Congress's conclusion that information obtained from employment applications and interviews

was often used to exclude individuals with disabilities—particularly those with so-called hidden disabilities such as epilepsy, diabetes, emotional illness, heart disease and cancer—before their ability to perform the job was even evaluated.†

Under Title I, Section 102(c)(2), safety and health professionals should be aware that conducting pre-employment physical examinations of applicants and asking prospective employees if they are qualified individuals with disabilities is prohibited. Employers are further prohibited from inquiring as to the nature or severity of the disability, even if the disability is visible or obvious. Safety professionals should also be aware that individuals may ask whether any candidates for transfer or promotion who have a known disability can perform the required tasks of the new position if the tasks are job-related and consistent with business necessity. An employer is also permitted to inquire about the applicant's ability to perform the essential job functions prior to employment. The employer should use the written job description as evidence of the essential functions of the position.‡

Safety and health professionals may require medical examinations of employees only if the medical examination is specifically job-related and is consistent with business necessity. Medical examinations are permitted only after the applicant

---

* Id.
† S. Comm. on Labor and Human Resources rep. at 38; H. Comm. on Jud. Rep. at 42.
‡ Id.

with a disability has been offered the job position. The medical examination may be given before the applicant starts the particular job, and the job offer may be contingent upon the results of the medical examination if all employees are subject to the medical examinations and information obtained from the medical examination is maintained in separate, confidential medical files. Companies or organizations are permitted to conduct voluntary medical examinations for current employees as part of an ongoing medical health program, but again, the medical files must be maintained separately and in a confidential manner. The ADA does not prohibit safety and health professionals or their medical staff from making inquiries or requiring medical or "fit for duty" examinations when there is a need to determine whether or not an employee is still able to perform the essential functions of the job, or where periodic physical examinations are required by medical standards or federal, state, or local law.*

In the area of medical testing, safety and health professionals should pay careful attention to the area of controlled substance testing. Under the ADA, the company or organization is permitted to test job applicants for alcohol and controlled substances prior to an offer of employment under Section 104(d). The testing procedure for alcohol and illegal drug use is not considered a medical examination as defined under the ADA. Companies or organizations may additionally prohibit the use of alcohol and illegal drugs in the workplace and may require that employees not be under the influence while on the job. Companies and organizations are permitted to test current employees for alcohol and controlled substance use in the workplace, subject to the limits permitted by current federal and state law. The ADA requires all employers to conform to the requirements of the Drug-Free Workplace Act of 1988. Thus, safety and health professionals should be aware that most existing pre-employment and post-employment alcohol and controlled substance programs which are not part of the pre-employment medical examination or ongoing medical screening program will be permitted in their current form.* Individual employees who choose to use alcohol and illegal drugs are afforded no protection under the ADA. However, employees who have successfully completed a supervised rehabilitation program and are no longer using or addicted to drugs or alcohol are offered the protection of a qualified individual with a disability under the ADA.†

Safety and health professionals should note that Title III also requires that "auxiliary aids and services" be provided for the qualified individual with a disability including, but not limited to, interpreters, readers, amplifiers, and other devices (not limited or specified under the ADA) to provide that individual with an equal opportunity for employment, promotion, etc.‡ Congress did, however, provide that auxiliary aids and services do not need to be offered to customers, clients, and other members of the public if the auxiliary aid or service creates an undue hardship on the business. Safety professionals may want to consider alternative methods of accommodating the qualified individual with a disability. This section

---

* ADA Section 102(c).
† ADA Section 511(b).
‡ ADA Section 3(1).

also addresses the modification of existing facilities to provide access to the individual, and requires that all new facilities be readily accessible and usable by the individual.

Under Title V, safety professionals should note that the ADA does not limit or invalidate other federal or state laws that provide equal or greater protection for the rights of individuals with disabilities. Safety professionals should become knowledgeable with regards to their individual state laws addressing disability or handicap discrimination in the workplace which may be more restrictive than the ADA.

Congress wrote the ADA in an all-encompassing manner and it is substantially broad in nature. Safety and health professionals should note that the ADA provides protections to all individuals associated with or having a relationship to the qualified individual with a disability. This inclusion is unlimited in nature, including family members, individuals living together, and an unspecified number of others. The ADA extends coverage to all "individuals," legal or illegal, documented or undocumented, living within the boundaries of the United States, regardless of their status.*

Under Section 102(b)(4), unlawful discrimination includes "excluding or otherwise denying equal jobs or benefits to a qualified individual because of the known disability of the individual with whom the qualified individual is known to have a relationship or association." Therefore, the protections afforded under this section are not limited to only familial relationships. There appear to be no limits regarding the kinds of relationships or associations that are afforded protection. Of particular note is the inclusion of unmarried partners of persons with AIDS or other qualified disabilities.[†]

Safety and health professionals should note that, similar to the OSHA requirements, the ADA requires that employers post notices of the pertinent provisions of the ADA in an accessible format in a conspicuous location within the employer's facilities. To further ensure ADA compliance, safety professionals may wish to consider providing additional notification on job applications and other pertinent documents.

Under the ADA, safety and health professionals should be aware that it is unlawful for an employer to "discriminate on the basis of disability against a qualified individual with a disability" in all areas, including:

- *recruitment, advertising, and job application procedures;*
- *hiring, upgrading, promoting, awarding tenure, demotion, transfer, layoff, termination, the right to return from layoff, and rehiring;*
- *rate of pay or other forms of compensation and changes in compensation;*
- *job assignments, job classifications, organization structures, position descriptions, lines of progression, and seniority lists;*
- *leaves of absence, sick leave, or other leaves;*

---

* H. Rep. 101-485, Part 2, 51.
[†] *Id.*

- *fringe benefits available by virtue of employment, whether or not administered by the employer;*
- *selection and financial support for training, including apprenticeships, professional meetings, conferences and other related activities, and selection for leave of absence to pursue training;*
- *activities sponsored by the employer, including social and recreational programs;*
- *any other term, condition, or privilege of employment.**

Safety and health professionals should be aware that the enforcement procedures adopted by the ADA mirror those of Title VII of the Civil Rights Act. A claimant under the ADA must file a claim with the EEOC within 180 days from the alleged discriminatory event, or within 300 days in states with approved enforcement agencies such as the Human Rights Commission. These are commonly called dual agency states or Section 706 agencies. The EEOC has 180 days to investigate the allegation and sue the employer, or to issue a right-to-sue notice to the employee. The employee will have 90 days to file a civil action from the date of this notice.†

The governing federal agency for the ADA is the Equal Employment Opportunity Commission. Enforcement of the ADA by the Attorney General or by private lawsuit is also permitted. Remedies, as identified below, can included the ordered modification of a facility, and civil penalties. Section 505 permits reasonable attorney fees and litigation costs for the prevailing party in an ADA action but, under section 513, Congress encourages the use of arbitration to resolve disputes arising under the ADA.‡

Compensatory and punitive damages may be awarded in cases involving intentional discrimination based on a person's race, color, national origin, sex (including pregnancy), religion, disability, or genetic information.

Compensatory damages pay victims for out-of-pocket expenses caused by the discrimination (such as costs associated with a job search or medical expenses) and compensate them for any emotional harm suffered (such as mental anguish, inconvenience, or loss of enjoyment of life).

Punitive damages may be awarded to punish an employer who has committed an especially malicious or reckless act of discrimination.

There are limits on the amount of compensatory and punitive damages a person can recover. These limits vary depending on the size of the employer:
For employers with 15–100 employees, the limit is $50,000.
For employers with 101–200 employees, the limit is $100,000.
For employers with 201–500 employees, the limit is $200,000.
For employers with more than 500 employees, the limit is $300,000.

---

* *EEOC Interpretive Guidelines, supra.* note 11.
† S. Rep. 101–116, 21; H.Rep. 101-485; Part 2, 51; Part 3, 28. Also see EEOC website at www.eeoc.gov.
‡ ADA Section 505 and 513.

As noted above, the remedies provided under the ADA were modified from their original version with the passage of the Civil Rights Act of 1991. Employment discrimination (whether intentional or by mistake) that has a discriminatory effect on qualified individuals may include hiring, reinstatement, promotion, back pay, front pay, reasonable accommodation, or other actions that will make an individual "whole." Payment of attorney fees, expert witness fees, and court fees are still permitted, and jury trials are also allowed. Compensatory and punitive damages were also made available if intentional discrimination is found. Damages may be available to compensate for actual monetary losses, future monetary losses, mental anguish, and inconvenience. Punitive damages are also available if an employer acted with malice or reckless indifference. The total amount of punitive and compensatory damages for future monetary loss and emotional injury for each individual is limited, and is based upon the size of the employer.

As can be seen, the potential entanglements and interactions between the safety and health function and the ADA can be extensive. Safety and health professionals should acquire a working understanding of the ADA and identify the areas in which the ADA intersects with the safety and health function. Safety and health professionals should carefully listen to employees and others and direct the employee or individual to human resources or the appropriate ADA representative for the company or organizations. Safety and health professionals are usually not ADA experts; however, it is important to be able to identify when the ADA and safety and health functions intersect to avoid potential risk and liability.

Case Study: this case may be modified for the purposes of this text.

122 S.Ct. 681
Supreme Court of the United States

TOYOTA MOTOR MANUFACTURING, KENTUCKY, INC., Petitioner,

v.

Ella WILLIAMS, Respondent.

No. 00–1089.

|

Argued Nov. 7, 2001.

|

Decided Jan. 8, 2002.

SYNOPSIS

Former employee sued former employer under Americans with Disabilities Act (ADA). The United States District Court for the Eastern District of Kentucky, Henry R. Wilhoit, Jr., Chief Judge, entered summary judgment for former employer. Former employee appealed. The United States Court of Appeals for the Sixth Circuit, 224 F.3d 840, reversed. Certiorari was granted. The Supreme Court, Justice O'Connor, held that: (1) to be substantially limited in performing manual tasks, an individual must have an impairment that prevents or severely restricts the individual from doing activities that are of central importance to most people's daily lives; (2) it is insufficient for individuals attempting to prove disability status to merely submit evidence of a medical diagnosis of an impairment; (3) it is not required that a class of manual activities be implicated for an impairment to substantially limit the major life activity of performing manual tasks; (4) Court of Appeals applied wrong standard in determining whether employee's carpal tunnel syndrome caused her to be substantially limited in major life activity of performing manual tasks, when it focused on her inability to perform manual tasks associated only with her job; (5) employee's inability to do repetitive work with hands and arms extended at or above shoulder levels was not sufficient proof that she was substantially limited in major life activity of performing manual tasks; and (6) medical conditions that caused employee to restrict certain activities did not constitute manual-task disability under ADA.

Reversed and remanded.

**684 Syllabus*

Claiming to be unable to perform her automobile assembly line job because she was disabled by carpal tunnel syndrome and related impairments, respondent sued petitioner, her former employer, for failing to provide her with a reasonable accommodation as required by the Americans with Disabilities Act of 1990(ADA), 42 U.S.C. § 12112(b)(5)(A). The District Court granted petitioner summary judgment, holding that respondent's impairment did not qualify as a "disability" under the ADA because it had not "substantially limit[ed]" any "major life activit[y]," § 12102(2)(A), and that there was no evidence that respondent had had a record of a substantially limiting impairment or that petitioner had regarded her as having such an impairment. The Sixth Circuit reversed, finding that the impairments substantially limited respondent in the major life activity of performing manual tasks. In order to demonstrate that she was so limited, said the court, respondent had to show that her manual disability involved a "class" of manual activities affecting the ability to perform tasks at work. Respondent satisfied this test, according to the court, because her ailments prevented her from doing the tasks associated with certain types of manual jobs that require the gripping of tools and repetitive work with hands and arms extended at or above shoulder levels for extended periods of time. In reaching this conclusion, the court found that evidence that respondent could tend to her personal hygiene and carry out personal or household chores did not affect a determination that her impairments substantially limited her ability to perform the range of manual tasks associated with an assembly line job. The court granted respondent partial summary judgment on the issue of whether she was disabled under the ADA.

---

* The syllabus constitutes no part of the opinion of the Court but has been prepared by the Reporter of Decisions for the convenience of the reader. See *United States v. Detroit Timber & Lumber Co.*, 200 U.S. 321, 337, 26 S.Ct. 282, 50 L.Ed. 499.

Held: The Sixth Circuit did not apply the proper standard in determining that respondent was disabled under the ADA because it analyzed only a limited class of manual tasks and failed to ask whether respondent's impairments prevented or restricted her from performing tasks that are of central importance to most people's daily lives. Pp. 689–694.

(a) The Court's consideration of what an individual must prove to demonstrate a substantial limitation in the major life activity of performing manual tasks is guided by the ADA's disability definition. ***185** "Substantially" in the phrase "substantially limits" suggests "considerable" or "to a large degree," and thus clearly precludes impairments that interfere in only a minor way with performing manual tasks. Cf. *Albertson's, Inc. v. Kirkingburg*, 527 U.S. 555, 565, 119 S.Ct. 2162, 144 L.Ed.2d 518. Moreover, because "major" means important, "major life activities" refers to those activities that are of central importance to daily life. In order for performing manual tasks to fit into this category, the tasks in question must be central to daily life. To be substantially limited in the specific major life activity of performing manual ****685** tasks, therefore, an individual must have an impairment that prevents or severely restricts the individual from doing activities that are of central importance to most people's daily lives. The impairment's impact must also be permanent or long term. See 29 CFR §§ 1630.2(j)(2)(ii)–(iii).

It is insufficient for individuals attempting to prove disability status under this test to merely submit evidence of a medical diagnosis of an impairment. Instead, the ADA requires them to offer evidence that the extent of the limitation caused by their impairment in terms of their own experience is substantial. 527 U.S., at 567, 119 S.Ct. 2162. That the ADA defines "disability" "with respect to an individual," § 12102(2), makes clear that Congress intended the existence of a disability to be determined in such a case-by-case manner. See, *e.g., Sutton v. United Air Lines, Inc.*, 527 U.S. 471, 483, 119 S.Ct. 2139, 144 L.Ed.2d 450. An individualized assessment of the effect of an impairment is particularly necessary when the impairment is one such as carpal tunnel syndrome, in which symptoms vary widely from person to person. Pp. 690–692.

(b) The Sixth Circuit erred in suggesting that, in order to prove a substantial limitation in the major life activity of performing manual tasks, a plaintiff must show that her manual disability involves a "class" of manual activities, and that those activities affect the ability to perform tasks at work. Nothing in the ADA's text, this Court's opinions, or the regulations suggests that a class-based framework should apply outside the context of the major life activity of working. While the Sixth Circuit addressed the different major life activity of performing manual tasks, its analysis erroneously circumvented *Sutton, supra,* at 491, 119 S.Ct. 2139, by focusing on respondent's inability to perform manual tasks associated only with her job. Rather, the central inquiry must be whether the claimant is unable to perform the variety of tasks central to most people's daily lives. Also without support is the Sixth Circuit's assertion that the question of whether an impairment constitutes a disability is to be answered only by analyzing the impairment's effect in the workplace. That the

ADA's "disability" definition applies not only to the portion of the ADA dealing with employment, but also to the other provisions dealing with public transportation and public accommodations, ***186** demonstrates that the definition is intended to cover individuals with disabling impairments regardless of whether they have any connection to a workplace. Moreover, because the manual tasks unique to any particular job are not necessarily important parts of most people's lives, occupation-specific tasks may have only limited relevance to the manual task inquiry. In this case, repetitive work with hands and arms extended at or above shoulder levels for extended periods, the manual task on which the Sixth Circuit relied, is not an important part of most people's daily lives. Household chores, bathing, and brushing one's teeth, in contrast, are among the types of manual tasks of central importance to people's daily lives, so the Sixth Circuit should not have disregarded respondent's ability to do these activities. Pp. 692–694.

224 F.3d 840, reversed and remanded.

O'CONNOR, J., delivered the opinion for a unanimous Court.

OPINION

**686 *187 Justice O'CONNOR delivered the opinion of the Court.

Under the Americans with Disabilities Act of 1990 (ADA or Act), 104 Stat. 328, 42 U.S.C. § 12101 *et seq.* (1994 ed. and Supp. V), a physical impairment that "substantially limits one or more ... major life activities" is a "disability." 42 U.S.C. § 12102(2)(A) (1994 ed.). Respondent, claiming to be disabled because of her carpal tunnel syndrome and other related impairments, sued petitioner, her former employer, for failing to provide her with a reasonable accommodation as required by the ADA. See § 12112(b)(5)(A). The District Court granted summary judgment to petitioner, finding that respondent's impairments did not substantially limit any of her major life activities. The Court of Appeals for the Sixth Circuit reversed, finding that the impairments substantially limited respondent in the major life activity of performing manual tasks, and therefore granting partial summary judgment to respondent on the issue of whether she was disabled under the ADA. We conclude that the Court of Appeals did not apply the proper standard in making this determination because it analyzed only a limited class of manual tasks and failed to ask whether respondent's impairments prevented or restricted her from performing tasks that are of central importance to most people's daily lives.

I

Respondent began working at petitioner's automobile manufacturing plant in Georgetown, Kentucky, in August 1990. She was soon placed on an engine fabrication assembly line, where her duties included work with pneumatic tools. Use of these tools eventually caused pain in respondent's hands, wrists, and arms. She sought treatment at petitioner's in-house medical service, where she was diagnosed with bilateral carpal tunnel syndrome and bilateral tendinitis. Respondent consulted a personal physician who placed her on permanent work restrictions that precluded her from lifting more than 20 pounds or from "frequently lifting or *188 carrying ... objects weighing up to 10 pounds," engaging in "constant repetitive ... flexion or extension of [her] wrists or elbows," performing "overhead work," or using "vibratory or pneumatic tools." Brief for Respondent 2; App. 45–46.

In light of these restrictions, for the next two years petitioner assigned respondent to various modified duty jobs. Nonetheless, respondent missed some work for medical leave, and eventually filed a claim under the Kentucky Workers' Compensation Act. Ky.Rev.Stat. Ann. § 342.0011 *et seq.* (1997 and Supp.2000). The parties settled this claim, and respondent returned to work. She was unsatisfied by petitioner's efforts to accommodate her work restrictions, however, and responded by bringing an action in the United States District Court for the Eastern District of Kentucky alleging that petitioner had violated the ADA by refusing to accommodate her disability. That suit was also settled, and as part of the settlement, respondent returned to work in December 1993.

Upon her return, petitioner placed respondent on a team in Quality Control Inspection Operations (QCIO). QCIO is responsible for four tasks: (1) "assembly paint"; (2) "paint second inspection"; (3) "shell body audit"; and (4) "ED surface repair." App. 19. Respondent was initially placed on a team that performed only the first two of these tasks, and for a couple of years, she rotated on a weekly basis between them. In assembly paint, respondent visually inspected painted cars moving slowly down a conveyor. She scanned for scratches, dents, chips, or any other flaws that may have occurred during the assembly or painting process, at a rate of one car every 54 seconds. When respondent began working in assembly paint, inspection team members were required to open and shut the doors, trunk, and/or hood of each passing car. Sometime during respondent's tenure, however, the position **687 was modified to include only visual inspection with few or no manual tasks. Paint second inspection required team members to use their hands to wipe each *189 painted car with a glove as it moved along a conveyor. *Id.*, at 21–22. The parties agree that respondent was physically capable of performing both of these jobs and that her performance was satisfactory.

During the fall of 1996, petitioner announced that it wanted QCIO employees to be able to rotate through all four of the QCIO processes. Respondent therefore received training for the shell body audit job, in which team members apply a highlight oil to the hood, fender, doors, rear quarter panel, and trunk of passing cars at a rate of approximately one car per minute. The highlight oil has the viscosity of salad oil, and employees spread it on cars with a sponge attached to a block of wood. After they wipe each car with the oil, the employees visually inspect it for flaws. Wiping the cars required respondent to hold her hands and arms up around shoulder height for several hours at a time.

A short while after the shell body audit job was added to respondent's rotations, she began to experience pain in her neck and shoulders. Respondent again sought care at petitioner's in-house medical service, where she was diagnosed with myotendinitis bilateral periscapular, an inflammation of the muscles and tendons around both of her shoulder blades; myotendinitis and myositis bilateral forearms with nerve compression causing median nerve irritation; and thoracic outlet compression, a condition that causes pain in the nerves that lead to the upper extremities. Respondent requested that petitioner accommodate her medical conditions by allowing her to return to doing only her original two jobs in QCIO, which respondent claimed she could still perform without difficulty.

The parties disagree about what happened next. According to respondent, petitioner refused her request and forced her to continue working in the shell body audit job, which caused her even greater physical injury. According to petitioner, respondent simply began missing work on a regular basis. Regardless, it is clear that on December 6, 1996, the ***190** last day respondent worked at petitioner's plant, she was placed under a no-work-of-any-kind restriction by her treating physicians. On January 27, 1997, respondent received a letter from petitioner that terminated her employment, citing her poor attendance record.

Respondent filed a charge of disability discrimination with the Equal Employment Opportunity Commission (EEOC). After receiving a right to sue letter, respondent filed suit against petitioner in the United States District Court for the Eastern District of Kentucky. Her complaint alleged that petitioner had violated the ADA and the Kentucky Civil Rights Act, Ky.Rev.Stat. Ann. § 344.010 et seq. (1997 and Supp.2000), by failing to reasonably accommodate her disability and by terminating her employment. Respondent later amended her complaint to also allege a violation of the Family and Medical Leave Act of 1993 (FMLA), 107 Stat. 6, as amended, 29 U.S.C. § 2601 et seq. (1994 ed. and Supp. V).

Respondent based her claim that she was "disabled" under the ADA on the ground that her physical impairments substantially limited her in (1) manual tasks; (2) housework; (3) gardening; (4) playing with her children; (5) lifting; and (6) working, all of which, she argued, constituted major life activities under the Act. Respondent also argued, in the alternative, that she was disabled under the ADA because she had a record of a substantially limiting impairment and because she was regarded as having such an impairment. See 42 U.S.C. §§ 12102(2)(B)–(C) (1994 ed.).

After petitioner filed a motion for summary judgment and respondent filed a motion for partial summary judgment on her disability claims, the District Court granted summary judgment to petitioner. ****688** Civ. A. No. 97–135 (Jan. 26, 1999), App. to Pet. for Cert. A–23. The court found that respondent had not been disabled, as defined by the ADA, at the time of petitioner's alleged refusal to accommodate her, and that she had therefore not been covered by the Act's protections or by the Kentucky Civil Rights Act, which is construed ***191** consistently with the ADA. Id., at A–29, A–34 to A–47. The District Court held that respondent had suffered from a physical impairment, but that the impairment did not qualify as a disability because it had not "substantially limit[ed]" any "major life activit[y]," 42 U.S.C. § 12102(2)(A). App. to Pet. for Cert. A–34 to A–42. The court rejected respondent's arguments that gardening, doing housework, and playing with children are major life activities. *Id.*, at A–35 to A–36. Although the court agreed that performing manual tasks, lifting, and working are major life activities, it found the evidence insufficient to demonstrate that respondent had been substantially limited in lifting or working. *Id.*, at A–36 to A–42. The court found respondent's claim that she was substantially limited in performing manual tasks to be "irretrievably contradicted by [respondent's] continual insistence that she could perform the tasks in assembly [paint] and paint [second] inspection without difficulty." *Id.*, at A–36. The court also found no evidence that respondent had had a record of a substantially limiting impairment, *id.*, at A–43, or that petitioner had regarded her as having such an impairment, *id.*, at A–46 to A–47.

The District Court also rejected respondent's claim that her termination violated the ADA and the Kentucky Civil Rights Act. The court found that even if it assumed that respondent was disabled at the time of her termination, she was not a "qualified individual with a disability," 42 U.S.C. § 12111(8) (1994 ed.), because, at that time, her physicians had restricted her from performing work of any kind, App. to Pet. for Cert. A–47 to A–50. Finally, the court found that respondent's FMLA claim failed, because she had not presented evidence that she had suffered any damages available under the FMLA. *Id.*, at A–50 to A–54.

Respondent appealed all but the gardening, housework, and playing-with-children rulings. The Court of Appeals for the Sixth Circuit reversed the District Court's ruling on whether respondent was disabled at the time she sought an ***192** accommodation, but affirmed the District Court's rulings on respondent's FMLA and wrongful termination claims. 224 F.3d 840 (2000). The Court of Appeals held that in order for respondent to demonstrate that she was disabled due to a substantial limitation in the ability to perform manual tasks at the time of her accommodation request, she had to "show that her manual disability involve[d] a 'class' of manual activities affecting the ability to perform tasks at work." *Id.*, at 843. Respondent satisfied this test, according to the Court of Appeals, because her ailments "prevent[ed] her from doing the tasks associated with certain types of manual assembly line jobs, manual product handling jobs and manual building trade jobs (painting, plumbing, roofing, etc.) that require the gripping of tools and repetitive work with hands and arms extended at or above shoulder levels for extended periods of time." *Ibid.* In reaching this conclusion, the court disregarded evidence that respondent could "ten[d] to her personal hygiene [and] carr[y] out personal or household chores," finding that such evidence "does not affect a determination that her impairment substantially limit[ed] her ability to perform the range of manual tasks associated with an assembly line job," *ibid.* Because the Court of Appeals concluded that respondent had been substantially limited in performing manual tasks and, for that reason, was entitled to partial summary judgment on the issue of whether she was disabled under the Act, it found that it did not need to determine whether respondent had been substantially limited in the major life activities of lifting or ****689** working, *ibid.*, or whether she had had a "record of" a disability or had been "regarded as" disabled, *id.*, at 844.

We granted certiorari, 532 U.S. 970, 121 S.Ct. 1600, 149 L.Ed.2d 466 (2001), to consider the proper standard for assessing whether an individual is substantially limited in performing manual tasks. We now reverse the Court of Appeals' decision to grant partial summary judgment to respondent on the issue of whether she was substantially limited in performing manual tasks at the *193 time she sought an accommodation. We express no opinion on the working, lifting, or other arguments for disability status that were preserved below, but which were not ruled upon by the Court of Appeals.

II

The ADA requires covered entities, including private employers, to provide "reasonable accommodations to the known physical or mental limitations of an otherwise qualified individual with a disability who is an applicant or employee, unless such covered entity can demonstrate that the accommodation would impose an undue hardship." 42 U.S.C. § 12112(b)(5)(A) (1994 ed.); see also § 12111(2) ("The term 'covered entity' means an employer, employment agency, labor organization, or joint labor-management committee"). The Act defines a "qualified individual with a disability" as "an individual with a disability who, with or without reasonable accommodation, can perform the essential functions of the employment position that such individual holds or desires." § 12111(8). In turn, a "disability" is:

(A) "a physical or mental impairment that substantially limits one or more of the major life activities of such individual;

(B) "a record of such an impairment; or

(C) "being regarded as having such an impairment." § 12102(2).

[1] There are two potential sources of guidance for interpreting the terms of this definition—the regulations interpreting the Rehabilitation Act of 1973, 87 Stat. 361, as amended, 29 U.S.C. § 706(8)(B) (1988 ed.), and the EEOC regulations interpreting the ADA. Congress drew the ADA's definition of disability almost verbatim from the definition of "handicapped individual" in the Rehabilitation Act, § 706(8)(B), and Congress' repetition of a well-established term generally implies that Congress intended the term to be construed *194 in accordance with pre-existing regulatory interpretations. *Bragdon v. Abbott*, 524 U.S. 624, 631, 118 S.Ct. 2196, 141 L.Ed.2d 540 (1998); *FDIC v. Philadelphia Gear Corp.*, 476 U.S. 426, 437–438, 106 S.Ct. 1931, 90 L.Ed.2d 428 (1986); *ICC v. Parker*, 326 U.S. 60, 65, 65 S.Ct. 1490, 89 L.Ed. 2051 (1945). As we explained in *Bragdon v. Abbott, supra*, at 631, 118 S.Ct. 2196, Congress did more in the ADA than suggest this construction; it adopted a specific statutory provision directing as follows:

"Except as otherwise provided in this chapter, nothing in this chapter shall be construed to apply a lesser standard than the standards applied under title V of the Rehabilitation Act of 1973 (29 U.S.C. § 790 et seq.) or the regulations issued by Federal agencies pursuant to such title." 42 U.S.C. § 12201(a) (1994 ed.).

[2] The persuasive authority of the EEOC regulations is less clear. As we have previously noted, see *Sutton v. United Air Lines, Inc.*, 527 U.S. 471, 479, 119 S.Ct. 2139, 144 L.Ed.2d 450 (1999), no agency has been given authority to issue regulations interpreting the term "disability" in the ADA. Nonetheless, the EEOC has done so. See 29 CFR §§ 1630.2(g)–(j) (2001). Because both parties accept the EEOC regulations as reasonable, we assume without deciding that they are, and we have no occasion to decide what level of deference, if any, they are due. See ****690** *Sutton v. United Air Lines, Inc., supra*, at 480, 119 S.Ct. 2139; *Albertson's, Inc. v. Kirkingburg*, 527 U.S. 555, 563, n. 10, 119 S.Ct. 2162, 144 L.Ed.2d 518 (1999).

[3] To qualify as disabled under subsection (A) of the ADA's definition of disability, a claimant must initially prove that he or she has a physical or mental impairment. See 42 U.S.C. § 12102(2)(A). The Rehabilitation Act regulations issued by the Department of Health, Education, and Welfare (HEW) in 1977, which appear without change in the current regulations issued by the Department of Health and Human Services, define "physical impairment," the type of impairment relevant to this case, to mean "any physiological disorder or condition, cosmetic disfigurement, or anatomical loss affecting one or more of the following body systems: neurological; ***195** musculoskeletal; special sense organs; respiratory, including speech organs; cardiovascular; reproductive, digestive, genito-urinary; hemic and lymphatic; skin; and endocrine." 45 CFR § 84.3(j)(2)(i) (2001). The HEW regulations are of particular significance because at the time they were issued, HEW was the agency responsible for coordinating the implementation and enforcement of § 504 of the Rehabilitation Act, 29 U.S.C. § 794 (1994 ed. and Supp. V), which prohibits discrimination against individuals with disabilities by recipients of federal financial assistance. *Bragdon v. Abbott, supra*, at 632, 118 S.Ct. 2196 (citing *Consolidated Rail Corporation v. Darrone*, 465 U.S. 624, 634, 104 S.Ct. 1248, 79 L.Ed.2d 568 (1984)).

[4] Merely having an impairment does not make one disabled for purposes of the ADA. Claimants also need to demonstrate that the impairment limits a major life activity. See 42 U.S.C. § 12102(2)(A) (1994 ed.). The HEW Rehabilitation Act regulations provide a list of examples of "major life activities" that includes "walking, seeing, hearing," and, as relevant here, "performing manual tasks." 45 CFR § 84.3(j)(2)(ii) (2001).

To qualify as disabled, a claimant must further show that the limitation on the major life activity is "substantia[l]." 42 U.S.C. § 12102(2)(A). Unlike "physical impairment" and "major life activities," the HEW regulations do not define the term "substantially limits." See Nondiscrimination on the Basis of Handicap in Programs and Activities Receiving or Benefiting from Federal Financial Assistance, 42 Fed.Reg. 22676, 22685 (1977) (stating HEW's position that a definition of "substantially limits" was not possible at that time). The EEOC, therefore, has created its own definition for purposes of the ADA. According to the EEOC regulations, "substantially limit[ed]" means "[u]nable to perform a major life activity that the average person in the general population can perform"; or "[s]ignificantly restricted as to the condition, manner or duration under which an individual can perform a particular major life activity as compared to ***196** the condition, manner, or duration under which the average person in the general population can perform that same major life activity." 29 CFR § 1630.2(j) (2001). In determining whether an individual is substantially limited in a major life activity, the regulations instruct that the following factors should be considered: "[t]he nature and severity of the impairment; [t]he duration or expected duration of the impairment; and [t]he permanent or long-term impact, or the expected permanent or long-term impact of or resulting from the impairment." §§ 1630.2(j)(2)(i)–(iii).

III

The question presented by this case is whether the Sixth Circuit properly determined that respondent was disabled under subsection (A) of the ADA's disability definition at the time that she sought an accommodation from petitioner. 42 U.S.C. § 12102(2) (A). The parties do not dispute that respondent's medical conditions, which include carpal tunnel syndrome, myotendinitis, and thoracic outlet compression, **691 amount to physical impairments. The relevant question, therefore, is whether the Sixth Circuit correctly analyzed whether these impairments substantially limited respondent in the major life activity of performing manual tasks. Answering this requires us to address an issue about which the EEOC regulations are silent: what a plaintiff must demonstrate to establish a substantial limitation in the specific major life activity of performing manual tasks.

[5] [6] Our consideration of this issue is guided first and foremost by the words of the disability definition itself. "[S]ubstantially" in the phrase "substantially limits" suggests "considerable" or "to a large degree." See *Webster's Third New International Dictionary* 2280 (1976) (defining "substantially" as "in a substantial manner" and "substantial" as "considerable in amount, value, or worth" and "being that specified to a large degree or in the main"); see also 17 *Oxford English Dictionary* 66–67 (2d ed.1989) ("substantial": "[r]elating to *197 or proceeding from the essence of a thing; essential"; "[o]f ample or considerable amount, quantity, or dimensions"). The word "substantial" thus clearly precludes impairments that interfere in only a minor way with the performance of manual tasks from qualifying as disabilities. Cf. *Albertson's, Inc. v. Kirkingburg*, 527 U.S., at 565, 119 S.Ct. 2162 (explaining that a "mere difference" does not amount to a "significant restric[tion]" and therefore does not satisfy the EEOC's interpretation of "substantially limits").

[7] [8] [9] "Major" in the phrase "major life activities" means important. See Webster's, *supra*, at 1363 (defining "major" as "greater in dignity, rank, importance, or interest"). "Major life activities" thus refers to those activities that are of central importance to daily life. In order for performing manual tasks to fit into this category—a category that includes such basic abilities as walking, seeing, and hearing—the manual tasks in question must be central to daily life. If each of the tasks included in the major life activity of performing manual tasks does not independently qualify as a major life activity, then together they must do so.

[10] That these terms need to be interpreted strictly to create a demanding standard for qualifying as disabled is confirmed by the first section of the ADA, which lays out

the legislative findings and purposes that motivate the Act. See 42 U.S.C. § 12101. When it enacted the ADA in 1990, Congress found that "some 43,000,000 Americans have one or more physical or mental disabilities." § 12101(a)(1). If Congress intended everyone with a physical impairment that precluded the performance of some isolated, unimportant, or particularly difficult manual task to qualify as disabled, the number of disabled Americans would surely have been much higher. Cf. *Sutton v. United Air Lines, Inc.*, 527 U.S., at 487, 119 S.Ct. 2139 (finding that because more than 100 million people need corrective lenses to see properly, "[h]ad Congress intended to include all persons with corrected physical limitations among those covered by the Act, it undoubtedly would have cited a much ***198** higher number [than 43 million] disabled persons in the findings").

[11] We therefore hold that to be substantially limited in performing manual tasks, an individual must have an impairment that prevents or severely restricts the individual from doing activities that are of central importance to most people's daily lives. The impairment's impact must also be permanent or long term. See 29 CFR §§ 1630.2(j) (2)(ii)–(iii) (2001).

[12] [13] It is insufficient for individuals attempting to prove disability status under this test to merely submit evidence of a medical diagnosis of an impairment. Instead, the ADA requires those "claiming the Act's protection ... to prove a disability by offering evidence that the extent of the limitation [caused by their impairment] in terms of their own experience ... is ****692** substantial." *Albertson's, Inc. v. Kirkingburg, supra*, at 567, 119 S.Ct. 2162 (holding that monocular vision is not invariably a disability, but must be analyzed on an individual basis, taking into account the individual's ability to compensate for the impairment). That the Act defines "disability" "with respect to an individual," 42 U.S.C. § 12102(2), makes clear that Congress intended the existence of a disability to be determined in such a case-by-case manner. See *Sutton v. United Air Lines, Inc., supra*, at 483, 119 S.Ct. 2139; *Albertson's, Inc. v. Kirkingburg, supra*, at 566, 119 S.Ct. 2162; cf. *Bragdon v. Abbott*, 524 U.S., at 641–642, 118 S.Ct. 2196 (relying on unchallenged testimony that the respondent's HIV infection controlled her decision not to have a child, and declining to consider whether HIV infection is a *per se* disability under the ADA); 29 CFR pt. 1630, App. § 1630.2(j) (2001) ("The determination of whether an individual has a disability is not necessarily based on the name or diagnosis of the impairment the person has, but rather on the effect of that impairment on the life of the individual"); *ibid.* ("The determination of whether an individual is substantially limited in a major life activity must be made on a case-by-case basis").

[14] [15] ***199** An individualized assessment of the effect of an impairment is particularly necessary when the impairment is one whose symptoms vary widely from person to person. Carpal tunnel syndrome, one of respondent's impairments, is just such a condition. While cases of severe carpal tunnel syndrome are characterized by muscle atrophy and extreme sensory deficits, mild cases generally do not have either of these effects and create only intermittent symptoms of numbness and tingling. Carniero, Carpal Tunnel Syndrome: The Cause Dictates the Treatment, 66 *Cleveland*

Clinic J. Medicine 159, 161–162 (1999). Studies have further shown that, even without surgical treatment, one quarter of carpal tunnel cases resolve in one month, but that in 22 percent of cases, symptoms last for eight years or longer. See DeStefano, Nordstrom, & Uierkant, Long-term Symptom Outcomes of Carpal Tunnel Syndrome and its Treatment, 22A *J. Hand Surgery* 200, 204–205 (1997). When pregnancy is the cause of carpal tunnel syndrome, in contrast, the symptoms normally resolve within two weeks of delivery. See Ouellette, Nerve Compression Syndromes of the Upper Extremity in Women, 17 *J. of Musculoskeletal Medicine* 536 (2000). Given these large potential differences in the severity and duration of the effects of carpal tunnel syndrome, an individual's carpal tunnel syndrome diagnosis, on its own, does not indicate whether the individual has a disability within the meaning of the ADA.

IV

The Court of Appeals' analysis of respondent's claimed disability suggested that in order to prove a substantial limitation in the major life activity of performing manual tasks, a "plaintiff must show that her manual disability involves a 'class' of manual activities," and that those activities "affec[t] the ability to perform tasks at work." See 224 F.3d, at 843. Both of these ideas lack support.

[16] The Court of Appeals relied on our opinion in *Sutton v. United Air Lines, Inc.*, for the idea that a "class" of manual *200 activities must be implicated for an impairment to substantially limit the major life activity of performing manual tasks. 224 F.3d, at 843. But *Sutton* said only that "*[w]hen the major life activity under consideration is that of working*, the statutory phrase 'substantially limits' requires ... that plaintiffs allege they are unable to work in a broad class of jobs." 527 U.S., at 491, 119 S.Ct. 2139 (emphasis added). Because of the conceptual difficulties inherent in the argument that working could be a major life activity, we have been hesitant to hold as much, and we need not decide this difficult question today. In **693 *Sutton*, we noted that even assuming that working is a major life activity, a claimant would be required to show an inability to work in a "broad range of jobs," rather than a specific job. *Id.*, at 492, 119 S.Ct. 2139. But *Sutton* did not suggest that a class-based analysis should be applied to any major life activity other than working. Nor do the EEOC regulations. In defining "substantially limits," the EEOC regulations only mention the "class" concept in the context of the major life activity of working. 29 CFR § 1630.2(j)(3) (2001) ("With respect to the major life activity of *working*[,] [t]he term *substantially limits* means significantly restricted in the ability to perform either a class of jobs or a broad range of jobs in various classes as compared to the average person having comparable training, skills and abilities"). Nothing in the text of the Act, our previous opinions, or the regulations suggests that a class-based framework should apply outside the context of the major life activity of working.

[17] [18] While the Court of Appeals in this case addressed the different major life activity of performing manual tasks, its analysis circumvented *Sutton* by focusing on respondent's inability to perform manual tasks associated only with her job. This was error. When addressing the major life activity of performing manual tasks, the

central inquiry must be whether the claimant is unable to perform the variety of tasks central to most people's daily lives, not whether the *201 claimant is unable to perform the tasks associated with her specific job. Otherwise, *Sutton's* restriction on claims of disability based on a substantial limitation in working will be rendered meaningless, because an inability to perform a specific job always can be recast as an inability to perform a "class" of tasks associated with that specific job.

[19] There is also no support in the Act, our previous opinions, or the regulations for the Court of Appeals' idea that the question of whether an impairment constitutes a disability is to be answered only by analyzing the effect of the impairment in the workplace. Indeed, the fact that the Act's definition of "disability" applies not only to Title I of the Act, 42 U.S.C. §§ 12111–12117 (1994 ed.), which deals with employment, but also to the other portions of the Act, which deal with subjects such as public transportation, §§ 12141–12150, 42 U.S.C. §§ 12161–12165 (1994 ed. and Supp. V), and privately provided public accommodations, §§ 12181–12189, demonstrates that the definition is intended to cover individuals with disabling impairments regardless of whether the individuals have any connection to a workplace.

[20] [21] Even more critically, the manual tasks unique to any particular job are not necessarily important parts of most people's lives. As a result, occupation-specific tasks may have only limited relevance to the manual task inquiry. In this case, "repetitive work with hands and arms extended at or above shoulder levels for extended periods of time," 224 F.3d, at 843, the manual task on which the Court of Appeals relied, is not an important part of most people's daily lives. The court, therefore, should not have considered respondent's inability to do such manual work in her specialized assembly line job as sufficient proof that she was substantially limited in performing manual tasks.

[22] At the same time, the Court of Appeals appears to have disregarded the very type of evidence that it should have focused upon. It treated as irrelevant "[t]he fact that [respondent] can ... ten[d] to her personal hygiene [and] carr[y] *202 out personal or household chores." *Ibid.* Yet household chores, bathing, and brushing one's teeth are among the types of manual tasks of central importance to people's daily lives, and should have been part of the assessment of whether respondent was substantially limited in performing manual tasks.

**694 [23] The District Court noted that at the time respondent sought an accommodation from petitioner, she admitted that she was able to do the manual tasks required by her original two jobs in QCIO.App. to Pet. for Cert. A–36. In addition, according to respondent's deposition testimony, even after her condition worsened, she could still brush her teeth, wash her face, bathe, tend her flower garden, fix breakfast, do laundry, and pick up around the house. App. 32–34. The record also indicates that her medical conditions caused her to avoid sweeping, to quit dancing, to occasionally seek help dressing, and to reduce how often she plays with her children, gardens, and drives long distances. *Id.*, at 32, 38–39. But these changes in her life did not amount to such severe restrictions in the activities that are of central importance to most people's daily lives that they establish a manual task disability as a matter of law. On this record, it was therefore inappropriate for the Court of Appeals to grant partial summary judgment to respondent on the issue of whether she was substantially limited in performing manual tasks, and its decision to do so must be reversed.

[24] In its brief on the merits, petitioner asks us to reinstate the District Court's grant of summary judgment to petitioner on the manual task issue. In its petition for certiorari, however, petitioner did not seek summary judgment; it argued only that the Court of Appeals' reasons for granting partial summary judgment to respondent were unsound. This Court's Rule 14.1(a) provides: "Only the questions set out in the petition, or fairly included therein, will be considered by the Court." The question of whether petitioner was entitled to summary judgment on the manual task issue is *203 therefore not properly before us. See *Irvine v. California*, 347 U.S. 128, 129–130, 74 S.Ct. 381, 98 L.Ed. 561 (1954).

Accordingly, we reverse the Court of Appeals' judgment granting partial summary judgment to respondent and remand the case for further proceedings consistent with this opinion.

So ordered.

ALL CITATIONS

534 U.S. 184, 122 S.Ct. 681, 151 L.Ed.2d 615, 200 A.L.R. Fed. 667, 70 USLW 4050, 67 Cal. Comp. Cases 60, 12 A.D. Cases 993, 22 NDLR P 97, 02 Cal. Daily Op. Serv. 149, 2002 Daily Journal D.A.R. 197, 15 Fla. L. Weekly Fed. S 39

NOTES

1. 42 Fed. Reg. 22686 (May 4, 1977); S. Rep. 101-116; H.Rep. 101-485, Part 2, 51.
2. 28 C.F.R. § 41.31.
3. See, *Jasany v. U.S. Postal Service*, 755 F.2d 1244 (6th Cir. 1985).
4. S. Rep. 101-116,23; H.Rep. 101-485, Part 2, 52–53.
5. ADA, Title I, Section 101(8).
6. EEOC Interpretive Guidelines, EEOC, 1994.
7. *Id.*
8. 56 Fed. Reg. 35,745 (July 26, 1991).
9. EEOC Interpretive Guidelines, 56 Fed. Reg. 35,751 (July 26, 1991).

12 Federal Wage and Hour Laws

******

Chapter Objectives:

1. Acquire an understanding of the Fair Labor Standards Act (FLSA).
2. Understand minimum wage and the difference between states.
3. Acquire an understanding of the exemptions to the FLSA.
4. Acquire an understanding of the "working time" and "overtime" requirements.
5. Acquire an understanding of the child labor laws.

******

Safety and health professionals should possess a working knowledge of the Fair Labor Standards Act (hereinafter "FLSA") in order to avoid "pitfalls" when encountering overtime issues, age requirements, and work time issues. Although the human resource function within most organizations is tasked with the management of the FLSA requirements within the organization, safety and health professionals often encounter safety and health-related issues which possess components which are governed by the FLSA.

The original Fair Labor Standards Act was enacted in 1938;[*] however, the original Act was amended in 1947 by the Portal-to-Portal Act,[†] and the Equal Pay Act in 1963,[‡] and again in 1977.[§] In general, the FLSA and its amendments addressed the different types of employees and how they are paid, payment for time spent on activities not directly related to the work functions, forbidding different wages and benefits based upon sex, tightened statutory exemption from coverage, reduced maximum weekly hours, and generally attempted to protect against discrimination and equalize the coverage for workers under the FLSA.

One of the most recent issues impacting safety and health professionals under the FLSA is the determination of whether a worker is an "employee" with coverage under the FLSA or an independent contractor who is not afforded coverage under the FLSA. Historically, in assessing this situation under the FLSA, the test was whether a worker was dependent upon the employer's business as a means of livelihood,[¶]

---

[*] See, 29 U.S.C.A. § 201, *et seq.*
[†] 29 U.S.C.A. § 251, *et seq.*
[‡] 29 U.S.C.A. § 206(d).
[§] P.L. 95-151, 91 Stat. 1245 (Nov. 1977).
[¶] *Walling v. Portland Terminal Co.*, 330 U.S. 148 (1947).

229

rather than the traditional common law employee categorization.* The test included whether a worker was "engaged in commerce" (closely related to interstate commerce) and "in the production of goods for commerce."† Safety and health professionals should also be aware that in 1974, an amendment to the FLSA extended coverage to employees of the federal, state, county, and municipal governments.‡

The FLSA defines an "employer" as the entity or individual who acts directly or indirectly in the interest of the employer.§ To make the determination as to the "employer," the enterprise test (extending coverage to fellow employees covered under the Act), minimum dollar volume test (specific yearly business sales),¶ and geographical test (outside U.S. territory)** were adopted. Of particular importance to safety and health professionals is the fact that certain positions or jobs are specifically excluded from the FLSA and thus the minimum wage and overtime provisions. In general, executive positions (paid on salary), administrative positions (paid on salary and exercise independent judgment), professional positions (salary, discretion, and judgment, specialized study), outside salespersons (work away from business), agricultural workers (farming, dairy), and specific retail or service positions. Additionally, FLSA provides for a number of specific exemptions for taxi drivers, babysitters, railroad employees, airline employees, and airline employees.

WAGE AND HOUR DIVISION (WHD)

(Revised July 2008)

FACT SHEET 13: EMPLOYMENT RELATIONSHIP UNDER THE FAIR LABOR STANDARDS ACT (FLSA)

This fact sheet provides general information concerning the meaning of "employment relationship" and the significance of that determination in applying provisions of the FLSA.

Characteristics

An employment relationship under the FLSA must be distinguished from a strictly contractual one. Such a relationship must exist for any provision of the FLSA to apply to any person engaged in work which may otherwise be subject to the Act. In the application of the FLSA an employee, as distinguished from a person who is engaged in a business of his or her own, is one who, as a matter of economic reality, follows the usual path of an employee and is dependent on the business which he or she serves. The employer-employee relationship

---

* *Weisal v. Singapore Joint Venture Inc.*, 602 F.2d 1185 (CA-5, 1979).
† 29 CFR Section 776.15(b) and 16(b); Also see, *Batts v. Professional Bldg. Inc.*, 276 F.Supp. 356 (DC W.VA. 1967).
‡ 29 U.S.C.A. § 203(5)(6).
§ 29 U.S.C.A. § 203(5)(6).
¶ 29 U.S.C.A. § 779.203.
** 29 U.S.C.A. § 203(c).

under the FLSA is tested by "economic reality" rather than "technical concepts." It is not determined by the common law standards relating to master and servant.

The U.S. Supreme Court has on a number of occasions indicated that there is no single rule or test for determining whether an individual is an independent contractor or an employee for purposes of the FLSA. The Court has held that it is the total activity or situation which controls. Among the factors which the Court has considered significant are:

1. The extent to which the services rendered are an integral part of the principal's business.
2. The permanency of the relationship.
3. The amount of the alleged contractor's investment in facilities and equipment.
4. The nature and degree of control by the principal.
5. The alleged contractor's opportunities for profit and loss.
6. The amount of initiative, judgment, or foresight in open market competition with others required for the success of the claimed independent contractor.
7. The degree of independent business organization and operation.

There are certain factors which are immaterial in determining whether there is an employment relationship. Such facts as the place where work is performed, the absence of a formal employment agreement, or whether an alleged independent contractor is licensed by State/local government are not considered to have a bearing on determinations as to whether there is an employment relationship. Additionally, the Supreme Court has held that the time or mode of pay does not control the determination of employee status.

Requirements

When it has been determined that an employer-employee relationship does exist, and the employee is engaged in work that is subject to the Act, it is required that the employee be paid at least the Federal minimum wage of $5.85 per hour effective July 24, 2007; $6.55 per hour effective July 24, 2008; and $7.25 per hour effective July 24, 2009, and in most cases overtime at time and one-half his/her regular rate of pay for all hours worked in excess of 40 per week. The Act also has youth employment provisions which regulate the employment of minors under the age of eighteen, as well as recordkeeping requirements.

Typical Problems

(1) One of the most common problems is in the construction industry where contractors hire so-called independent contractors, who in reality should be considered employees because they do not meet the tests for independence, as stated above. (2) Franchise arrangements can pose problems in this area as well. Depending on the level of control the franchisor has over the franchisee,

employees of the latter may be considered to be employed by the franchisor. (3) A situation involving a person volunteering his or her services for another may also result in an employment relationship. For example, a person who is an employee cannot "volunteer" his/her services to the employer to perform the same type of service performed as an employee. Of course, individuals may volunteer or donate their services to religious, public service, and non-profit organizations, without contemplation of pay, and not be considered employees of such organization. (4) Trainees or students may also be employees, depending on the circumstances of their activities for the employer. (5) People who perform work at their own home are often improperly considered as independent contractors. The Act covers such homeworkers as employees and they are entitled to all benefits of the law.

Where to Obtain Additional Information

For additional information, visit our Wage and Hour Division Website: http:// www.dol.gov/whd/ and/or call our toll-free information and helpline, available 8 a.m. to 5 p.m. in your time zone, 1-866-4USWAGE (1-866-487-9243).

This publication is for general information and is not to be considered in the same light as official statements of position contained in the regulations.*

Safety and health professionals should be aware that the FLSA previously prescribed a minimum wage for specific employees; however, the FLSA now provides a universal minimum wage level for all covered employees.† With this minimum level being established under the FLSA, safety and health professionals should be aware that individual states (and even municipalities) can establish minimum wage levels above the minimum wage level within their state or jurisdiction. For example, the minimum wage at the time of this writing in New York state is $13.00 per hour while in Kentucky, the minimum wage correlated with the FLSA level of $7.25. Safety and health professionals with operations in several states should be aware of the differences in the minimum hourly payment requirements between the states (as well as municipalities).

Of particular importance for safety and health professionals is the issue of "working time" or the time in which an employee is entitled to compensation under the FLSA.‡ In general, "working time" includes all time the employee is required to be on duty at the employer's property or prescribed work place.§ in 1947, Congress enacted the "portal-to-portal exclusion excluding "preliminary" and "postliminary" activities—before or after work activities—from the definition of "working time."¶

---

* See, U.S. Department of Labor website located at www.dol.gov.
† 29 U.S.C.A. § 207(a)(1).
‡ *Skidmore v. Swift & Co.*, 323 U.S. 134 (1944).
§ 29 C.F.R. § 778.223.
¶ 29 U.S.C.A. § 251 *et seq.*

At that time, such activities as clothes changing time and donning and doffing safety equipment were excluded from "working time" and thus were not paid. However, in recent years, various courts, including the U.S. Supreme Court, have addressed various issues involving whether various activities, such as changing work clothing, donning and doffing personal protective equipment (PPE), and other activities constituted "working time" and thus would be compensable under the FLSA. In the recent case of *Sandler v. United States Steel Corp.*,* the U.S. Supreme Court held that steelworkers' donning and doffing of certain items of required protective gear constituted "changing clothes" and thus was not compensable. However, safety and health professionals should be aware that in *IBP, Inc. v. Alvarez*,† the U.S. Supreme Court found that donning and doffing of personal protective equipment in a meat packing facility was "integral and indispensable" to the work activities and thus was compensable. Other cases, such as *Steiner v. Mitchell*‡ found that clothes changing time in a battery plant was "integral and indispensable" and thus compensable.

Safety and health professionals should be aware that the U.S. Department of Labor has adopted a regulation to assist in defining what does and what does not constitute "working time."§ In short, this regulation identifies that any activities that are primarily for the benefit of the employer and that are permitted by the employer constitute work time (and are thus compensable). In interpreting this regulation, the courts have found, as a general rule, that the employer is liable for off-the-clock work time if the employer knew or should have known that the employee was working.¶ Prudent safety and health professionals should carefully review the issue of donning and doffing personal protective safety equipment with your human resource or legal counsel when adding or changing PPE requirements.

Although safety and health professionals are not FLSA experts, today's workplace is changing and safety and health professionals should be cognizant that such activities as transporting safety equipment to the worksite, travel to different worksites, time during security screenings, repairing tools, waiting and walking time, readying machinery for work, booting computers, and related activities could constitute working time. As the American workplace changes and adapts to new technologies and practices, prudent safety and health professionals should be aware of the issues of working time and the potential that the activity could be compensable for the participating employees.

Another requirement under the FLSA of which safety and health professionals should be aware is that of overtime compensation. In general, employees working more than forty hours in any one workweek should be paid at the rate of not less than one and one-half times the employee's normal hourly rate. Although there are many exceptions and the calculation can be impacted by such issues as piece-work rates, collective bargaining agreements, different designations of workweeks, and other issues, it is important that safety and health professionals be aware of the overtime

---

* No. 12-417, 201 WL 273241 (January 27, 2014).
† 546 U.S. 21 (2005).
‡ 350 U.S. 247 (1956)
§ 29 C.F.R. § 785.11
¶ *Id.*

calculation for their hourly personnel and factor this into safety and health training and other required safety activities which are conducted before and after the normal working hours.

Safety and health professionals should also be aware that the FLSA governs issues involving child labor. The FLSA restricts the use of child labor to certain types of employment, certain ages, and, most importantly, the degree of hazards involved in particular jobs.* Safety and health professionals should be aware that the FLSA has a proscribed listing of hazardous jobs, such as roofing, excavation, operating saws, and numerous other activities which employees under the age of eighteen are prohibited from performing. Although the human resource department for most organizations will verify the age for employment, safety and health professionals should be cognizant of the age requirements for internships, apprentice programs, and related activities where employees may be below the requisite age limits.

As with virtually all laws, the FLSA requires extensive documentation and recordkeeping to insure compliance. Additionally, the FLSA requires a poster displaying the requirements of the FLSA. Safety and health professionals should be aware that the FLSA is regulated by the U.S. Department of Labor, and inspections and investigations are conducted by the Wage and Hour Division of the U.S. Department of Labor. Under the FLSA, violations discovered by the Wage and Hour Division can result in civil actions, with damages ranging from unpaid minimum wages, unpaid overtime, and liquidated damages. The FLSA also prohibits retaliation against employees who file a complaint.[†]

Safety and health professionals should be aware that in addition to the FLSA, there are several other correlating labor laws which address various aspects of the employer–employee relationship and the workplace. The Walsh–Healey Act of 1936 requires minimum wages and overtime for government contracts.[‡] The Davis–Bacon Act of 1931 requires minimum wage and overtime on federal public works contracts.[§] The Copeland Act (often referred to as the "Anti-kickback" Act) prohibited government contractors from inducing employees to give back an amount of their compensation to the employer.[¶]

Additionally, the Miller Act required contractors on government projects to post a bond to ensure that employees are paid their wages,[**] and the Contract Work Hours and Safety Standards Act required overtime and safe and healthful working conditions on construction sites financed by the federal government.[††] Safety and health professionals should also be aware that Title III of the Consumer Credit Protection Act addresses garnishments and the amounts which can be garnished from a worker's wages.[‡‡]

In summation, although safety and health professionals do not need to be wage and hour experts, awareness of these important requirements is important so not

---

* 29 U.S.C.A. § 203.
† 29 U.S.C.A. § 15(a)(3).
‡ 29 U.S.C.A. § 35.
§ 40 U.S.C.A. § 276 a(a).
¶ 41 U.S.C.A. § 874.
** 40 U.S.C.A. § 270a(a)(2)
†† 40 U.S.C.A. § 329.
‡‡ 15 U.S.C.A. § 1671 *et seq.*

to create issues for employees or the human resource function. Within the safety and health function, such activities as training before or after a work shift, required weekend work on safety projects and other safety and health-related activities may create overtime issues. Safety and health professionals can be assured that if an employee has not been paid correctly for his/her efforts, there will be an issue!

Case Study: this case may be modified for the purposes of this text.

814 F.Supp.2d 17
United States District Court, D. Massachusetts.

Barbatine NORCEIDE, Narces Norceide, and Jack Walsh on behalf of themselves
and all others similarly situated, Plaintiffs,

v.

CAMBRIDGE HEALTH ALLIANCE, d/b/a Cambridge Hospital, Somerville
Hospital, and Whidden Hospital, Defendant.

Civil Action No. 10cv11729–NG.
|
Aug. 28, 2011.

SYNOPSIS

Background: Current and former employees brought action against employer, alleging violations of the Fair Labor Standards Act (FLSA), Massachusetts Wage Act (MWA), and their employment contracts. Employer filed motion to dismiss, and employees filed motions to amend and for conditional certification.

Holdings: The District Court, Gertner, J., held that:

[1] calculation of whether an employee was paid minimum wage required by FLSA was to be done pursuant to an hour-by-hour method;

[2] allegations were sufficient to state a claim for a FLSA minimum wage violation;

[3] allegations were sufficient to state a claim for a FLSA overtime wage violation;

[4] claim under Massachusetts statute mandating that non-exempt hourly employees be paid their hourly wage for all time worked was precluded;

[5] employment at hospital precluded claims under Massachusetts law for overtime;

[6] granting motion to amend employees' complaint to include breach of contract claim was warranted; and

[7] employees put forth sufficient evidence that putative class members were similarly situated.

Defendant's motion granted in part and denied in part; plaintiffs' motions granted.

MEMORANDUM AND ORDER RE: MOTION TO DISMISS, MOTION TO AMEND, MOTION FOR CONDITIONAL CERTIFICATION

GERTNER, District Judge:

This is a wage-and-hour class action in which current and former employees of Defendant Cambridge Health Alliance ("CHA") allege that CHA deprived them of compensation for time worked during their 30-minute meal breaks and before and after their shifts. They claim these practices were in violation of the Fair Labor Standards Act ("FLSA"), the Massachusetts Wage Act, and their employment contracts with CHA. The violations allegedly resulted from CHA's practice of pressuring its employees not to record time worked outside of their shifts, thereby routinely stripping its employees of wages to which they were entitled.

CHA has moved to dismiss most of Plaintiffs' state and federal law claims (document # 16). Plaintiffs have moved to amend their complaint (document # 29) and for conditional certification of their FLSA claims (document # 15). For the foregoing reasons, CHA's motion to dismiss is **GRANTED IN PART** and **DENIED IN PART**, Plaintiffs' motion to amend is **GRANTED**, and Plaintiffs' motion for conditional certification is **GRANTED**.

I. BACKGROUND

Taking all well-pled allegations as true and making all reasonable inferences in Plaintiffs' favor, *Gargano v. Liberty Int'l Underwriters, Inc.*, 572 F.3d 45, 48–49 (1st Cir.2009), the facts are as follows:

Formed in 1996 and comprised of Cambridge Hospital, Somerville Hospital, and Whidden Hospital, along with more than 30 clinics, primary care centers, and community health centers, CHA is a healthcare system that serves the residents of Cambridge, Somerville and Boston's Metro–North region. Compl. 16, 17 (document # 8); Bennett Aff. 3 (document # 21–1). CHA employs approximately 4,200 people, including nurses, certified nursing assistants, secretaries, pharmacy technicians, dietary food workers, and housekeeping staff. *Id.*; Gilyan Aff. 7 (document # 21–15). The three named Plaintiffs—Barbatine Norceide ("Barbatine"), Narces Norceide ("Narces"), and Jack Walsh ("Walsh")—are current and former, non-exempt, hourly employees of these hospitals. Compl. 19. Barbatine is a former unit secretary for Cambridge Hospital; Narces, a member of the Massachusetts Nurses Association ("MNA"), is a currently-employed registered nurse who practices at both Cambridge Hospital and Whidden Hospital; and Walsh is a former employee who worked both in the pharmacy department and as a Counselor and Patient Access Representative at Cambridge Hospital. *Id.*

From October 12, 2007, through October 12, 2010 (the "class period"), Barbatine, Narces, Walsh, and the similarly-situated CHA workers they represent (collectively "Plaintiffs") routinely worked before and after their shifts and during their 30-minute meal breaks, often at the request of their supervisors. *Id.* 20–22. They were required to carry cell phones and pagers on them during their meals breaks and instructed to respond immediately to all calls. *Id.* 20. As a result, employees often took "meals at their desks or in locations convenient so that they [could] be contacted by their supervisors and co-workers for work-related matters during *20 their meal breaks." *Id.* "[N]early all of [Barbatine's] lunches were interrupted with work-related matters; a colleague, a doctor or a nurse or a patient always needed something," and she was regularly asked to stay an extra 30 to 60 minutes after her shifts ended. Barbatine Decl. 3, 6 (document # 8–1). Narces' meal breaks "were interrupted more than half of the time," and he often completed preparatory tasks for about 10 minutes before his shifts began and was "expected" to stay late in order to "transition to the next shift." Narces Decl. 4 (document # 8–2). Walsh "seldom" got a meal break and was often "directed" by his managers to work an additional 1–3 hours each week, either before or after his shifts. Walsh Decl. 4 (document # 8–3).

CHA employed three different time-keeping systems during the class period. Up until late 2007, all non-exempt CHA employees were individually responsible for entering their time into an electronically maintained timesheet located on CHA's intranet, known as "Staffnet." Gilyan Aff. 4. After employees input their time, their managers reviewed the Staffnet entries and approved the timesheets. *Id.* 6. Beginning in late November 2007, CHA transitioned to a new software product called "ANSOS," which scheduled staff working in CHA's "costs centers" at the three hospital campuses. *Id.* 7, 11. The ANSOS system allowed managers to schedule the employees necessary to fill each shift and then print a paper schedule, which was posted in each clinical unit of the three hospitals. *Id.* 8. According to CHA, employees could make notations on this sheet when their time exceeded their scheduled hours. *Id.* 10. All employees who did not work in "costs centers"—i.e., pharmacy technicians, housekeepers, laboratory technicians—continued to use Staffnet. *Id.* 7.

Then, in April 2008, CHA implemented a project to add an electronic timekeeping component, known as the McKesson Time and Attendance System, to the ANSOS system. *Id.* 12. This system allowed employees to record the time they began and ended their shifts by either swiping a card or entering the time into a webclock. *Id.* 15. The McKesson Time and Attendance system would then automatically deduct 30 minutes from the compensable working time, unless a manager informed payroll otherwise. *Id.* 15–16. The McKesson system was first rolled out in August 2009 for employees working in the laboratory at Cambridge Hospital and, then in October 2009, for employees covered by the collective bargaining agreement between CHA and the Massachusetts Nurses Association ("MNA"). According to CHA, Barbatine used McKesson Time and Attendance for a few weeks, Narces used ANSOS prior to the implementation of McKesson Time and Attendance, and Walsh only used Staffnet. *Id.* 17–18.

Regardless of what timekeeping system was used, CHA employees were routinely paid only for the time worked during their pre-scheduled shifts—which allotted 30–minutes for an unpaid meal break—rather the actual time during which they performed compensable labor. Compl. 9, 22–24. As a result, Walsh, Narces, and Barbatine claimed that they did not receive wages for about two to four hours of work each week. *See* Walsh Decl. 4; Narces Decl. 1, 4; Barbatine Decl. 2, 3, 6. While all three time-keeping systems theoretically enabled its employees to report time worked during a meal break or before/after a shift, CHA "fail[ed] and refused/discouraged payment of time outside of schedule shifts, creating an atmosphere w[h]ere time worked outside of the schedule shift was 'swept under the rug' and not paid." Compl. 23. Accordingly to *21 Narces, even when he complained that he worked through his break, "[t]here was no override of the payroll system's automatic deduction." Narces Decl. 3. Barbatine alleges that "[t]here was no avenue for [her] to complain that [she] did not get a meal break," and she could not "override ... the payroll system's automatic deduction." Barbatine Decl. 5. According to Walsh, "[t]here was ... an atmosphere whereby if you complained about not getting a meal break or not being paid for time worked outside of your scheduled shift, management would ask you: 'Do you still want your job?'" Walsh Decl. 5.

II. Discussion

There are three pending motions in this case: Defendant's motion to dismiss, Plaintiffs' motion to amend, and Plaintiffs' motion for conditional class certification. I will address each in turn.

A. Defendant's Motion to Dismiss

CHA moves to dismiss on the following grounds: (1) Barbatine and Narces' federal and state overtime claims fail because they do not allege working more than 40 hours in a week; (2) Plaintiffs have failed to state a claim for a violation of the FLSA's minimum wage provision; (3) Plaintiffs are precluded from recouping unpaid wages pursuant to Mass. Gen. Laws ch. 148, § 149 because they have failed to exhaust their administrative remedies; and (4) as hospital employees, Plaintiffs are exempt from recouping unpaid overtime under state law. Addressing each, I must accept all well-pleaded facts as true and make all reasonable inferences in the Plaintiffs' favor. *Gargano*, 572 F.3d at 48. I may take into account documents "central to plaintiffs' claim" or "sufficiently referred to in the complaint" without converting a motion to dismiss into one for summary judgment. *Watterson v. Page*, 987 F.2d 1, 3 (1st Cir.1993). To survive CHA's motion to dismiss, Plaintiffs' complaint must allege "enough facts to state a claim to relief that is plausible on its face." *Bell Atl. Corp. v. Twombly*, 550 U.S. 544, 570, 127 S.Ct. 1955, 167 L.Ed.2d 929 (2007).

1. FLSA (Count I).

Enacted in 1938, the FLSA ensures basic worker rights for non-exempt employees. It guarantees that every non-exempt employee be paid at least the minimum wage for each hour worked and receive overtime for time worked in excess of 40 hours in a workweek at a rate of not less than time and one-half their regular rates of pay. 29 U.S.C. §§ 206, 207. As explained by the FLSA's governing regulations, "[w]ork not requested but suffered or permitted is work time." 29 C.F.R. § 785.11 ("For example, an employee may voluntarily continue to work at the end of the shift."). Plaintiffs allege that, by failing to compensate them for time worked during meal breaks and outside of their scheduled shifts, CHA violated these two fundamental worker protections.*

---

* Section 216 of the FLSA holds an employer who violates the provisions of § 206 or § 207 liable to the employee or employees affected in the amount of their unpaid minimum wages or their unpaid overtime compensation and an additional equal amount as liquidated damages. 29 U.S.C. § 216(b).

a. *Unpaid Minimum Wage.* [1] According to CHA, Plaintiffs' minimum wage claims should be dismissed because they do not allege that CHA ever paid them less than the operative minimum wage. Specifically, CHA argues that, if the total wages paid to any given plaintiff in a week were divided by the total hours worked in the week, then the average hourly wage would be greater ***22** than the minimum wage. For instance, suppose that one week Barbatine was scheduled to work 26 hours at a rate of $10.00 an hour and was paid accordingly, meaning she earned $260. However, in fact, she worked an additional 4 hours during her breaks and before/after her shifts and was not paid for this time. According to CHA, Barbatine has no claim for a minimum wage violation, since $260 divided by 30 hours is an average hourly wage of $8.67, which still exceeds the minimum wage.*

In reality, Plaintiffs counter, Barbatine was being paid at a rate of $0 per hour for her additional 4 hours. CHA intended for its payment of $260 to cover her scheduled shifts and nothing more. Barbatine's payment statement for the week in question on its face would indicate that she was getting paid for 26 hours of work, not 30. I agree with Plaintiffs.

The weekly average wage measuring rod that CHA argues should be utilized when assessing minimum wage violations stems from the Second Circuit's decision in *United States v. Klinghoffer Bros. Realty Corp.*, 285 F.2d 487 (2d Cir.1960). In that case, due to some financial difficulties their employer faced, security guard employees agreed to work an additional six hours per week but not be paid for this time until some later date. *Id.* at 489–90. The employer, however, never provided compensation for this extra work. The federal government charged the company with violating the FLSA. The Second Circuit dismissed the government's minimum wage claim on the basis of the weekly average wage theory. *Id.* at 490. Articulating the purpose of the FLSA only as "guarantee[ing] a minimum livelihood to the employees," the court found that "the Congressional purpose is accomplished so long as the total weekly wage paid by an employer meets the minimum weekly requirements of the statute, such minimum weekly requirement being equal to the number of hours actually worked that week multiplied by the minimum hourly statutory requirement." *Id.* at 490 (citing H.R.Rep. No. 75–2738, at 28 (1938); Sen. Rep. No. 75–884, at 1–3 (1937); H.R.Rep. No. 75–1452, at 8–9 (1937)).

Since the court's decision in 1960, several other circuits have adopted the Second Circuit's approach—what has come to be known as the *Klinghoffer* rule. However, they have mostly done so by citing to *Klinghoffer* without any further analysis of whether, in fact, the weekly average rule effectuates the legislative intent of the FLSA's minimum wage law. *See, e.g., U.S. Dep't of Labor v. Cole Enter., Inc.*, 62 F.3d 775, 780 (6th Cir.1995) (simply noting what "several courts have held"); *Hensley v. MacMillan Bloedel Containers, Inc.*, 786 F.2d 353, 357 (8th Cir.1986) (citing to *Klinghoffer* without analysis); *Blankenship v. Thurston Motor Lines*, 415 F.2d 1193, 1198 (4th Cir.1969) (stating without explanation that *Klinghoffer* "contains a correct statement of the law"). The D.C. Circuit is a notable exception, accepting the weekly average wage rule in *Dove v. Coupe* only after its own analysis. 759 F.2d 167, 171–72

---

* For work performed from July 24, 2007 to July 23, 2008, the federal minimum wage was $5.85 per hour. For work performed from July 24, 2008 to July 23, 2009, the federal minimum wage was $6.55 per hour. Since July 24, 2009, the minimum wage has been $7.25 per hour.

(D.C.Cir.1985).* The ***23** First Circuit, however, has yet to address whether to use the hour-by-hour or the *Klinghoffer* weekly average measure for evaluating minimum wage law compliance. In my view, as explained below, the *Klinghoffer* weekly average method ignores the plain language of the minimum wage provision and undermines the FLSA's primary purpose of ensuring a fair wage for workers.

My review of the FLSA is guided by principles of statutory construction; my interpretation "depends upon reading the whole statutory text, considering the purpose and context of the statute, and consulting any precedents or authorities that inform the analysis." *Dolan v. Postal Service*, 546 U.S. 481, 486, 126 S.Ct. 1252, 163 L.Ed.2d 1079 (2006).

I begin by looking to the language of the statute. *See Kasten v. Saint-Gobain Performance Plastics Corp.*, ––– U.S. –––––, 131 S.Ct. 1325, 1331, 179 L.Ed.2d 379 (2011). The FLSA's minimum wage provision mandates that an employer pay to each non-exempt employee "wages at the following rates: (1) except as other provided ... not less than—(A) \$5.85 an hour, beginning on the 60th day after May 25, 2007; (B) \$6.55 an hour, beginning 12 months after that 60th day; and (C) \$7.25 an hour, beginning 24 months after that 60th day." 29 U.S.C. § 206(a). As the other courts to have considered this language concede, it speaks only of an *hourly* wage. Thus, while it is does not explicitly state how to calculate what an employee has been paid for an hour's worth of work, the statute's text is explicit that, with respect to the minimum wage, the only metric Congress envisioned was the hour, with each hour having its own discrete importance.†

To be sure, other parts of the FLSA speak of a "workweek." But, this unit of time is used for determining worker entitlement to other protections, most importantly overtime, not for assessing violations of the minimum wage law. *See, e.g.,* 29 U.S.C. § 207(a) (2) ("[N]o employer shall employ any of his employees who in any workweek is engaged

---

* Claiming that the purpose of the FLSA's minimum wage law is to "'protect certain groups of the population from sub-standard wages ... due to ... unequal bargaining power,'" *Dove,* 759 F.2d at 171 (quoting *Brooklyn Sav. Bank v. O'Neil,* 324 U.S. 697, 706, 65 S.Ct. 895, 89 L.Ed. 1296 (1945)), the D.C. Circuit first recognized *Klinghoffer's* finding that this purpose "is accomplished so long as the total weekly wage paid by an employer meets the minimum weekly requirements." *Dove* 759 F.2d at 171. The court continued its analysis by noting that while the "minimum wage laws logically could be construed as requiring hour-by-hour compliance," it followed that "both administrative and judicial decisions established the workweek as the measuring rod for compliance at a very early date." *Id.* (citing *Travis v. Ray,* 41 F.Supp. 6, 9 (D.C.Ky.1941). In support, the court cited to a 1942 manual published by the Department of Labor's Wage and Hour Division, Wage & Hour Release No. R–609 (Feb. 5, 1940), *reprinted in* 1942 Wage Hour Manual (BNA) 185); various Wage and Hour Division Opinion Letters; and the *Klinghoffer* line of cases. *Dove,* 759 F.2d at 171–72. Finally, the court held that, "[i]n keeping with the view of [its] sister courts, and giving respectful consideration to the position of the agency charged with administration of the Fair Labor Standards Act" the average workweek wage is the proper measure of compliance with minimum wage law. *Id.* at 172.

† In 1961, Congress amended the minimum wage provision to include the phrase "in any workweek" to the following opening sentence: "Every employer shall pay to each of his employees who in any workweek is engaged in commerce or in the production of good for commerce ...." An Act to Amend the Fair Labor Standards Act, Pub. L. No. 87–30, § 6(a), 75 Stat. 69 (1961). However, as the Act's purpose statement makes explicit, the phrase "workweek" was not intended to change the discrete importance of each hour. *See id.* 75 Stat. 65. Rather, the amendment was part of Congress' attempt to "to provide coverage for employees of large enterprises engaged in retail trade or service and of other employers engaged in commerce or in the production of goods for commerce." *Id.*

in commerce ... for a workweek longer than forty hours ... unless such employee receives compensation *24 for his employment in excess of the hours above specified at a rate not less than one and one-half times the regular rate at which he is employed."). In fact, the other provisions of the FLSA support the conclusion that, for the purpose of determining a minimum wage violation, the use of any unit of time other than an hour is a contrivance. When Congress meant to use the word "workweek," it did so. When it meant to use "hour," that was the word it used.

The FLSA's legislative history does not explicitly address whether an hour-by-hour or weekly-average method should be employed when determining compliance with the minimum wage law. However, it does makes clear that Congress' overriding purpose when enacting the FLSA was to ensure, as the bill's name implies, fairness for workers. "The principal congressional purpose in enacting the [FLSA] was to protect all covered workers from substandard wages and oppressive working hours, 'labor conditions [that are] detrimental to the maintenance of the minimum standard of living necessary for health, efficiency and general well-being of workers.'" *Barrentine v. Arkansas–Best Freight Sys. Inc.*, 450 U.S. 728, 739, 101 S.Ct. 1437, 67 L.Ed.2d 641 (1981) (citing 29 U.S.C. § 202(a)). One way Congress attempted to effectuate this somewhat amorphous goal through the FLSA was by "guarantee[ing] a minimum livelihood to the employees covered by the Act," *Klinghoffer*, 285 F.2d at 490 (citing H.R.Rep. No. 75–2738, at 28 (1938); Sen. Rep. No. 75–884, at 1–3 (1937); H.R.Rep. No. 75–1452, at 8–9 (1937)). While the Senate and House reports do not indicate whether Congress had in mind a formula for determining the amount necessary for "a minimum livelihood," they do reveal that Congress considered the test to be whether a worker received "'a fair day's pay for a fair day's work,'" *Overnight Motor Transp. Co. v. Missel*, 316 U.S. 572, 578, 62 S.Ct. 1216, 86 L.Ed. 1682 (1942) (quoting 81 Cong. Rec. 4983 (1937) (message of President Roosevelt)); *see also Barrentine*, 450 U.S. at 739, 101 S.Ct. 1437.*

Congress' primary concern with protecting workers—not employers—buttresses the above conclusion that the plain language of the minimum wage provision should be read as an endorsement of the hour-by-hour method. When a statute is susceptible to two opposing interpretations—here, the hour-by-hour and weekly average methods—it must be read "in the manner which effectuates rather than frustrates the major purpose of the legislative draftsmen." *Shapiro v. United States*, 335 U.S. 1, 31–32, 68 S.Ct. 1375, 92 L.Ed. 1787 (1948). While the weekly method does ensure that workers earn a base amount after working a certain number of hours in a week, it frustrates the overall purpose of promoting fairness for workers.

Take the Barbatine example above. There, CHA intended for the $260 to compensate for only the 26 hours she was scheduled to work. CHA, therefore, got four

---

* One of the ways that Congress articulated it overriding concern for worker fairness was by imposing strict liability on employers, signaling that even well-intentioned and negligent employees were liable. *See* 29 U.S.C. §§ 206, 207; *see also* 29 U.S.C. § 255(a). In 1966, Congress amended the statute of limitations provision of the FLSA to extend the two-year limit to three years for causes of action "arising out of a willful violation." See 29 U.S.C. § 255(a) (1966). This indicates that, while Congress wanted to impose strict liability, it was also concerned with the employer's intent. Put another way, an important aspect of fairness is whether an employer is intentionally taking advantage of an employee. By purposefully paying $0 for certain hours of work, an employer is arguably acting intentionally to exploit a worker.

free hours of work from Barbatine, *25 while Barbatine received the same amount of compensation after working 30 hours as she would have for working 26 hours. Such a compensation scheme does not promote an environment in which a worker is ensured "a fair day's pay for a fair day's work."' See Travis, 41 F.Supp. at 9 ("[I]f the act is given a very strict construction[,] averaging is probably not permitted."); see also Dove, 759 F.2d at 171.

[2] Taken together, the plain language of the minimum wage provision, the remaining parts of the FLSA, and the Congress' primary goal of protecting workers buttresses the conclusion that Congress intended for the hour-by-hour method to be used for determining a minimum wage violation.* Here, Plaintiffs have alleged that CHA knew the Plaintiffs were working more hours than reported on their time sheet and that it was not compensating its employees for this time. In other words, Plaintiffs have alleged that CHA intentionally paid its workers $0 for each *26 unrecorded

---

* To the extent that any—albeit minimal—ambiguity remains, under the familiar Chevron framework, this Court would generally defer to the interpretation of the Department of Labor ("DOL") as reflected in its regulations. Chevron, U.S.A., Inc. v. Natural Res. Def. Council, Inc., 467 U.S. 837, 842–45, 104 S.Ct. 2778, 81 L.Ed.2d 694 (1984); see Christensen v. Harris Cnty., 529 U.S. 576, 586–88, 120 S.Ct. 1655, 146 L.Ed.2d 621 (2000); Duckworth v. Pratt & Whitney, Inc., 152 F.3d 1, 5–6 (1st Cir.1998). However, the DOL has not promulgated regulations governing the calculation method that should be used for determining a minimum wage violation. See 29 C.F.R. § 776.0, et seq.; see Dove, 759 F.2d at 172 & n. 7. In 29 C.F.R. § 776.4(a) (1984), the DOL explains that the "workweek is to be taken as the standard in determining the applicability of the Act," meaning that "if in any workweek an employee is engaged in both covered and noncovered work he is entitled to both the wage and hours benefits of the Act for all the time worked in that week, unless exempted there from by some specific provision of the Act." However, this regulation speaks simply to the "applicability" of the FLSA, not how to determine the hourly wage for purpose of the minimum wage. See Dove, 759 F.2d at 172 n. 7.

The DOL has produced a few opinion letters and manuals that endorse the use of the weekly average methods. For instance, in 1940, the General Counsel of the Wage and Hour Division of the Department of Labor stated in a press release later incorporated in a manual that "[for] enforcement purposes, the Wage and Hour Division is at present adopting the workweek as the standard period of time over which wages may be averaged to determine whether the employer has paid the equivalent of the [the minimum wage]." Dove, 759 F.2d at 171 (quoting Wage Hour Release No. R–609 (Feb. 5, 1940), reprinted in 1942 Wage and Hour Manual (BNA) 185). And, in 2008, the DOL relied upon Klinghoffer when responding to a school's request for an opinion regarding its compliance with the minimum wage and overtime provision of the FLSA. U.S. Dept't of Labor, Opinion Letter FLSA 2008–5 (May 30, 2008), available at www.dol.gov/whd/opinion/flsa.htm (last visited Aug. 12, 2011).

However, these opinion letters and manual are entitled to only "respect," not deference, Christensen, 529 U.S. at 587, 120 S.Ct. 1655 (quoting Skidmore v. Swift & Co., 323 U.S. 134, 140, 65 S.Ct. 161, 89 L.Ed. 124 (1944)), and "only to the extent that th[e]se interpretations have the 'power to persuade.' " I do not find them persuasive for several reasons. Most importantly, they do not engage with the FLSA's purpose of ensuring a fair wage for employees. Furthermore, the DOL's 1940 press release was "mindful" that the FLSA defined the minimum wage as an hourly rate, noting that its opinion did not bind the court and, in fact, that the courts might hold to be a violation of the law [a case] where the employer does not pay anything for hours properly considered to be hours worked ... but until directed otherwise by authoritative ruling of the courts, the Division will take the workweek as the standard for determining whether there has been compliance with the law.

Dove, 759 F.2d at 172 (quoting Wage Hour Release No. R–609 (Feb. 5, 1940)). Likewise, the DOL's May 2008 letter illustrated that the agency was simply following the Second Circuit's ruling in Klinghoffer, in the absence of conflicting judicial interpretation, rather than offering their own independent conclusion.

hour worked during their meal breaks and before/after their shifts.* This allegation is sufficient to state a claim for a minimum wage violation at this stage, and CHA's motion to dismiss Plaintiffs' FLSA minimum wage claim is **DENIED**.

b. Overtime (Norceides only) [3] Defendants argue that, to the extent that the Norceides (Barbatine and Narces) seek to recover overtime pay, their claims should be dismissed because their declarations do not provide sufficient support for their claim that they worked more than 40 hours in any given work week. CHA is placing too high an evidentiary burden on Plaintiffs at the pleading stage. In their complaint, Plaintiffs claim that they have worked in excess of 40 hours in a week.† That is sufficient at this point.‡ Accordingly, CHA's motion to dismiss Plaintiffs' FLSA overtime claim is **DENIED**.

2. State Law Claims (Count II)

Plaintiffs allege violations of Mass. Gen. Law ch. 149, §§ 148, 150, which mandates that non-exempt hourly employees be paid their hourly wage (also known as their straight-time wage) for all time worked; and Mass. Gen. Law ch. 151, §§ 1A, 1B, which requires overtime for all hours worked in excess of 40 in a week.§ Defendant *27 argues that 1) Plaintiffs' claim for unpaid straight-time wages fails because they did not exhaust their administrative remedies; and 2) Plaintiffs' overtime claim fail because CHA employee are subject to the hospital exemption. I will address each in turn.

---

* Indeed, Defendants have repeatedly claimed that, had the employees reported that they had worked extra hours, they would have received additional compensation.

† In fact, Narces' affidavit details how he worked over 40 hours. Narces says that he worked for CHA as a Registered Nurse "approximately 3 shifts per week, always 12.5 hour shifts," from October 2009 to August 2010. Narces Decl. 1. And, since August 2010, he has had "one 12.5 hour [shift] at Cambridge hospital and 8.5 hour shifts at Whidden hospital for 32 hours a week." *Id.* He was supposed to get a 30 minute meal break during each shift, but his "meal breaks were interrupted more than half of the time." *Id.* 1–2. Nonetheless, his paycheck always showed a 30 minute deduction for the full meal break. *Id.* 3. Similarly, he was "paid based only on [his] scheduled shift, as opposed the times [he] beg [an] and end[ed]" work, even though he routinely started early and worked late. *Id.* 4. By defendant's math, between October 2009 through August 2010, Narces worked at most 40 hours, but not more. However, Narces says that he worked "approximately" three shifts. *Id.* 1. As a result, there certainly could have been weeks that he worked four shifts, and during those weeks, he would have worked more than 40 hours, even if he always got a meal break and never started early or left late.

‡ This factual dispute is particularly unsuited for the pleading stage given the lack of payroll records in this case. According to Plaintiffs, CHA lacks records of every hour worked by Plaintiffs, Compl. 10, in violation of 29 C.F.R. § 516.2(7), which mandates that every employer keep and preserve a payroll or other records containing, amongst other things, information on "[h]ours worked each workday and total hours worked each workweek." As the Supreme Court made clear in *Anderson v. Mt. Clemens Pottery,* this lack of records encumbers the employer, not the employee. 328 U.S. 680, 686–87, 66 S.Ct. 1187, 90 L.Ed. 1515 (1946). At trial, if the evidence shows that CHA did not maintain the legally required records and an employee has shown that he has performed some amount of work for which he was improperly compensated, then the burden "shifts to the employer to come forward with evidence of the precise amount of work performed or with evidence to negative the reasonableness of the inference to be drawn from the employee's evidence." *Id.* Extending *Anderson*'s burden-shifting doctrine to the pleading stage, in the absence of records, a plaintiff need do no more than allege that she has worked in excess of 40 hours in a week in order to state a claim for an FLSA overtime violation.

§ Currently, Plaintiffs have dropped their state law claims on behalf of any unionized employees. (Narces is the only named unionized Plaintiff.) Accordingly, this court need not address Defendant arguments concerning preemption by the Labor Management Relations Act.

a. Unpaid Straight-Time Wages—Exhaustion [4] Sections 148 and 150 of Mass. Gen. Laws ch. 149 work in tandem. Section 148 provides that all non-exempt workers have the right to receive straight-time wages for each hour worked, while Section 150 grants an aggrieved employee the right to bring a civil action for a violation of Section 148. Under Section 150, an employee's private right of action is conditioned on the filing of a complaint with the Massachusetts Attorney General.* See, e.g., Sterling Research, Inc. v. Pietrobono, No. 02–40150–FDS, 2005 WL 3116758, at *11 (D.Mass. Nov. 21, 2005); Daly v. Norton Co., No. 99452B, 1999 WL 1204011, at *2 (Mass.Super. Nov. 15, 1999) (noting that the provision "mandates that all compensation claims must be filed with the Office of the Massachusetts Attorney General before any court may adjudicate such claims").

[5] According to Joan Bennett, a senior vice president at CHA, neither Narces Norceide nor Barbatine Norceide has filed a complaint with the Massachusetts Attorney General's Office regarding claims for unpaid wages. Bennett Aff. 11 (document # 16–1). Plaintiffs did not address this issue in their complaint, but they have since stated in their briefing and at oral argument that each of the three named plaintiffs filed complaints with the Attorney General's office for unpaid wages. Their oral representations notwithstanding, Plaintiffs are obliged to seek to amend the complaint to make this claim. CHA's motion to dismiss Plaintiffs state-law claim for unpaid straight-time wages is **GRANTED WITHOUT PREJUDICE**.

b. Overtime [6] Under state law, a worker employed in a "hospital, sanatorium, convalescent or nursing home, infirmary, rest home or charitable home for the aged" is not protected by the overtime provision. Mass. Gen. Laws ch. 151, § 1A(16). Claiming that Plaintiffs are all "hospital" employees, CHA moves to dismiss Plaintiffs' state-law overtime claims.† Plaintiffs respond that this is "not a case about hospital employees only," noting that Defendant have more than 30 clinical locations "both in the community and at its hospital campuses." Pls.' Mot. Amend Compl. 3 (document # 29). While Plaintiffs ***28** may indeed be correct that CHA employees in the community centers are entitled to overtime under state-law, Narces, Barbatine, and Walsh, by their own admissions, are hospital employees. See Barbatine Decl. 1; Narces Decl. 1;

---

* Mass. Gen. Law. Ch. 149, § 150 states:
  > An employee claiming to be aggrieved by a violation of sections 33E, 148, 148A, 148B, 150C, 152, 152A or 159C or section 19 of chapter 151 may, 90 days after the filing of a complaint with the attorney general, or sooner if the attorney general assents in writing, and within 3 years after the violation, institute and prosecute in his own name and on his own behalf, or for himself and for others similarly situated, a civil action for injunctive relief, for any damages incurred, and for any lost wages and other benefits.

† CHA also attacks Plaintiffs' state-law overtime claim on the same grounds that they challenged the FLSA overtime claim, namely that some of the Plaintiffs have not sufficiently alleged that they worked more than 40 hours per week without receiving overtime pay. CHA's argument with respect for the state-law claim is rejected for the same reasons as it was for the FLSA overtime claim. *See Swift v. AutoZone, Inc.,* 441 Mass. 443, 447, 806 N.E.2d 95 (2004) (quoting *Valerio v. Putnam Assocs. Inc.,* 173 F.3d 35, 40 (1st Cir.1999)) (finding that the state overtime laws are "intended to be 'essentially identical'" to the FLSA's overtime provision).

Walsh Decl. 1. Accordingly, Narces, Barbatine, and Walsh's claims for overtime are **DISMISSED**.*

B. Plaintiffs' Motion to Amend

Plaintiffs have moved to amend their complaint in two ways: first, to clarify that they are bringing their state-law claims (Count II) on behalf of non-union, non-exempt employees only; and, second, to add a claim for breach of contract on behalf of non-union, non-exempt employees (Count III). CHA does not contest Plaintiffs' effort to limit its state-law claims to non-union employees, but it does object to the addition of the breach of contract claim.

[7] [8] Since Plaintiffs have already amended their complaint once as a matter of course, their request to add a breach of contract claim requires leave of court, which I "should freely give ... when justice so requires," Fed.R.Civ.P. 15(a)(2), unless the amendment is futile, rewards undue delay, or causes grave injustice to CHA, *see Adorno v. Crowley Towing And Transp. Co.*, 443 F.3d 122, 126 (1st Cir.2006); *Patton v. Guyer*, 443 F.2d 79, 86 (10th Cir.1971). In this case, justice so requires Plaintiffs' amendment because federal and state wage-and-hour laws do not provide for all of the potential relief to which Plaintiffs may be entitled under a breach of contract claim.† Nor does this amendment subject CHA to "a grave injustice." *See Patton*, 443 F.2d at 86. This case is still at the pleading stage, and Plaintiffs' breach of contract claim stems for the same general set of alleged practices employed by CHA as do the FLSA and state-law claims, meaning CHA is not being forced to defend itself on a whole new front. Accordingly, Plaintiffs' motion to amend (document # 29) is **GRANTED**.

C. Class Certification

Plaintiffs have moved for conditional certification of their FLSA claims. Under the FLSA, an employee may bring an action on behalf of him/herself and other "employees similarly situated." 29 U.S.C. § 216(b). This Court has generally used a "two-tiered" approach for determining whether putative class members are similarly situation for purposes of class certification. *See, e.g., O'Donnell v. Robert Half Int'l, Inc.*, 429 F.Supp.2d 246, 249 (D.Mass.2006); *Trezvant v. Fidelity Emp'r Servs. Corp.*, 434 F.Supp.2d 40, 42–45 (D.Mass.2006). At the first step (also known as the notice stage), the Court relies upon the pleadings and affidavits to determine whether the plaintiffs have made "a modest factual showing" that the putative class members were together subject to "'a single decision, policy, or plan that

---

* This finding, however, does not necessarily dismiss the overtime claim for all class members.
† Despite CHA's assertion to the contrary, Plaintiffs breach of contract claim is not necessarily pre-empted by the FLSA. In certain situations, the FLSA may preempt causes of action in which a party seeks to enforce rights guaranteed by the FLSA, *see, e.g., Anderson v. Sara Lee Corp.*, 508 F.3d 181, 194 (4th Cir.2007) (noting that plaintiffs cannot enforce their FLSA rights by way of a § 1983 action). In this case, however, Plaintiffs' breach of contract claim may extend to rights not guaranteed by the FLSA. As CHA repeatedly emphasizes, Plaintiffs have no right to recoup unpaid straight-time wages under the FLSA, though they may have such a remedy under their employment contracts.

violated the law.'" *Trezvant*, 434 F.Supp.2d at 43 (quoting ***29** *Thiessen v. Gen. Elec. Capital*, 267 F.3d 1095, 1102 (10th Cir.2001)). At the second step, the court revisits the certification issue at the completion of discovery using a stricter standard. *Kane v. Gage Merch. Servs., Inc.*, 138 F.Supp.2d 212, 214 (D.Mass.2001).

[9] Plaintiffs' motion for conditional class certification is at the first step. At this notice stage, given the minimal evidence available, Plaintiffs' burden of proving that putative class members are similarly situated is "fairly lenient," typically resulting in conditional certifications of the representative class. *Trezvant*, 434 F.Supp.2d at 43. Plaintiffs satisfy their burden if they put forth "some evidence that the legal claims and factual characteristics are similar." *Id.* at 44. Here, Plaintiffs have done so by making a modest factual showing that the putative class members were all subject to CHA's practice of discouraging its workers from recording time worked before and after their shifts and during their meal breaks, thereby paying its employees based on their theoretical schedules rather than actual time worked.

To be sure, CHA assigned its employees to various different time-keeping systems, some of which theoretically allowed employees to record the full time they worked. However, regardless of the timekeeping systems used, Plaintiffs were all allegedly either explicitly or implicitly "discouraged" by their managers from reporting any time worked during meal breaks or before/after their shifts. Compl. 23; *see* Narces Decl.; Barbatine Decl.; Walsh Decl. As described by Walsh, "[t] here was ... an atmosphere whereby if you complained about not getting a meal break or not being paid for time worked outside of your scheduled shift, management would ask: 'Do you still want your job?'" Walsh Compl. 5. Barbatine and Narces describe the ways that CHA made it difficult for them to report having worked additional time. According to Narces, his complaints that he had worked through his break fell on deaf ears. Narces Decl. 3. Barbatine likewise alleges that "[t]here was no avenue for [her] to complain that [she] did not get a meal break[.]" Barbatine Decl. 5.

[10] While not a formal policy, CHA's practice was uniform, effectively depriving workers of compensation for time worked beyond their shifts. As mentioned above, the FLSA is a strict-liability statute, meaning than an employer is obliged to ensure that an employee who could have but did not record all time on the job is properly compensated with a minimum wage and overtime for all time *actually* worked. *See* 29 U.S.C. §§ 206, 207; *see also* 29 U.S.C. § 255(a). Employers and employees do not have equal bargaining power. Employees, often fearing that they may lose their jobs if they do not oblige, regularly succumb to employer pressure to perform uncompensated labor.

Accordingly, Plaintiffs' motion for conditional certification (document # 15) is **GRANTED**, and Plaintiffs' proposed opt-in notice and form (document # 15–1) are approved.

III. CONCLUSION

For the foregoing reasons, CHA's Partial Motion to Dismiss (**document # 16**) is **DENIED IN PART** and **GRANTED IN PART**, Plaintiffs' Motion to Amend (**document # 29**) is **GRANTED**, and Plaintiffs' Motion for Conditional Certification of FLSA Claims (**document # 15**) is **GRANTED**.   ·

SO ORDERED.

ALL CITATIONS

814 F.Supp.2d 17, 161 Lab.Cas. P 35,941

13 National Labor Relations Act and Unfair Labor Practices

******

Chapter Objectives:

1. Acquire an understanding of the National Labor Relations Act and other labor laws.
2. Acquire an understanding of the National Labor Relations Board (NLRB).
3. Acquire an understanding of unfair labor practices.
4. Understand the correlation between safety and labor.

******

The current labor relations landscape was created by a number of different laws enacted during this turbulent period of beginning in the 1930s. The Norris–LaGuardia Act of 1932; the National Labor Relations Act (also known as the Wagner Act of 1935); the Labor Management Relations Act of 1947 (also known as the Taft–Hartley Act); and the Labor-Management Reporting and Disclosure Act of 1959 (also known as the Landrum–Griffin Act) were enacted during this time period. There were also a number of earlier laws, including the Sherman Antitrust Act of 1890, the Clayton Act of 1914, the Railroad Labor Act of 1926, and the National Industrial Recovery Act of 1933, which laid the groundwork and the framework for today's labor-management relations, and the collective bargaining process was established by the combination of the National Labor Relations Act, the Labor Management Relations Act, and the Labor Management Reporting and Disclosure Act. Although there is a myriad of other laws which impact the labor-management relations area in various ways today, including such laws as the Occupational Safety and Health Act and the Civil Rights Act, the basic foundation and framework of labor law in the United States remains in the Wagner Act, the Taft–Hartley Act and the Landrum–Griffin Act.

Safety and health professionals usually are not involved directly in the relationship between labor and management; however, a base level knowledge of the important aspects which impact the safety and health function is essential. A labor organization, also commonly called a union, can be the representative of the employees who choose or are required to become members for the purposes of collective bargaining or negotiating with the company or organization over mandatory subjects, including wages, hours, and conditions of employment. In addition to the federal labor laws, many states have passed "right-to-work" laws which outlaw compulsory union membership and the required dues paid by employees.

In general, and following the specific labor laws identified above, as well as the procedures established by the National Labor Relations Board, employees can form or join a labor organization. If elected, the labor organization becomes the employee's representative. The labor organization and the company or organization are required to negotiate to impasse over the wages, hours, and conditions of employment (which includes safety and health). These negotiations often lead to a collective bargaining agreement, or union contract, which contractually establishes the rules of the workplace. Members of the labor organization often pay a membership fee to join, and a periodic payment, usually referred to as dues, to maintain membership.

Safety and health professionals should be aware that there are different types of union security arrangements which affect the payment of dues by employees. First, safety and health professionals should be aware of the Taft–Hartley Act which made "closed shops" or workplaces requiring membership in the union prior to employment, illegal. However, several other types of union security arrangements are permitted to be included within a collective bargaining agreement. In a "union shop," the company or organization is free to hire whomever they choose; however, employees hired for positions covered by the collective bargaining agreement are required to join the labor organization after a specified probationary period. An "agency shop" is a union security arrangement where nonunion employees are required to pay the union monies equal to the union dues or fees as a condition of continuing employment with the company. The purpose of this payment of union dues, despite not being a union member, is to compensate the union for their collective bargaining work and the company's desire to make union membership voluntary. The most common and the least desirable form of union security arrangement is an "open shop." In an open shop, membership in the union is voluntary, and employees choosing not to belong to the union are not required to pay dues.

In very general terms, safety and health professionals should be aware of the process governed by the National Labor Relations Board through which employees can vote to elect or not to elect a labor organization to represent them in the workplace. This general process includes the following steps:

1. Employees contact the labor organization.
2. The labor organization files a petition with the National Labor Relations Board.
3. Bargaining unit determined by the NLRB.
4. Union authorization cards signed by employees. (Note: If over 50%, union can ask for voluntary recognition of the union without an election).
5. Company notified. *Excelsior* list of employees usually requested.
6. NLRB established "laboratory conditions" in the workplace.
7. Very specific rules during pre-election period.
8. Picketing, solicitation, boycotts, and campaigning.
9. Unfair labor practice charges filed by company and union.

10. NLRB hold the election.
11. If the labor organization wins the election, collective bargain negotiations begin.
12. If the company wins, the union is barred for a period of time.

The foundational element of which safety and health professionals should be aware in the Labor Management Relations Act and the Wagner Act are employees' Section 7 (of the Labor Management Relations Act) rights which provide protection for employees to form, join, or assist a labor organization; to bargain collectively through representatives of their choosing; and to engage in concerted activities for mutual aid and protection. Section 7 also provides that employees have the right to refrain from any or all such activities except to the extent that such right may be affected by an agreement requiring membership in a labor organization as a condition of employment. Most of the provisions of Section 7 of the Labor Management Relations Act are designed to protect the rights of the employee identified above.

Safety and health professionals should pay particular attention to the actions and/ or inactions which can be an unfair labor practice under Section 8 of the LMRA. Safety and health professionals, as agents for the company or organization, can be charged with an unfair labor practice against the company or organization if he/she should:

1. interfere with, restrain or coerce employees in the exercise of the rights guaranteed under Section 7;
2. dominate or interfere with the formation or administration of any labor organization or contribute financial or other support to it;
3. discriminate in regards to hire or tenure of employment or any term or condition of employment to encourage or discourage membership in any labor organization;
4. discharge or otherwise discriminate against an employee because he has fled charges or given testimony under the Act; and/or
5. refuse to bargain collectively with representatives of his employees.[1]

Safety and health professionals should be aware of the normal safety and health activities that can constitute an unfair labor practice. Even if the safety and health professional is acting in good faith and it is in the performance of a safety activity, the action or inaction can often place the safety and health professional is a position where an unfair labor practice charge may be filed by the employee or labor organization.

Safety and health professionals should be aware that many companies or organizations often go to great lengths to avoid unionization. Safety and health professionals, as agents of the company or organization, can easily be "caught in the middle" during a union organizing campaign with unfair labor practice charges being filed against the company or organization for his/her actions or inactions. Some of the common unfair labor practice charges files by employees or labor organizations include, but are not limited to: interrogation of employees; polling employees; investigative interviews (in violation of the *Weingarten* rule); threats, promises, and reprisals; granting of benefits; and spying. Some of the activities and statements which the NLRB could find to be in violation of Section 8(a) (1) include the following:

THREATS

- A safety and health professional threatens an employee with violence because of the employee's union activities.
- A safety and health professional tell an employee "the company will never sign a contact with that union."
- A safety and health professional tells an employee that the company knows who signed the authorization cards and will "get them."

INTERROGATION

- A safety and health professional asks an employee who is attending the union meeting tonight.
- A safety and health professional asks an employee how she feels about the union.
- A safety and health professional asks an employee how other employees feel about the union.

PROMISES

- A safety and health professional states that an employee will get a raise if she votes against the union.
- A safety and health professional tells an employee he will get a promotion if he votes against the union.

SPYING

- A safety and health professional implies surveillance when she tells an employee that he had a lot to say at the union meeting last evening.
- A safety and health professional visits an employee's home for the purposes of ascertaining her union support.

The above are only examples of possible violations of Section 8(a)(1) and should not be construed in any way as all-inclusive. Safety and health professionals with responsibilities within the areas of workers' compensation and security as well as safety and health should exercise caution and acquire legal guidance before embarking on any activities or conversations which could constitute an unfair labor practice.

Safety and health professionals should be aware that most private sector safety professionals fall within Section 2 (11) as "supervisors" and are considered management and thus outside of the collective bargaining unit. A "supervisor" means

> any individual having authority, in the interest of the employer, to hire, transfer, suspend, layoff, recall, promote, discharge, assign, reward, or discipline other employees, or responsible to direct them, or to adjust their grievances, or effectively to recommend

such action, if in connection with the foregoing exercise of such authority is not of a merely routine or clerical nature, but requires the use of independent judgment.[2]

Safety and health professionals should exercise caution whenever a union organizing campaign is underway. Guidance should always be sought from the human resource department of legal counsel. Generally, safety and health professionals should not look at any lists of employees provided by the labor organization, or look at any authorization cards or letters from the labor organization. This could constitute recognition of the labor organization. Additionally, safety and health professionals should not accept any registered mail or any written documents attempted to be handed to the safety professional from the labor representative.

A labor organizing campaign can be very disruptive to safety and health program efforts. Safety and health professionals should be aware that the labor organization must acquire signed authorization cards from 30% of the employees in the identified collective bargaining unit for the NLRD to order an election. The NLRB conducts a secret ballot election, usually onsite, and employees who signed the authorization cards are not bound by the card to vote for the union. If the majority of the employees vote for a union, the NLRB would recognize the union and order collective bargaining negotiations with the company or organization to begin. If the union does not receive a majority of the votes from the collective bargaining unit employees, the NLRB has the ability to recognize the union and order the company or organization to bargain with the union. However, before the NLRB can order bargaining, the union must show that the employer improperly denied the union's pre-election bargaining requests or the union lost its majority as a result of unfair labor practices by the company or organization. If the company or organization committed serious unfair labor practices, the NLRB can order the company or organization to bargain with the union if the union obtained a majority of the authorization cards without even holding a secret ballot election.

Given the visibility and functions of the safety and health professional within most organizations, the safety and health professional can often be a focus point for the filing of unfair labor practices by the labor organization. One of the first tactics often employed is to contact OSHA to initiate an inspection in order that any alleged violations could potentially be utilized during the labor organization's organizing campaign. The unfair labor practice charges, which usually are filed by both the labor organization and the company, play an important role in the organization process. If the labor organization can prove unfair labor practices by the company or organization, the NLRB can order the company to bargain with the union even if the union lost the election. Additionally, in the event of a strike or lockout situation, unfair labor practices play a key role in determining whether the strike was an economic strike or an unfair labor practice strike and thus whether the striking employees or replacement employees return to the jobs.

A labor organizing campaign can be a mine field for the safety and health professional. Safety and health programs, or lack thereof, may be one of the reasons why the employees initially contacted the labor organization. The organizing campaign can be very disruptive to the overall safety and health efforts within the company or organizations while the company or organization and the union battle for the hearts and minds of the voting employees. Safety and health professionals, as members of the management team, should exercise caution in their daily job activities and be extremely cautious in their interaction and communications with employees and the labor organization representatives. The safety and health professional's actions or inactions directly reflect on and impact the company or organization. Prudent safety and health professionals should always receive their direction and advice from their human resource or legal counsel when encountering a labor organizing campaign.

Safety and health professionals should be aware that the National Labor Relations Board is an independent federal agency vested with the power to safeguard employees' rights to organize and to determine whether to have unions as their bargaining representative. The agency also acts to prevent and remedy unfair labor practices committed by private sector employers and unions."[3] The National Labor Relations Board was established by the Wagner Act in 1935 to administer and enforce the provisions of the National Labor Relations Act.

The NLRB consists of three members serving five-year terms as well as a General Counsel, administrative law judges, an executive secretary, and regional directors. There are currently thirty-three regional NLRB offices and several sub-regional offices located throughout the United States. The Labor Management Relations Act bifurcated the authority of the NLRB between the board and the General Counsel. The NLRB is primarily a judicial body hearing unfair labor practice charges and representation matters; however, the NLRB does have limited investigative and prosecutorial authority. The Office of General Counsel has absolute authority over the issuance of unfair labor practice complaints. Section 6 of the NLRA provides the authority to the NLRB to issue rules and regulations related to the NLRA. Safety and health professionals should note that an actual "labor dispute" must exist before the NLRB possesses jurisdiction and thus can become involved in the situation. Unlike OSHA and other governmental agencies, the NLRB cannot become involved until there is a controversy involving the

> terms, tenure or conditions of employment, or concerning the association or representation of persons in negotiating, fixing, maintaining, changing or seeking to arrange terms or conditions of employment, regardless of whether the disputants stand in the proximate relation of employer and employee.[4]

In addition to a "labor dispute," the NLRB is required to identify that the labor dispute "affects commerce."[5]

The NLRB is responsible for conducting labor organization representation elections in a "fair and impartial manner."[6] To conduct this election, the NLRB as adopted the "laboratory conditions" standard which the NLRB imposes on the workplace and both the company or organization and the labor union. Laboratory conditions are usually imposed before and during the election on both the labor union and the company or organization. Safety and health professionals should be aware of when the NLRB's laboratory conditions are in effect, given the fact that if the laboratory conditions are violated, the NLRB can set aside the election and conduct it again. The NLRB usually evaluates the conduct of the labor union and company representatives from the time the petition is filed by the labor organization up to and including the election.[7]

The environment in which the safety and health professional functions in the workplace and the activities which the safety and health professional can perform are often substantially modified during the laboratory conditions leading up to an NLRB election. Safety and health professionals should take their lead and direction as to which activities or functions should or should not be performed during this period of time from their human resource department or legal counsel. Safety and health professionals should be aware that activities or functions, such as job safety observations, which may have constituted the "norm" prior to the organizing campaign can possibly constitute an unfair labor practice. Given the importance of safety and health in the workplace and the visibility of the safety and health professional in the operation, safety and health professionals should be prepared for program or personal challenges and be knowledgeable as to prescribed methods to address these challenges.

Given the laws and regulations of the NLRB in a labor organizing campaign, it is imperative that the safety and health professional seek guidance from their human resource department or legal counsel from the initial onset. Safety and health professionals may also want to research the most recent NLRB decisions on any particular issue which may arise on the NLRB website located at www.NLRB.gov. Safety and health professionals should be aware that the "rules of the road" in performing safety and health program activities may change dramatically during an organizing campaign leading to an NLRB election, and prudent safety and health professionals do not want to jeopardize their company's position with their workers through errors and missteps within the safety and health function.

Case Study: This case may be modified for the purposes of this text.

808 F.3d 1013
United States Court of Appeals,
Fifth Circuit.

MURPHY OIL USA, INCORPORATED, Petitioner/Cross-Respondent

v.

NATIONAL LABOR RELATIONS BOARD, Respondent/Cross-Petitioner.

No. 14–60800.

|

Oct. 26, 2015.

SYNOPSIS

Background: Employer filed petition for review of order of the National Labor Relations Board, 361 NLRB No. 72, 2014 WL 5465454, finding that it had unlawfully required employees to sign arbitration agreement waiving their right to pursue class and collective actions.

Holdings: The Court of Appeals, Leslie H. Southwick, Circuit Judge, held that:

[1] employer waived claims that employee's unfair labor practice charge was untimely and that Board was collaterally estopped from considering whether it was lawful to enforce arbitration agreement;

[2] employer committed no unfair labor practice by requiring employees to sign arbitration agreements;

[3] Board's refusal to follow Court of Appeals' decision did not warrant finding of contempt;

[4] employer's earlier arbitration agreement constituted unfair labor practice;

[5] employer's revised arbitration agreement did not constitute unfair labor practice; and

[6] employer did not engage in unfair labor practice by filing motion to dismiss and compel arbitration in employees' collective action.

Petition granted in part and denied in part.

OPINION

LESLIE H. SOUTHWICK, Circuit Judge:

The National Labor Relations Board concluded that Murphy Oil USA, Inc., had unlawfully required employees at its Alabama facility to sign an arbitration agreement waiving their right to pursue class and collective actions. Murphy Oil, aware that this circuit had already held to the contrary, used the broad venue rights governing the review of Board orders to file its petition with this circuit. The Board, also aware, moved for en banc review in order to allow arguments that the prior decision should be overturned. Having failed in that motion and having the case instead heard by a three-judge panel, the Board will not be surprised that we adhere, as we must, to our prior ruling. We GRANT Murphy Oil's petition, and hold that the corporation did not commit unfair labor practices by requiring employees to sign its arbitration agreement or seeking to enforce that agreement in federal district court.

We DENY Murphy Oil's petition insofar as the Board's order directed the corporation to clarify language in its arbitration agreement applicable to employees hired prior to March 2012 to ensure they understand they are not barred from filing charges with the Board.

FACTS AND PROCEDURAL BACKGROUND

Murphy Oil USA, Inc., operates retail gas stations in several states. Sheila Hobson, the charging party, began working for Murphy Oil at its Calera, Alabama facility in November 2008. She signed a "Binding Arbitration Agreement and Waiver of Jury Trial" (the "Arbitration Agreement"). The Arbitration Agreement provides that, "[e]xcluding claims which must, by ... law, be resolved in other forums, [Murphy Oil] and Individual agree to resolve any and all disputes or claims ... which relate ... to Individual's employment ... by binding arbitration." The Arbitration Agreement further requires employees to waive the right to pursue class or collective claims in an arbitral or judicial forum.

In June 2010, Hobson and three other employees filed a collective action against Murphy Oil in the United States District Court for the Northern District of Alabama alleging violations of the Fair Labor Standards Act ("FLSA"). Murphy Oil moved to dismiss the collective action and compel individual arbitration pursuant to the Arbitration Agreement. The employees opposed the motion, contending that the FLSA prevented enforcement of the Arbitration Agreement because that statute grants a substantive right to collective action that cannot be waived. The employees also argued that the Arbitration Agreement interfered with their right under the National Labor Relations Act *1016 ("NLRA") to engage in Section 7 protected concerted activity.

While Murphy Oil's motion to dismiss was pending, Hobson filed an unfair labor charge with the Board in January 2011 based on the claim that the Arbitration Agreement interfered with her Section 7 rights under the NLRA. The General Counsel for the Board issued a complaint and notice of hearing to Murphy Oil in March 2011.

In a separate case of first impression, the Board held in January 2012 that an employer violates Section 8(a)(1) of the NLRA by requiring employees to sign an arbitration agreement waiving their right to pursue class and collective claims in all forums. *D.R. Horton, Inc.*, 357 N.L.R.B. 184 (2012). The Board concluded that such agreements restrict employees' Section 7 right to engage in protected concerted activity in violation of Section 8(a)(1). *Id.* The Board also held that employees could reasonably construe the language in the *D.R. Horton* arbitration agreement to preclude employees from filing an unfair labor practice charge, which also violates Section 8(a)(1). *Id.* at *2, *18.

Following the Board's decision in *D.R. Horton*, Murphy Oil implemented a "Revised Arbitration Agreement" for all employees hired after March 2012. The revision provided that employees were not barred from "participating in proceedings to adjudicate unfair labor practice[ ] charges before the" Board. Because Hobson and the other employees involved in the Alabama lawsuit were hired before March 2012, the revision did not apply to them.

In September 2012, the Alabama district court stayed the FLSA collective action and compelled the employees to submit their claims to arbitration pursuant to the Arbitration Agreement.* One month later, the General Counsel amended the complaint before the Board stemming from Hobson's charge to allege that Murphy Oil's motion to dismiss and compel arbitration in the Alabama lawsuit violated Section 8(a)(1) of the NLRA.

Meanwhile, the petition for review of the Board's decision in *D.R. Horton* was making its way to this court. In December 2013, we rejected the Board's analysis

---

* The employees never submitted their claims to arbitration. In February 2015, the employees moved for reconsideration of the Alabama district court's order compelling arbitration. The district court denied their motion and ordered the employees to show cause why their case should not be dismissed with prejudice for failing to adhere to the court's order compelling arbitration. The district court ultimately dismissed the case with prejudice for "willful disregard" of its instructions in order to "gain [a] strategic advantage." *Hobson v. Murphy Oil USA, Inc.*, No. CV–10–S–1486–S, 2015 WL 4111661, at *3 (N.D.Ala. July 8, 2015), *appeal docketed*, No. 15–13507 (11th Cir. Aug. 5, 2015). The employees timely appealed. The case is pending before the Eleventh Circuit.

of arbitration agreements. *D.R. Horton, Inc. v. NLRB*, 737 F.3d 344 (5th Cir.2013). We held: (1) the NLRA does not contain a "congressional command overriding" the Federal Arbitration Act ("FAA");* and (2) "use of class action procedures ... is not a substantive right" under Section 7 of the NLRA. *Id.* at 357, 360–62. This holding means an employer does not engage in unfair labor practices by maintaining and enforcing an arbitration agreement prohibiting employee class or collective actions and requiring employment-related claims to be resolved through individual arbitration. *Id.* at 362.

In analyzing the specific arbitration agreement at issue in *D.R. Horton*, however, we held that its language could be "misconstrued" as prohibiting employees from filing an unfair labor practice charge, ***1017** which would violate Section 8(a)(1). *Id.* at 364. We enforced the Board's order requiring the employer to clarify the agreement. *Id.* The Board petitioned for rehearing en banc, which was denied without a poll in April 2014.

The Board's decision as to Murphy Oil was issued in October 2014, ten months after our initial *D.R. Horton* decision and six months after rehearing was denied. The Board, unpersuaded by our analysis, reaffirmed its *D.R. Horton* decision. It held that Murphy Oil violated Section 8(a)(1) by "requiring its employees to agree to resolve all employment-related claims through individual arbitration, and by taking steps to enforce the unlawful agreements in [f]ederal district court." The Board also held that both the Arbitration Agreement and Revised Arbitration Agreement were unlawful because employees would reasonably construe them to prohibit filing Board charges.

The Board ordered numerous remedies. Murphy Oil was required to rescind or revise the Arbitration and Revised Arbitration agreements, send notification of the rescission or revision to signatories and to the Alabama district court, post a notice regarding the violation at its facilities, reimburse the employees' attorneys' fees incurred in opposing the company's motion to dismiss and compel arbitration in the Alabama litigation, and file a sworn declaration outlining the steps it had taken to comply with the Board order.

Murphy Oil timely petitioned this court for review of the Board decision.

DISCUSSION

[1] [2] Board decisions that are "reasonable and supported by substantial evidence on the record considered as a whole" are upheld. *Strand Theatre of Shreveport Corp. v. NLRB*, 493 F.3d 515, 518 (5th Cir.2007) (citation and quotation marks omitted); *see also* 29 U.S.C. § 160(e). "Substantial evidence is such relevant evidence as a reasonable mind would accept to support a conclusion." *J. Vallery Elec., Inc. v. NLRB*, 337 F.3d 446, 450 (5th Cir.2003) (citation and quotation marks omitted). This court reviews the Board's legal conclusions *de novo,* but "[w]e will enforce the Board's order if its construction of the statute is reasonably defensible." *Strand Theatre*, 493 F.3d at 518 (citation and quotation marks omitted).

---

* 9 U.S.C. § 1 *et seq.*

I. Statute of Limitations and Collateral Estoppel

Murphy Oil asserts that Hobson filed her charge too late after the execution of the Arbitration Agreement and the submission of Murphy Oil's motion to compel in the Alabama litigation. By statute, "no complaint shall issue based upon any unfair labor practice occurring more than six months prior to the filing of the charge with the Board." 29 U.S.C. § 160(b). Murphy Oil also contends that the Board is collaterally estopped from considering whether it was lawful to enforce the Arbitration Agreement because the district court had already decided that issue in the Alabama litigation.

[3] Both of these arguments were raised in Murphy Oil's answer to the Board's complaint. They were not, though, discussed in its brief before the Board. "No objection that has not been urged before the Board ... shall be considered by the court ... ." 29 U.S.C. § 160(e), (f). Similarly, we have held that "[a]ppellate preservation principles apply equally to petitions for enforcement or review of NLRB decisions." *NLRB v. Catalytic Indus. Maint. Co. (CIMCO)*, 964 F.2d 513, 521 (5th Cir.1992). While Murphy Oil may have properly pled its statute of limitations and collateral estoppel defenses, it did not sufficiently press those arguments ***1018** before the Board. Thus, they are waived. *See* 29 U.S.C. § 160(e), (f).

II. D.R. Horton and Board Nonacquiescence

[4] The Board, reaffirming its *D.R. Horton* analysis, held that Murphy Oil violated Section 8(a)(1) of the NLRA by enforcing agreements that "requir[ed] ... employees to agree to resolve all employment-related claims through individual arbitration." In doing so, of course, the Board disregarded this court's contrary *D.R. Horton* ruling that such arbitration agreements are enforceable and not unlawful. *D.R. Horton*, 737 F.3d at 362.* Our decision was issued not quite two years ago; we will not repeat its analysis here. Murphy Oil committed no unfair labor practice by requiring employees to relinquish their right to pursue class or collective claims in all forums by signing the arbitration agreements at issue here. *See id.*

[5] Murphy Oil argues that the Board's explicit "defiance" of *D.R. Horton* warrants issuing a writ or holding the Board in contempt so as to "restrain [it] from continuing its nonacquiescence practice with respect to this [c]ourt's directive." The Board, as far as we know, has not failed to apply our ruling in *D.R. Horton* to the parties in that case. The concern here is the application of *D.R. Horton* to new parties and agreements.

An administrative agency's need to acquiesce to an earlier circuit court decision when deciding similar issues in later cases will be affected by whether the new

---

* Several of our sister circuits have either indicated or expressly stated that they would agree with our holding in *D.R. Horton* if faced with the same question: whether an employer's maintenance and enforcement of a class or collective action waiver in an arbitration agreement violates the NLRA. *See Walthour v. Chipio Windshield Repair, LLC*, 745 F.3d 1326, 1336 (11th Cir.2014), *cert. denied*, ---- U.S. ----, 134 S.Ct. 2886, 189 L.Ed.2d 836 (2014); *Richards v. Ernst & Young, LLP*, 744 F.3d 1072, 1075 n. 3 (9th Cir.2013), *cert. denied*, ---- U.S. ----, 135 S.Ct. 355, 190 L.Ed.2d 249 (2014); *Owen v. Bristol Care, Inc.*, 702 F.3d 1050, 1053–55 (8th Cir.2013); *Sutherland v. Ernst & Young LLP*, 726 F.3d 290, 297 n. 8 (2d Cir.2013).

decision will be reviewed in that same circuit. *See* Samuel Estreicher & Richard L. Revesz, Nonacquiescence by Federal Administrative Agencies, 98 *YALE L.J.* 679, 735–43 (1989). Murphy Oil could have sought review in (1) the circuit where the unfair labor practice allegedly took place, (2) any circuit in which Murphy Oil transacts business, or (3) the United States Court of Appeals for the District of Columbia. 29 U.S.C. § 160(f). The Board may well not know which circuit's law will be applied on a petition for review. We do not celebrate the Board's failure to follow our *D.R. Horton* reasoning, but neither do we condemn its nonacquiescence.

III. The Agreements and NLRA Section 8(a)(1)

The Board also held that Murphy Oil's enforcement of the Arbitration Agreement and Revised Arbitration Agreement violated Section 8(a)(1) of the NLRA because employees could reasonably believe the contracts precluded the filing of Board charges. Hobson and the other employees involved in the Alabama litigation were subject to the Arbitration Agreement applicable to employees hired before March 2012. The Revised Arbitration Agreement contains language that sought to correct the possible ambiguity.

A. The Arbitration Agreement in Effect Before March 2012

Section 8(a) of the NLRA makes it unlawful for an employer to commit unfair labor practices. 29 U.S.C. § 158(a). For ***1019** example, an employer is prohibited from interfering with employees' exercise of their Section 7 rights. *Id.* § 158(a)(1). Under Section 7, employees have the right to self-organize and "engage in other concerted activities for the purpose of collective bargaining or other mutual aid or protection." *Id.* § 157.

[6] The Board is empowered to prevent unfair labor practices. This power cannot be limited by an agreement between employees and the employer. *See id.* § 160(a). "Wherever private contracts conflict with [the Board's] functions, they ... must yield or the [NLRA] would be reduced to a futility." *J.I. Case Co. v. NLRB*, 321 U.S. 332, 337, 64 S.Ct. 576, 88 L.Ed. 762 (1944). Accordingly, as we held in *D.R. Horton*, an arbitration agreement violates the NLRA if employees would reasonably construe it as prohibiting filing unfair labor practice charges with the Board. 737 F.3d at 363.

[7] Murphy Oil argues that Hobson's choice to file a charge with the Board proves that the pre-March 2012 Arbitration Agreement did not state or suggest such charges could not be filed. The argument misconstrues the question. "[T]he actual practice of employees is not determinative" of whether an employer has committed an unfair labor practice. *See Flex Frac Logistics, L.L.C. v. NLRB*, 746 F.3d 205, 209 (5th Cir.2014). The Board has said that the test is whether the employer action is "likely to have a chilling effect" on employees' exercise of their rights. *Id.* (citing *Lafayette Park Hotel*, 326 N.L.R.B. 824, 825 (1998)). The possibility that employees will misunderstand their rights was a reason we upheld the Board's rejection of a similar provision of the arbitration agreement in *D.R. Horton*. We explained that the FAA and NLRA have "equal importance in our review" of employment arbitration contracts. *D.R. Horton*, 737 F.3d. at 357. We held that even though requiring

arbitration of class or collective claims in all forums does not "deny a party any statutory right," an agreement reasonably interpreted as prohibiting the filing of unfair labor charges would unlawfully deny employees their rights under the NLRA. *Id.* at 357–58, 363–64.

[8] Murphy Oil's Arbitration Agreement provided that "any and all disputes or claims [employees] may have ... which relate in any manner ... to ... employment" must be resolved by individual arbitration. Signatories further "waive their right to ... be a party to any group, class or collective action claim in ... any other forum." The problem is that broad "any claims" language can create "[t]he reasonable impression ... that an employee is waiving not just [her] trial rights, but [her] administrative rights as well." *D.R. Horton*, 737 F.3d at 363–64 (citing *Bill's Electric, Inc.*, 350 N.L.R.B. 292, 295–96 (2007)).

We do not hold that an express statement must be made that an employee's right to file Board charges remains intact before an employment arbitration agreement is lawful. Such a provision would assist, though, if incompatible or confusing language appears in the contract. *See id.* at 364.

We conclude that the Arbitration Agreement in effect for employees hired before March 2012, including Hobson and the others involved in the Alabama case, violates the NLRA. The Board's order that Murphy Oil take corrective action as to any employees that remain subject to that version of the contract is valid.

B. The Revised Arbitration Agreement in Effect After March 2012

In March 2012, following the Board's decision in *D.R. Horton*, Murphy Oil added the following clause in the Revised Arbitration Agreement: "[N]othing in this ***1020** Agreement precludes [employees] ... from participating in proceedings to adjudicate unfair labor practice[ ] charges before the [Board]." The Board contends that Murphy Oil's modification is also unlawful because it "leaves intact the entirety of the original Agreement," including employees' waiver of their right "to commence or be a party to any group, class or collective action claim in ... any other forum." This provision, the Board said, could be reasonably interpreted as prohibiting employees from pursuing an administrative remedy "since such a claim could be construed as having 'commence[d]' a class action in the event that the [Board] decides to seek classwide relief."

[9] We disagree with the Board. Reading the Murphy Oil contract as a whole, it would be unreasonable for an employee to construe the Revised Arbitration Agreement as prohibiting the filing of Board charges when the agreement says the opposite. The other clauses of the agreement do not negate that language. We decline to enforce the Board's order as to the Revised Arbitration Agreement.

IV. Murphy Oil's Motion to Dismiss and NLRA Section 8(a)(1)

Finally, the Board held that Murphy Oil violated Section 8(a)(1) by filing its motion to dismiss and compel arbitration in the Alabama litigation. As noted above, Section 8(a) prohibits employers from engaging in unfair labor practices. 29 U.S.C. § 158(a). Section 8(a)(1) provides that an employer commits an unfair labor practice by "interfer[ing] with, restrain[ing], or coerc[ing] employees in the exercise" of their Section 7 rights, including engaging in protected concerted activity. *Id.* §§ 157, 158(a)(1).

[10] The Board said that in filing its dispositive motion and "eight separate court pleadings and related [documents] ... between September 2010 and February 2012," Murphy Oil "acted with an illegal objective [in] ... 'seeking to enforce an unlawful contract provision'" that would chill employees' Section 7 rights, and awarded attorneys' fees and expenses incurred in "opposing the ... unlawful motion." We disagree and decline to enforce the fees award.

The Board rooted its analysis in part in *Bill Johnson's Restaurants, Inc. v. NLRB*, 461 U.S. 731, 103 S.Ct. 2161, 76 L.Ed.2d 277 (1983). That decision discussed the balance between an employer's First Amendment right to litigate and an employee's Section 7 right to engage in concerted activity. In that case, a waitress filed a charge with the Board after a restaurant terminated her employment; she believed she was fired because she attempted to organize a union. *Id.* at 733, 103 S.Ct. 2161. After the Board's General Counsel issued a complaint, the waitress and several others picketed the restaurant, handing out leaflets and asking customers to boycott eating there. *Id.* In response, the restaurant filed a lawsuit in state court against the demonstrators alleging that they had blocked access to the restaurant, created a threat to public safety, and made libelous statements about the business and its management. *Id.* at 734, 103 S.Ct. 2161. The waitress filed a second charge with the Board alleging that the restaurant initiated the civil suit in retaliation for employees' engaging in Section 7 protected concerted activity, which violated Section 8(a)(1) and (4) of the NLRA. *Id.* at 734–35, 103 S.Ct. 2161.

The Board held that the restaurant's lawsuit constituted an unfair labor practice because it was filed for the purpose of discouraging employees from seeking relief with the Board. *Id.* at 735–37, 103 S.Ct. 2161. The Supreme Court remanded the case for further consideration, stating: "The right to litigate is an important one," *1021 but it can be "used by an employer as a powerful instrument of coercion or retaliation." *Id.* at 740, 744, 103 S.Ct. 2161. To be enjoinable, the Court said the lawsuit prosecuted by the employer must (1) be "baseless" or "lack[ing] a reasonable basis in fact or law," and be filed "with the intent of retaliating against an employee for the exercise of rights protected by" Section 7, or (2) have "an objective that is illegal under federal law." *Id.* at 737 n. 5, 744, 748, 103 S.Ct. 2161.

We start by distinguishing this dispute from that in *Bill Johnson's*. The current controversy began when three Murphy Oil employees filed suit in Alabama. Murphy Oil defended itself against the employees' claims by seeking to enforce the Arbitration Agreement. Murphy Oil was not retaliating as *Bill Johnson's* may have been. Moreover, the Board's holding is based solely on Murphy Oil's enforcement of an agreement that the Board deemed unlawful because it required employees to individually arbitrate employment-related disputes. Our decision in *D.R. Horton* forecloses that argument in this circuit. 737 F.3d at 362. Though the Board might not need to acquiesce in our decisions, it is a bit bold for it to hold that an employer who followed the reasoning of our *D.R. Horton* decision had no basis in fact or law or an "illegal objective" in doing so. The Board might want to strike a more respectful balance between its views and those of circuit courts reviewing its orders.

Moreover, the timing of Murphy Oil's motion to dismiss when compared to the timing of the *D.R. Horton* decisions counsels against finding a violation of Section 8(a)(1). The relevant timeline of events is as follows:

1. July 2010: Murphy Oil filed its motion to dismiss and sought to compel arbitration in the Alabama litigation;
2. January 2012: the Board in *D.R. Horton* held it to be unlawful to require employees to arbitrate employment-related claims individually, and the *D.R. Horton* agreement violated the NLRA because it could be reasonably construed as prohibiting the filing of Board charges;
3. October 2012: the Board's General Counsel amended the complaint against Murphy Oil to allege that Murphy Oil's motion in the Alabama litigation violated Section 8(a)(1); and
4. December 2013: this court granted *D.R. Horton's* petition for review of the Board's order and held that agreements requiring individual arbitration of employment-related claims are lawful but that the specific agreement was unlawful because it could be reasonably interpreted as prohibiting the filing of Board charges.

In summary, Murphy Oil's motion was filed a year and a half before the Board had even spoken on the lawfulness of such agreements in light of the NLRA. This court later held that such agreements were generally lawful. Murphy Oil had at least a colorable argument that the Arbitration Agreement was valid when its defensive motion was made, as its response to the lawsuit was not "lack[ing] a reasonable basis in fact or law," and was not filed with an illegal objective under federal law. *See Bill Johnson's*, 461 U.S. at 737 n. 5, 744, 748, 103 S.Ct. 2161. Murphy Oil's motion to dismiss and compel arbitration did not constitute an unfair labor practice because it was not "baseless." We decline to enforce the Board's order awarding attorneys' fees and expenses.
* * *

The Board's order that Section 8(a)(1) has been violated because an employee would reasonably interpret the Arbitration Agreement in effect for employees hired before March 2012 as prohibiting the filing of an unfair labor practice charge is ENFORCED. ***1022** Murphy Oil's petition for review of the Board's decision is otherwise GRANTED.

ALL CITATIONS

808 F.3d 1013, 204 L.R.R.M. (BNA) 3489, 166 Lab.Cas. P 10,823

NOTES

1. NLRA Section 8(a).
2. NLRA Section 2(11).
3. NLRB website located at www.nlrb.gov.
4. 29 U.S.C.A. § 152(9).
5. 29 U.S.C.A. § 159 (c).
6. *Collins & Aikman Corp. v. NLRB*, 383 F.2d 722 (CA-4, 1967).
7. *NLRB v. Blades Mfg. Co.*, 344 F.2d 998 (CA-8, 1965).

14 State Workers' Compensation Laws

******

Chapter Objectives:

1. Acquire a general understanding of the concept of workers' compensation.
2. Acquire knowledge of the general benefits to injured workers.
3. Acquire an understanding of the structure of state workers' compensation laws.
4. Identify the correlation between the safety and health function and workers' compensation.

******

With many organizations, the workers' compensation function and the safety and health function are thought to be intertwined and synonymous when, in fact, these functions are separate and distinct functions and, to a great extent, polar opposites. The safety and health function is proactive, while the workers' compensation function is reactive. The safety and health function is primarily governed by federal laws and regulations, while the workers' compensation function is solely governed by individual state laws and regulations. However, in the eyes of many organizations, the injury, and the correlating workers' compensation costs, equate to the lack of prevention by the safety and health function and thus, the two functions are often housed within the same organizational unit or department. Additionally, it is far easier to track the reactive workers' compensation costs than to track the preventative savings within the proactive safety and health function. In essence, the thought process is: no injury or illness equals no workers' compensation costs.

The workers' compensation system is individual to each state and governed by the laws and regulation established by each state. Although the framework is generally the same, providing monetary payment for time loss, medical costs, and permanency of the injury or illness, the rates vary substantially among and between the states. Workers' compensation coverage is required in all states and generally companies or organizations acquire coverage from private sector insurance companies, state-operated insurance entities or, for large corporations, self-insurance. For larger companies electing self-insurance, a third party often manages the payments and functions and states usually require a substantial bond to ensure payment to injured workers. Companies provide payment to private sector insurance companies or state-operated insurance entities based upon their injury and illness record ("MOD rate") and periodic payments, not unlike your auto or home insurance, are provided to the insurer. The more injuries or illnesses incurred, the higher the workers' compensation coverage costs to the organization.

Safety and health professionals should be aware that under most state workers' compensation systems, the safety and health professional will have little or no control over the workers' compensation claim once filed. State workers' compensation systems usually are developed provide immediate payment to injured workers (referred to as "claimants") within the prescribed regulations and the attending physician determines if/when the injured employee can return to work and the percentage of loss the worker incurred. Most states possess prescribed forms to be completed and the amounts received often have a formula with a minimum and maximum amount. For employers, the benefit provided under virtually all workers' compensation systems is that the employee waives his/her right to sue the employer for the claim in a court of law (except in willful situations in some states).

With virtually all state workers' compensation system, there are prescribed state laws and regulations governing the system. All systems require the injury or illness to have occurred "arising out of or in the course of employment," or a showing or proof that the injury or illness was incurred in the course of the worker's employment. Given the fact that state laws can vary, it is important for safety and health professionals to be familiar with these state laws and regulations. If the claim can be proven not to be work-related, denial of the claim is possible. However, in most states, the laws and regulations are liberally interpreted or construed in favor of the claimant and thus payment of benefits. Safety and health professionals should be aware that denial of a claim in most states requires substantial proof which is often difficult to acquire.

Most state workers' compensation systems provide the injured worker with a percentage of the employee's wages (+/- 66.75% with a minimum and maximum amount based upon the employee's rate of pay), provides payment in full for all medical costs (no deductibles), and if a permanent disability resulted, a prescribed percentage of the loss of the use of the function. Once a worker files the claim for benefits and benefits are initiated, the worker usually has a short waiting period (3–7 days) before receiving time loss benefits, and payment of medical costs start immediately. In most states, the injured worker would continue to receive time loss benefits until such time as his/her attending physician identifies that the injured worker can return to restricted duty or has reached maximum medical recovery. In many states, this period of time in which benefits can be received extends seven years or longer before reaching the maximum benefit cap.

Safety and health professionals should be aware that virtually all states employ a specific internal administrative structure to hear disputes before permitting pursuing an action within the state court systems. In many state systems, disputes resolution is often attempted through an initial conference followed by a more formal administrative hearing before a state appointed administrative law judge (commonly known as an "ALJ"). Safety and health professionals should be aware that some states provide a final arbitrator or commission as the final step for appeals, while others permit appeal to the state court system. If the safety and health professionals have responsibility for the workers' compensation function, it is imperative that the safety and health professional acquires or already possesses a firm grasp not only on the law and regulations, but also the appeal process.

For many safety and health professionals, one of the major challenges is the education of your management team as to the requirements of your state's workers' compensation system. On many occasions, especially with strains, sprains, and soft tissue injuries, management team members will identify the injury as not "arising out of or in the course of employment" (i.e., it did not happen on the job) and want the safety and health professional to deny the worker's claim. In most states, the workers' compensation laws and regulations are liberally construed in favor of the claimant and, additionally, acquisition of the appropriate evidence to dispute the claim of injury is often non-existent. Safety and health professionals should be cognizant that most claims for workers' compensation claims are upheld and denial of a claim without substantial evidence that the injured workers was not your employee or the injury occurred outside of the workplace would be difficult.

Third party claims is an area of which safety and health professionals should be aware. With employers immune from tort-related actions, injured workers can pursue an action against a third party who may be involved in the injury or illness. These actions, such as a malpractice action against a physician or product liability action, permits the injured worker to bring an action in civil court against the third party. Safety and health professionals should be aware that their company or organization may be brought into the third party action and, depending on the state's subrogation laws, may permit recovery of workers' compensation costs if the injured worker is successful.

An area of the workers' compensation function which creates many issues for safety and health professionals is the issue of restricted or "light duty" programs. Conceptually, a company or organization possesses little or no control over the workers' compensation claim once filed. Time loss benefits and medical costs are fixed by statute. However, some companies and organizations have attempted to reduce or shift the cost of an injured worker's time loss benefits through working with the attending physician to acquire permission for the injured worker to return to work with specific medical restrictions until such time as the employee is fully recovered and can return to full duty. The injured worker would return to work and be paid their normal rate of pay and be provided restricted activities meeting with the approval of the physician. The positive aspects of a restricted duty program is the reduction of time loss benefits, maintaining the injured worker's daily routine, and achieving some productivity during the time of recovery. The negative aspects of a restricted duty program is the interference in the relationship between the physician and injured worker, impact on other workers, maintaining the injured worker's work activities within the restrictions, and the injured worker's attitude toward returning to restricted duty. The management of a restricted duty program can be difficult, time consuming, and tenuous. The relationship between the physician, injured worker, and safety and health professional is fragile and requires constant attention. Many companies and organizations have adopted the policy that injured workers remain on time loss benefits until such time as they are fully and completely released by their attending physician before returning to the workplace.

Of particular importance for safety and health professionals with workers' compensation responsibilities is the fact that once an employee is injured and workers' compensation has been established, other laws, such as the ADA, may come into play within the management of the claim. Additionally, issues regarding the confidentiality of medical records may require additional infrastructure and protocols when viewing or discussing workers' compensation claims and correlating medical information. OSHA has also addressed the issue of medical information and employee's rights in viewing specific information, and FMLA issues have an impact when an injured worker possesses a workers' compensation claim and is acquiring benefits. In short, just as the safety and health function does not work in a vacuum, the workers' compensation function has the potential to be impacted by a number of different laws and regulations.

The management of the safety and health function and the management of the workers' compensation function are substantially different. Additionally, in situations where the safety and health professionals also possess workers' compensation responsibilities, most safety and health professionals have found that the workers' compensation function requires far more of their time and efforts and the safety and health function is often neglected. This can create a vicious circle of more workers' compensation claims, because the risk and hazards have increased due to lack of time. Companies and organizations who understand the significant differences between the proactive safety and health function and the reactive workers' compensation function have separated these functions and thus permit each to achieve the full attention of the professionals. However, if you are a safety and health professional with workers' compensation as well as safety and health responsibilities, it is imperative that you acquire and become functionally knowledgeable regarding the workers' compensation laws and regulations for each state in which you have operations, and acquire appropriate guidance from your workers' compensation administrators and legal counsel, as necessary.

Case Study: This case may be modified for the purposes of this text.

UNITED STATES DISTRICT COURT

EASTERN DISTRICT OF KENTUCKY

NORTHERN DIVISION AT COVINGTON

CIVIL ACTION NO. 16-129-DLB-CJS

JOHN SCHWEITZER PLAINTIFF vs. WAL-MART STORES, INC. DEFENDANT

MEMORANDUM AND ORDER

1.   INTRODUCTION: This matter is before the Court upon Defendant Wal-Mart Stores, Inc.'s (Wal-Mart) motion for summary judgment (Doc. # 12). Plaintiff having responded to the motion (Doc. # 15), and Wal-Mart having filed its reply (Doc. # 18), the motion is ripe for review. For the reasons stated

below, the Court finds the motion to be well-taken, and will grant summary judgment in favor of Wal-Mart. II. FACTUAL AND PROCEDURAL BACKGROUND Plaintiff John Schweitzer worked for Advantage Sales and Marketing ("ASM"), representing sales of GlaxoSmithKline products to retail stores. (Doc. # 12-2 at 6). His job required him to go into retailers, such as Wal-Mart or Kroger, and ensure that the brands he represented were properly tagged on the shelves and were stocked on shelves in an appealing manner. Id. at 6–7. In addition, Plaintiff would construct displays for his products when necessary, and talk to department heads about brand volume, targeted advertising, and freshness. Id. at 7. Plaintiff indicated in his deposition that Wal-Mart's two employees engaged in the same type of work Plaintiff did, for different products. Id. at 7. The Wal-Mart in Maysville stocked approximately 150 of the brands Plaintiff represented, including Tums, Nicorette, and Abreva. Id. at 6. Plaintiff's work for ASM took him to that store once per month from 2002 until early 2015. Id. at 18. On March 2, 2015, Plaintiff arrived at the Wal-Mart in Maysville at about 9:00 a.m., coming from a previous stop at the Maysville Kroger. Id. at 18. In a time period spanning approximately three minutes, Plaintiff exited his car, started walking across the parking lot, and slipped on a patch of ice 15–20 feet from his car. Id. at 18, 21–22. As a result, Plaintiff filed a workers' compensation claim, indicating that he was injured doing work for ASM at Wal-Mart. Id. at 10; see also Doc. # 18-1 at 1 ("Describe how the injury occurred: Walking into Wal-Mart in Maysville, KY to deliver products, slipped and fell on ice injuring arm.") On February 29, 2016, Plaintiff filed a complaint against Wal-Mart in the Mason County Circuit Court, alleging that as the result of his fall, he has sustained a permanent and serious injury to his right shoulder. (Doc. # 1–3, at 1–2). Plaintiff alleges three counts against Wal-Mart: (1) negligent inspection and maintenance; (2) premises liability; and (3) negligent hiring and supervision. Id. at 2–3. Specifically, Plaintiff seeks compensatory damages and claims that as a result of Wal-Mart's negligence and recklessness, he has suffered permanent injuries, resulting in past, present, and future medical expenses, lost wages, and great mental and physical pain and suffering. Id. at 3–4. On August 2, 2017, Wal-Mart filed a Motion for Summary Judgment, arguing that Plaintiff had already been compensated for his injuries through the Kentucky Workers' Compensation program, and as an up-the-ladder employer, Wal-Mart was immune from Case: 2:16-cv-00129-DLB-CJS Doc #: 23 Filed: 12/20/17 Page: 2 of 10 – Page ID#: 3 tort liability for injuries incurred on its premises and subsequently compensated under a workers' compensation claim. (Docs. # 12 and 12-1). Plaintiff responded in opposition (Doc. # 15), to which Wal-Mart replied. (Doc. # 18).

2.   ANALYSIS A. Standard of Review Summary judgment is appropriate when "there is no genuine dispute as to any material fact and the movant is entitled to judgment as a matter of law." Fed. R. Civ. P. 56(a). The moving party has the initial burden of "showing the absence of any genuine issues of material fact." *Sigler v. Am. Honda Motor Co.*, 532 F.3d 469, 483 (6th

Cir. 2008). Once the moving party has met its burden, the nonmoving party must cite to evidence in the record upon which "a reasonable jury could return a verdict" in its favor; a mere "scintilla of evidence" will not do. *Anderson v. Liberty Lobby, Inc.*, 477 U.S. 242, 248–52 (1986). At the summary judgment stage, a court "views the evidence in the light most favorable to the nonmoving party and draws all reasonable inferences in that party's favor." *Slusher v. Carson*, 540 F.3d 449, 453 (6th Cir. 2008). B. Kentucky law provides up-the-ladder immunity for contractors. "The Kentucky Workers' Compensation Act is a legislative remedy which affords an injured worker a remedy without proof of the common law elements of fault." *General Electric Co. v. Cain*, 236 S.W.3d 579, 606 (Ky. 2007). On balance, two sections of this Act, when read in conjunction, give a premises owner immunity from tort liability with respect to tort-related injuries so long as the premises owner had workers' compensation coverage and the worker was injured performing work of the type that was a regular or recurring part of the premises owner's business. Cain, 236 S.W.3d at 585 (discussing Case: 2:16-cv-00129-DLB-CJS Doc #: 23 Filed: 12/20/17 Page: 3 of 10 – Page ID#: 4 Ky. Rev. Stat. Ann. §§ 342.690(1) and 342.610(2)). The premises owner asserting this up-the-ladder defense has the burden of both pleading and proving the affirmative defense. Id. "Regular or recurring" has been interpreted to mean that the type of work is performed as part of the usual, normal, or customary part of the particular business, which one would expect the employees to normally perform, and that it is repeated "with some degree of regularity." Id. at 588. The test to determine whether the type of work is that which would be normally performed by the business is a relative test, not an absolute one. Id. And proof of purchase of a workers' compensation policy, absent evidence that the policy was deficient, is a sufficient showing to "invoke the exclusive remedy provision 5 Wal-Mart has satisfied each of the three prongs of the "regular or recurring" inquiry. Plaintiff was hired to perform the work at Wal-Mart's Maysville store, among others. (Doc. # 12-2 at 6). Wal-Mart has shown that Plaintiff performed the same or similar work once per month for over a decade. Id. at 18. And Plaintiff testified that Wal-Mart's employees performed similar work as Plaintiff, but for different products. Id. at 7. Therefore, Wal-Mart has met its burden of showing that the work Plaintiff did was "regular or recurring." Similarly, in *Mueller v. 84 Lumber Co.*, No. 3:15-cv-838-JHM, 2016 WL 5868087 (W.D. Ky. Oct. 6, 2016), the court found that 84 Lumber had met each of these prongs in granting summary judgment on an action brought by the plaintiff for an injury incurred while on 84 Lumber's premises. The plaintiff was employed by a company contracted by 84 Lumber to deliver lumber and other goods to 84 Lumber's customers on a regular basis, and testimony showed that this type of work was provided by 84 Lumber through contracted delivery drivers (such as the plaintiff), or through 84 Lumber's own employees. *Mueller*, 2016 WL 5868087 at *2–3. Thus, 84 Lumber was an up-the-ladder employer and immune from liability for the plaintiff's injury. In another case, the Sixth Circuit found that that the employee of a contractor who had been injured while he was mowing grass was doing "regular and recurring work" when he was injured. *Himes v. United States*, 645 F.3d 771, 781–82 (6th Cir. 2011) (citing *Cain*, 236 S.W.3d at 588) ("[Plaintiff has] failed to come forward with any specific facts to dispute the facts that mowing by [the plaintiff] at the [Blue Grass Army Depot ("BGAD")] is regularly performed, that it is considered ordinary maintenance work, or that such work [is also] done by BGAD employees."). Case: 2:16-cv-00129-DLB-CJS Doc #: 23 Filed: 12/20/17 Page: 5 of 10 – Page ID#: 6 In contrast, Plaintiff's arguments regarding the "regular and recurring" analysis are not tenable. Broadly speaking, Plaintiff argues that he was not yet working for Wal-Mart at the time he was injured, removing Wal-Mart's up-the-ladder immunity. (Doc. # 15 at 5–9). But neither the case law he cites—*McMillen v. Ford Motor Co.*, No. 3:07-cv-309, 2009 WL 5169871 (W.D. Ky. Dec. 20, 2009)—nor his argument—that because he had not yet walked into Wal-Mart's building or written on the sign-in sheet, he was not yet working for Wal-Mart—are persuasive. First, McMillen is inapposite. In *McMillen*, the plaintiff was employed as an electrician and a superintendent by a company subcontracted by a contractor to the defendant. *McMillen*, 2009 WL 5169871 at *5. Although he spent most of his time at the defendant's worksite, had his own trailer at the worksite, and did electrical work almost exclusively for defendant, this did not guarantee that all work done on the worksite was for the defendant. Id. On the day that he was injured, the plaintiff had been evaluating a worksite for a proposed bid by his employer, which would be then submitted to and modified by the defendant's contractor before being submitted to the defendant. Id. There was no guarantee of acceptance by the defendant. Id. As a result, the court found that the work the plaintiff had been doing when he was injured was solely for his actual employer—the subcontractor—and did not create up-the-ladder immunity for the defendant. Id. at *7. In contrast, Plaintiff was on Wal-Mart's premises to do exactly the work that ASM was contracted by Wal-Mart to do. (Doc. # 12-2 at 6, 18). Second, Plaintiff's broader argument that he had not yet begun work for Wal-Mart because he was only in the parking lot is equally unpersuasive. Plaintiff argues that he "fell prior to ever entering [Wal-Mart's] store, [ ] had not signed the vendor log to begin his Case: 2:16-cv-00129-DLB-CJS Doc #: 23 Filed: 12/20/17 Page: 6 of 10 – Page ID#: 7 work and thus was not conducting any work in [Wal-Mart's] store at the time of the accident." (Doc. # 15 at 6). In addition to being contrary to Kentucky case law, this argument is illogical. The Court declines to adopt an arbitrary rule that requires presence inside a building, or the act of clocking or signing in, before workers' compensation coverage applies. Finally, Plaintiff's broader argument ignores the "purpose of Ky. Rev. Stat. Ann. 342.610(2) [, which] is not to shield owners or contractors from potential tort liability but to assure that contractors and subcontractors provide workers' compensation coverage." *Cain*, 236 S.W. 3d at 587. Plaintiff was on Wal-Mart's premises to do the work ASM had hired him to do (Doc. # 12-2 at 18), just as one of Wal-Mart's employees would be on the premises to do the work Wal-Mart had hired him or her to do. Where an employee sustains a work-related injury on premises controlled by the employer, that employee is eligible for workers' compensation benefits. *Jackson Purchase Med. Assoc. v. Crossett*, 412 S.W.3d 170, 172 (Ky. 2013); see also *Pierson v. Lexington Public Library*, 987 S.W.2d 316, 318 (Ky. 1999) (holding that an employer was liable for workers' compensation benefits to a worker who injured himself on the way to work, where the employee was injured in a garage that the employer had some control over). These workers' compensation benefits provide an exclusive remedy against the employer, precluding additional liability for the injury. Ky. Rev. Stat. Ann. § 342.690(1). If that same employee had sustained the same injury as Plaintiff, he would be eligible for workers' compensation. *Crossett*, 412 S.W.3d at 173. And under § 342.690(1), workers' compensation, once demanded, would be that employee's sole remedy for the employer's liability. Case: 2:16-cv-00129-DLB-CJS Doc #: 23 Filed: 12/20/17 Page: 7 of 10 – Page ID#: 8 Placing Plaintiff in the shoes of that employee should lead to the same result. Plaintiff was injured on Wal-Mart's premises, which Wal-Mart admits full control of (Doc. # 18-2), moving from his car into the confines of the store, with the sole purpose of performing the work ASM was subcontracted to do. (Doc. # 12-2 at 18). Plaintiff requested—and received—workers' compensation benefits for his injury (Docs. # 12-2 at 10 and 18-1 at 1), without needing to file suit. Because Plaintiff was compensated through the workers' compensation program, he is now precluded from seeking additional money on the grounds of tort liability. Because Plaintiff was engaged in "regular or recurring" work for Wal-Mart when he was injured, the sole remaining question in determining if Wal-Mart enjoys immunity as an up-the-ladder employer is whether Wal-Mart had workers'-compensation coverage at the time of the injury. It most certainly did. D. Wal-Mart had workers' compensation coverage at the time of the injury. Wal-Mart satisfies its burden of showing the coverage component of the up-the-ladder liability test. Wal-Mart has included with its Motion for Summary Judgment a document showing that it had workers' compensation insurance on the date of the injury, March 2, 2015. (Doc. 12-4). This document shows that an insurer—Marsh USA, Inc.—covered the Maysville, Kentucky store against workers' compensation claims under policy number 037083149, effective from September 15, 2014 through September 15, 2015. Id. This is sufficient to show that Wal-Mart has "secure[d] payment of compensation as required by [the workers' compensation statutes.]" Ky. Rev. Stat. Ann. § 342.690(1). Case: 2:16-cv-00129-DLB-CJS Doc #: 23 Filed: 12/20/17 Page: 8 of 10 – Page ID#: 9 Plaintiff argues that that the workers' compensation coverage was deficient because the Kentucky Supreme Court has since found that one section of the Kentucky Workers' Compensation Act is unconstitutional. (Doc. # 15 at 8-10). In support, Plaintiff cites the recent Supreme Court of Kentucky case *Parker v. Webster Cty. Coal, LLC*, 529 S.W.3d 759 (Ky. 2017). The majority in Parker found that one section of the Kentucky workers' compensation scheme—§ 342.730(4)—violated older workers' right to equal protection by treating "injured older workers who qualify for normal old-age Social Security retirement benefits differently than it treats injured older workers who do not qualify." *Parker*, 529 S.W.3d at 768. This section had provided for the reduction in number of weeks of benefits available to injured workers as those workers approached and then surpassed the age of Social Security eligibility. Id. at 767–68. Comparing the rights of these workers to those of another group—Kentucky teachers—the court determined that § 342.730(4) "invidiously discriminates against those who qualify for one type of retirement benefit (social security) from those who do not qualify for the that type of retirement benefit but do qualify for another type of retirement benefit (teacher retirement)." Id. at 769. Plaintiff argues that because he received fewer benefits than others under the now unconstitutional section, this has rendered Wal-Mart's insurance policy deficient. But by declaring Wal-Mart's workers' compensation coverage deficient because of *Parker*, Plaintiff erroneously conflates two schema—one that provides for employers to pay into the workers'-compensation system, and one that provides for employees by paying out of the workers-compensation system. They are not the same. Section 342.690(1) requires an employer to secure payment of compensation as required by the Kentucky Case: 2:16-cv-00129-DLB-CJS Doc #: 23 Filed: 12/20/17 Page: 9 of 10 – Page ID#: 10 Workers' Compensation Act, not to ensure that the Kentucky Workers' Compensation Act is constitutionally sufficient. Plaintiff's argument is unavailing, and Wal-Mart has established that it had a valid and covering workers' compensation insurance policy. To summarize, there is no genuine dispute as to any material fact which would preclude summary judgment here. Wal-Mart has met its burden of showing that it is an up-the-ladder employer for purposes of Plaintiff's injury, and is therefore immune from Plaintiff's tort claims arising out of the same injury. IV.

3.   CONCLUSION: Accordingly, for the reasons stated herein, IT IS ORDERED as follows: (1) Defendant's Motion for Summary Judgment (Doc. # 12) is GRANTED; and (2) Plaintiff's personal-injury claim is DISMISSED WITH PREJUDICE; and (3) A Judgment in favor of Defendant shall be filed contemporaneously herewith. This 20th day of December, 2017.

15 e-Workplace
Privacy Laws, Cybersecurity, and Social Media

******

Chapter Objectives:

1. Understand the privacy laws regarding e-communications.
2. Understand the safety issues regarding cellphones in the workplace.
3. Understand the risk and issues involving cybersecurity.
4. Understand the issues involved in social media.

******

In the evolving world of technology in our workplaces today with cellphones, computers, the cloud, social media, blogs, wikis, and microblogging sites, as well as other emerging technologies, safety and health professionals should become cognizant of the emerging laws governing communication in the workplace, as well as the numerous issues and risks these technologies can create in the workplace. Although safety and health professionals are not experts in e-communications, it is important to be able to recognize the issues involved in the new e-workplace and avoid the potential pitfalls, and to know the boundaries in this fast-moving environment.

The laws historically have lagged behind the technology. Under the Electronic Communications Privacy Act,* safety and health professionals, as agents for the employer, are regulated in their ability to intercept and review electronic communications. Under the Stored Communications Act,† employers are prevented from accessing stored electronic communications. The SCA prohibits employers from "intentionally accessing, without authorization, a facility through which an electronic communication is provided."‡ However, safety and health professionals should be aware that there is an exception for "the person or entity providing a wire or electronic communications service."§ In short, employers are permitted to access emails that are on an email server maintained by the company and provided to their employees for use. However, this emerging area can be a minefield. Employee email on web-based email accounts, offline communications, and employee-owned

---

* 18 U.S.C. § 2701.
† *Id.*
‡ *Id.*
§ *Id.*

phones create privacy concerns which prudent safety and health professionals should review with human resources and legal counsel. Many companies have developed specific policies to address such privacy concerns and variables as bringing your own electronic devices to work, overtime and wage issues, reimbursement for device expenses, and related issues. Safety and health professionals should avoid any invasion of an employee's electronic communications and ensure compliance with all company policies and procedures regarding employee electronic communications.

In today's society, the issue of data security has moved to the forefront. Although safety and health professionals are not IT professionals, it is important to ensure that employees' data, especially medical-related data, is secure and all company data privacy policies and protocols are followed. Safety and health professionals should be aware that such activities as taking your phone with company or employee data on it for repair and providing the password to a third party may constitute a data breach. Additionally, use of third party apps may provide a pathway for others to breach the system and acquire sensitive information.

In general, safety and health professionals using company-provided phones or using their personal phones for company business should consider using a strong password, use encryption software (if appropriate), install security updates, and have appropriate physical protection on the device. Additionally, safety and health professionals should report a lost or stolen phone or device to the employer, appropriately remove all data before replacing the phone or device, and install safeguards against malware and spyware. Safety and health professionals should additionally consider ensuring frequent backup of data, avoid using wifi hotspots in public places, and ensure that all company-provided devices are collected from employees leaving the company.

Although data protection is important, safety and health professionals should also provide attention to the risk created by the phone or device in the workplace. In 2010, Assistant Secretary for OSHA David Michaels, in an open letter, stated that

> *it is your responsibility and legal obligation to have a clear, unequivocal and enforced policy against texting while driving .... Companies are in violation of the Occupational Safety and Health Act if, by policy or practice, they require texting while driving, or create incentives which encourage or condone it, or they structure work so that texting is a practical necessity for workers to carry out their jobs. OSHA will investigate worker complaints, and employers who violate the law will be subject to citations and penalties.**

Safety and health professionals should be aware that OSHA can cite employers for texting and driving and other situations when the use of the phone or device creates a recognized hazard under the "General Duty Clause" or Section 5(a)(1) of the Occupational Safety and Health Act.†

In the area of social media, safety and health professionals should exercise caution. In the area of background checks for job applicants, the risk, and potential legal

---

* Mark A. Lies, II and Adam R. Young, Cell phones in the Workplace: Protecting Employee Safety. *Emp't. L. Lookout* (Oct. 13, 2016), www.laborandemploymentlawcounsel/2016/10/cell-phones-at-the-workplce-protecting-employee-safety/.
† *Id.*

liability, can include, but not limited to claims under Title VII of the Civil Rights Act, the Americans with Disabilities Act (ADA), the Age Discrimination in Employment Act (ADEA) as well as various state laws. Safety and health professionals should be aware that although there is no federal law explicitly prohibiting employers from requesting social media passwords and login information from applicants and employees, a number of states prohibit employers from asking for social media login information from applicants and employees.* Several other states are considering legislation addressing password protection for applicants and employees.† Safety and health professionals should be aware that company policy will vary among states; however, this policy usually addresses federal SCA as well as company electronic equipment use and other related policies.

Conversely, safety and health professionals should be cautious when posting information on the internet and especially on social media: "There is no judicially accepted social media privilege or blanket of privacy when it comes to social media sites."‡ Safety and health professionals should be aware that information posted on the internet and made available to the public is generally considered public information.§ Additionally, for safety and health professionals, this is an emerging area of the law and even with the use of privacy settings, information posted on social media and other locations on the internet may be discoverable in a legal action.

Another emerging area of which safety and health professionals should be aware is that of biometric information. Biometrics are measurements of individual characteristics or patterns, such as fingerprints, voiceprints, and eye scans which can quickly identify employees. Personal biometrics are biologically unique to the individual or employee; thus, the need for protection of this important data. Currently there are three states with statutes regulating the collection and storage of biometric information.¶ These laws, in general, require employee consent, protection of the data and correlating conversion to code, a standard of care when protecting the data, and the destruction of the data.

Although safety and health professionals are not expected to be IT gurus, the increased use in the types of technologies in the workplace has created new and unique opportunities within the safety and health arena. Nor unlike texting and driving, the use of technology within the workplace can create risk to employees operating equipment and create liabilities in the areas of discrimination and harassment, as well as create data storage and use issues. Prudent safety and health professionals should acquire guidance from their human resource or legal department as to their data storage, as well as exploration into applicants' or employees' social media or related areas. E-communication is an emerging area of the law, and safety and health professionals should tread cautiously in order to protect employees' data and adhere to new laws governing this area.

---

* Arkansas, California, Colorado, Connecticut, Delaware, Illinois, Maine, Maryland, Michigan, Montana, Nebraska, Nevada, New Hampshire, New Jersey, New Mexico, Oklahoma, Oregon, Rhode Island, Tennessee, Utah, Virginia, Washington and Wisconsin.
† Alaska, Florida, Georgia, Massachusetts, Minnesota, Missouri and West Virginia.
‡ See *Davenport v. State Farm Mut. Auto. Ins. Co.*, 2012 WL 555759 (M.D. Fla. Feb. 21,2012).
§ *Morena v. Handford Sentinel, Inc.* 172 Cal. App. 4th 1125 (2009).
¶ Illinois, Texas and Washington.

Case Study: This case may have been modified for the purpose of this text.

851 F.3d 332
United States Court of Appeals, Fourth Circuit.

Mark GRUTZMACHER, Plaintiff,
and
Kevin Patrick Buker, Plaintiff-Appellant,
v.
HOWARD COUNTY; Chief William F. Goddard, III; John Jerome; John S. Butler,
Defendants-Appellees.

No. 15-2066
|
Argued: December 7, 2016
|
Decided: March 20, 2017

SYNOPSIS

Background: Former county employee, a paramedic, brought § 1983 action against former county employer and county officials, alleging that he was discharged in retaliation for exercising his First Amendment free speech rights, and that the fire department's social media policy, which played a role in the employee's termination, was facially violative of the First Amendment. The United States District Court for the District of Maryland, Marvin J. Garbis, J., 2015 WL 3456750, granted summary judgment in favor of defendants on the First Amendment retaliation claim, and dismissed as moot the facial challenge. Former employee appealed.

Holdings: The Court of Appeals, Wynn, Circuit Judge, held that:

[1] certain comments by employee on his social media webpage amounted to protected speech;

[2] employer's interest in efficiency and preventing workplace disruption outweighed employee's free speech interest; and

[3] facial challenge to department's social media policy was moot.

Affirmed.

OPINION

Affirmed by published opinion. Judge Wynn wrote the opinion, in which Chief Judge Gregory and Judge Thacker joined.

WYNN, Circuit Judge:

Plaintiff Kevin Patrick Buker is a former Battalion Chief with the Howard County, Maryland Department of Fire and Rescue Services (the "Department").

Defendants are Howard County, Maryland; former Howard County Fire Chief William F. Goddard, III ("Chief Goddard"); former Howard County Deputy Chief John Butler ("Deputy Chief Butler");* and Howard County Assistant Chief John Jerome ("Assistant Chief Jerome," and collectively with Howard County, Chief Goddard, and Deputy Chief Butler, "Defendants").

Plaintiff brought this matter in the District Court for the District of Maryland, at Baltimore, alleging that Defendants retaliatorily fired him for exercising his First Amendment free-speech rights and, second, that the Department's social media policy, which played a role in Plaintiff's termination, was facially unconstitutional under the First Amendment. This appeal arises from the district court's orders granting summary judgment in favor of Defendants on Plaintiff's First Amendment retaliation claim and dismissing as moot Plaintiff's facial challenge to the social media policy. On review, we affirm the judgment of the district court.

I.

A.

The Department employed Plaintiff as a paramedic for the Howard County Fire Department from 1997 through 2012. In 2012, Chief Goddard promoted Plaintiff to the rank of battalion chief and assigned Plaintiff to the second battalion as its commander. According to Chief Goddard, as a battalion chief, Plaintiff was responsible for "manag[ing] the day-to-day operations of the field," as well as "ensur[ing] ... the policies and procedures as written in the department are complied with." J.A. 139.

*337 As a paramilitary-type organization, the Department executes the enforcement of its orders in a hierarchical manner that requires employees to strictly follow a chain-of-command. At the top of the Department's chain-of-command is the fire chief, followed by deputy fire chiefs, assistant chiefs, battalion chiefs, and, lastly, first responders. Although positioned at the lower end of the chain-of-command, Chief Goddard described the rank of battalion chief as "the most critical leadership position in the organization," as battalion chiefs directly supervise first responders. J.A. 138.

In 2011, Chief Goddard, along with the Department's public information officer, began drafting a social media policy for the Department, partially in response to national debate about the use of social media within fire and emergency services departments. The Department's decision to develop a social media policy also stemmed from an incident involving a Howard County volunteer firefighter posting to Facebook a photograph of a lynching, depicted by a noosed, brown beer bottle surrounded by white beer cans with paper cones for hoods. In a comment accompanying the photograph, the volunteer firefighter said that he "[w]ant[ed] to go fishing for mud sharks/there are way too many here in Maryland. They are not good to eat though, I hear they taste like decayed chicken." J.A. 835; Dist. Ct. Dkt. 40–4, at 5; Dist. Ct. Dkt.

---

* Butler was appointed as Fire Chief in January 2015, following Fire Chief Goddard's retirement.

40–7, at 3; Dist. Ct. Dkt. 40–24. Throughout the drafting process, the Department provided internal stakeholders—including Plaintiff, as well as all of the other battalion chiefs—opportunities to review and comment on the forthcoming policy.

On November 5, 2012, the Department issued General Order 100.21, entitled "Social Media Guidelines," which set forth the Department's policy regarding the use of social media by Department personnel. Under the Social Media Guidelines, the Department prohibited personnel "from posting or publishing any statements, endorsements, or other speech, information, images or personnel matters that could reasonably be interpreted to represent or undermine the views or positions of the Department, Howard County, or officials acting on behalf of the Department or County." J.A. 32. The Social Media Guidelines also barred Department employees "from posting or publishing statements, opinions or information that might reasonably be interpreted as discriminatory, harassing, defamatory, racially or ethnically derogatory, or sexually violent when such statements, opinions, or information may place the Department in disrepute or negatively impact the ability of the Department in carrying out its mission." J.A. 32. Additionally, the Social Media Guidelines prohibited Department personnel from "post[ing] any information or images involving off-duty activities that may impugn the reputation of the Department or any member of the Department." J.A. 32.

Further, on December 6, 2012, the Department issued General Order 100.22, entitled "Code of Conduct," which was "aimed at ensuring members of the Department maintain the highest level of integrity and ethical conduct both on and off duty." J.A. 34. In relevant part, the Code of Conduct prohibited Department personnel from "intentionally engag[ing] in conduct, through actions or words, which are disrespectful to, or that otherwise undermines the authority of, a supervisor or the chain of command" and "publicly criticiz[ing] or ridicul[ing] the Department or Howard County government or their policies." J.A. 38–39. The Code of Conduct also required "[m]embers [to] conduct themselves at all times, both on and off duty, in such a manner as to reflect favorably on the Department." J.A. 38. The *338 Code of Conduct further prohibited Department employees from engaging in "[c]onduct unbecoming" to the Department, which it defined as "any conduct that reflects poorly on an individual member, the Department, or County government, or that is detrimental to the public trust in the Department or that impairs the operation and efficiency of the Department." J.A. 38.

On January 20, 2013, Plaintiff was watching news coverage of a gun control debate in his office and posted the following statement to his Facebook page while on-duty*:

> My aide had an outstanding idea .. lets all kill someone with a liberal ... then maybe we can get them outlawed too! Think of the satisfaction of beating a liberal to death with another liberal ... its almost poetic ...

---

* We reproduce the Facebook posts and comments as they appear in the record and without the benefit of editing.

J.A. 82–83 (ellipses in original). Twenty minutes later, Mark Grutzmacher, a county volunteer paramedic, replied to Plaintiff's earlier post with the following comment:

> But.... was it an "assult liberal"? Gotta pick a fat one, those are the "high capacity" ones. Oh ... pick a black one, those are more "scary". Sorry had to perfect on a cool idea!

J.A. 84 (ellipses in original). Six minutes later, Plaintiff "liked" Grutzmacher's comment and replied, "Lmfao! Too cool Mark Grutzmacher!" J.A. 85.

Two Department employees subsequently forwarded Plaintiff's and Grutzmacher's Facebook posts to another battalion chief within the Department. On January 22, 2013, that battalion chief sent a screenshot of Plaintiff's initial Facebook post to Assistant Chief Jerome with a text message stating, "Chief, not sure this is something that should be displayed from one of our battalion chiefs." J.A. 82. Assistant Chief Jerome then contacted his direct supervisor, Deputy Chief Butler, along with another assistant chief, regarding Plaintiff's Facebook posts. Later that day, the three chiefs met to discuss whether Plaintiff's posts violated the Social Media Guidelines or Code of Conduct and, if so, what corrective measures the Department would take. Following their meeting, Assistant Chief Jerome emailed Plaintiff, directing him to review his recent Facebook posts and to remove anything inconsistent with the Department's social media policy. Though Plaintiff maintained that he was in compliance with the social media policy, Plaintiff removed the January 20 posts.

On January 23—a few hours after Plaintiff informed Assistant Chief Jerome that he had removed the posts—Plaintiff posted the following to his Facebook "wall":

> To prevent future butthurt and comply with a directive from my supervisor, a recent post (meant entirley in jest) has been deleted. So has the complaining party. If I offend you, feel free to delete me. Or converse with me. I'm not scared or ashamed of my opinions or political leaning, or religion. I'm happy to discuss any of them with you. If you're not man enough to do so, let me know, so I can delete you. That is all. Semper Fi! Carry On.

J.A. 96. One of Plaintiff's Facebook friends then replied, "As long as it isn't about the [Department], shouldn't you be able to express your opinions?" J.A. 96. Plaintiff responded:

> Unfortunately, not in the current political climate. Howard County, Maryland, *339 and the Federal Government are all Liberal Democrat held at this point in time. Free speech only applies to the liberals, and then only if it is in line with the liberal socialist agenda. County Governement recently published a Social media policy, which the Department then published it's own. It is suitably vague enough that any post is likely to result in disciplinary action, up to and including termination of employment, to include this one. All it took was one liberal to complain ... sad day. To lose the First Ammendment rights I fought to ensure, unlike the WIDE majority of the Government I serve.

J.A. 96 (ellipses in original). Another of Plaintiff's Facebook friends then commented, "Oh, your gonna get in trouble for saying that too." J.A. 96. "Probably ...," Plaintiff replied. J.A. 96.

The following day, January 24, a captain in the Department emailed Chief Goddard a screenshot of Plaintiff's January 23 Facebook posts. The captain also emailed Deputy Chief Butler and an assistant chief a "summary of the Buker issue," in which he noted the "racial overtones" of Grutzmacher's comment on Plaintiff's January 20 Facebook post. J.A. 101. The captain stated that by replying to the comment, Plaintiff "endorsed" Grutzmacher's racially charged statement. J.A. 101. The captain also characterized Plaintiff's January 23 posts as "insubordinate toward [management]." J.A. 101. The captain suggested treating the incidents "like any other investigation" and determining any disciplinary action "after the conclusion of the investigation." J.A. 101. The next day, the Department moved Plaintiff out of field operations to an administrative assignment pending the results of an internal investigation.

Approximately three weeks later, on February 17, 2013, Mike Donnelly, a member of a Department-affiliated volunteer company, posted to his own Facebook page a picture of an elderly woman with her middle finger raised. Overlaid across the picture was the following caption: "THIS PAGE, YEAH THE ONE YOU'RE LOOKING AT IT'S MINE[.] I'LL POST WHATEVER THE FUCK I WANT[.]" J.A. 100. Above the picture, Donnelly wrote, "for you Chief." J.A. 100. Plaintiff, who was one of Donnelly's Facebook friends, "liked" the photograph.

Chief Goddard served Plaintiff with charges of dismissal on February 25. The charges referenced Plaintiff's: (1) January 20 and January 23 Facebook posts; (2) "like" of and reply to Grutzmacher's January 20 comment; (3) replies to comments on Plaintiff's January 23 post; and (4) "like" of Donnelly's February 17 post.* The charges asserted that these posts violated the Department's Code of Conduct and Social Media Guidelines. In particular, the charging document asserted, among other things, that Plaintiff's Facebook activity improperly:

- "[A]dopted" and "approv[ed]" Grutzmacher's comment, which "had racial overtones and was insensitive and derogatory in nature";

  *340 • Reflected a "[f]ailure to grasp the impact and implications of [the] comments" on Plaintiff's "leadership position within the Department as a Battalion Chief," in which Plaintiff was "responsible for enforcing Department policies and taking appropriate action for violations of those policies by the people [he] supervise[d]";

---

* We observe that the act of "liking" a Facebook post makes the post attributable to the "liker," even if he or she did not author the original post. *See Bland v. Roberts*, 730 F.3d 368, 386 (4th Cir. 2013), *as amended* (Sept. 23, 2013) ("[C]licking on the 'like' button literally causes to be published the statement that the User 'likes' something, which is itself a substantive statement ... That a user may use a single mouse click to produce that message ... instead of typing the same message with several individual key strokes is of no constitutional significance."). Accordingly, for ease of reference, we refer to Plaintiff's various Facebook posts, comment replies, and "likes," collectively, as Plaintiff's "Facebook activity" or "speech."

- Demonstrated "repeated insolence and insubordination" by replacing the January 20 post "with another posting tirade mocking the Chain-of-Command, the Department, and the County"; and
- "[I]nterfered with Department operations" and caused "disruption [in] the Department's Chain-of-Command and authority."

J.A. 105.

Chief Goddard provided Plaintiff with an opportunity to rebut the specific charges at a pre-termination meeting held on March 8. Following that meeting, on March 14, 2013, Chief Goddard terminated Plaintiff's employment with the Department.

B.

On October 12, 2013, Plaintiff brought an action under 42 U.S.C. § 1983 in federal district court seeking reinstatement and damages. Plaintiff alleged that his Facebook posts were a substantial motivation for his termination and that, by terminating him, the Department impermissibly retaliated against Plaintiff for exercising his First Amendment rights. Plaintiff also alleged that the Department's Social Media Guidelines and Code of Conduct, as drafted and applied to Plaintiff, violated the First Amendment by impermissibly restricting Department employees' ability to speak on matters of public concern. The district court later construed the second of Plaintiff's claims as a facial challenge to the Department's Social Media Guidelines and Code of Conduct.

Following discovery, Defendants moved for summary judgment, arguing that Plaintiff's Facebook activity did not involve matters of public concern and that Plaintiff's interest in speaking did not outweigh the Department's interest in minimizing disruption. Defendants later filed a second motion for summary judgment as to Plaintiff's facial challenge claims, arguing that the Department's policies were not unconstitutionally overbroad or vague and did not constitute prior restraints.

The district court granted Defendants' first summary judgment motion on March 30, 2015. *Buker v. Howard County.*, Nos. MJG–13–3046, MJG–13–3747, 2015 WL 3456750 (D. Md. May 27, 2015). In doing so, the district court concluded that Plaintiff's January 20 Facebook posts and "like" were unprotected speech because they were "capable of impeding the [Fire Department]'s ability to perform its duties efficiently." *Id.* at *13 (alteration in original) (internal quotation marks omitted) (quoting *Duke v. Hamil*, 997 F.Supp.2d 1291, 1302 (N.D. Ga. 2014)). The district court further concluded that Plaintiff's January 23 posts and February 17 "like" similarly did not amount to protected speech because Plaintiff failed to show that he was speaking as a citizen on a matter of public concern. *Id.* at *13–14. The district court's memorandum decision and order did not, however, address Defendants' second motion for summary judgment, leaving unresolved Plaintiff's facial challenge.

On June 22, 2015, the Department replaced its Social Media Guidelines and Code of Conduct policies with revised versions. The revised version of the Social Media

Guidelines eliminated many of the earlier version's prohibitions on Department
341 personnel's private use of social media. And the revised Code of Conduct
did not include any of the provisions in the previous version that Plaintiff had challenged. Highlighting these changes, Defendants moved to dismiss Plaintiff's facial
challenge as moot, arguing that the Department's revised policies did not contain the
provisions Plaintiff challenged as overbroad, void for vagueness, or prior restraints.
The district court thus denied Defendants' earlier motion for summary judgment as
moot and granted Defendants' motion to dismiss on August 12, 2015.

Plaintiff timely appealed the district court's (1) award of summary judgment in favor
of Defendants on Plaintiff's First Amendment retaliation claim and (2) dismissal on
mootness grounds of Plaintiff's facial challenge to the Social Media Guidelines and
Code of Conduct.

II.

A.

[1] On appeal, Plaintiff first argues that the district court erred in granting summary
judgment in favor of Defendants on his First Amendment retaliation claim. Summary
judgment is appropriate "if the movant shows that there is no genuine dispute as to
any material fact and the movant is entitled to judgment as a matter of law." Fed. R.
Civ. P. 56(a). "We review a district court's decision to grant summary judgment de
novo, applying the same legal standards as the district court and viewing all facts and
reasonable inferences therefrom in the light most favorable to the nonmoving party."
Smith v. Gilchrist, 749 F.3d 302, 307 (4th Cir. 2014) (quoting *T-Mobile Ne. LLC v.
City Council of Newport News*, 674 F.3d 380, 384–85 (4th Cir. 2012)).

[2] [3] From the outset, we point out that "[t]he First Amendment 'was fashioned to
assure unfettered interchange of ideas for the bringing about of political and social
changes desired by the people.'" *Connick v. Myers*, 461 U.S. 138, 145, 103 S.Ct.
1684, 75 L.Ed.2d 708 (1983) (quoting *Roth v. United States*, 354 U.S. 476, 484, 77
S.Ct. 1304, 1 L.Ed.2d 1498 (1957)). "Protection of the public interest in having debate
on matters of public importance is at the heart of the First Amendment." *McVey v.
Stacy*, 157 F.3d 271, 277 (4th Cir. 1998) (citing *Pickering v. Bd. of Educ.*, 391 U.S.
563, 573, 88 S.Ct. 1731, 20 L.Ed.2d 811 (1968)).

[4] To resolve Plaintiff's appeal, we start by considering the First Amendment rights
of public employees. Public employees do not "relinquish First Amendment rights
to comment on matters of public interest by virtue of government employment."
Connick, 461 U.S. at 140, 103 S.Ct. 1684. To the contrary, the Supreme Court has
long recognized

> that public employees are often the members of the community who are likely to have
> informed opinions as to the operations of their public employers, operations which are
> of substantial concern to the public. Were they not able to speak on these matters, the
> community would be deprived of informed opinions on important public issues.

City of San Diego v. Roe, 543 U.S. 77, 82, 125 S.Ct. 521, 160 L.Ed.2d 410 (2004) (per curiam) (citing *Pickering*, 391 U.S. at 572, 88 S.Ct. 1731). To that end, the Supreme Court has repeatedly "underscored the 'considerable value' of 'encouraging, rather than inhibiting, speech by public employees. For government employees are often in the best position to know what ails the agencies for which they work.'" *Hunter v. Town of Mocksville*, 789 F.3d 389, 396 (4th Cir. 2015) (quoting ***342** *Lane v. Franks*, —— U.S. ——, 134 S.Ct. 2369, 2377, 189 L.Ed.2d 312 (2014)). As such, we do not take lightly "[o]ur responsibility ... to ensure that citizens are not deprived of fundamental rights by virtue of working for the government." *Connick*, 461 U.S. at 147, 103 S.Ct. 1684.

[5] [6] [7] "That being said, precedent makes clear that courts must also consider 'the government's countervailing interest in controlling the operation of its workplaces.'" *Hunter*, 789 F.3d at 397 (quoting *Lane*, 134 S.Ct. at 2377). Just as there is a "public interest in having free and unhindered debate on matters of public importance," *Pickering*, 391 U.S. at 573, 88 S.Ct. 1731, "[t]he efficient functioning of government offices is a paramount public interest," *Robinson v. Balog*, 160 F.3d 183, 189 (4th Cir. 1998). Therefore, a public employee "by necessity must accept certain limitations on his or her freedom." *Garcetti v. Ceballos*, 547 U.S. 410, 418, 126 S.Ct. 1951, 164 L.Ed.2d 689 (2006). In particular, under the balancing test developed by the Supreme Court in *Pickering* and *Connick*, "the First Amendment does not protect public employees when their speech interests are outweighed by the government's interest in providing efficient and effective services to the public." *Lawson v. Union Cty. Clerk of Court*, 828 F.3d 239, 247 (4th Cir. 2016).

[8] [9] Regarding Plaintiff's retaliation claim, "a public employer contravenes a public employee's First Amendment rights when it discharges ... '[the] employee ... based on the exercise of' that employee's free speech rights." *Ridpath v. Bd. of Governors Marshall Univ.*, 447 F.3d 292, 316 (4th Cir. 2006) (alteration in original) (quoting *Suarez Corp. Indus. v. McGraw*, 202 F.3d 676, 686 (4th Cir. 2000)). To state a claim under the First Amendment for retaliatory discharge, a plaintiff must satisfy the three-prong test set forth in *McVey v. Stacy*, 157 F.3d 271 (4th Cir. 1998). In particular, the plaintiff must show: (1) that he was a "public employee ... speaking as a citizen upon a matter of public concern [rather than] as an employee about a matter of personal interest;" (2) that his "interest in speaking upon the matter of public concern outweighed the government's interest in providing effective and efficient services to the public;" and (3) that his "speech was a substantial factor in the employer's termination decision." 157 F.3d at 277–78.

The district court found that Plaintiff's January 20 speech failed on the second prong of the *McVey* test, and that Plaintiff's January 23 and February 17 speech failed on the first *McVey* prong. *Buker*, 2015 WL 3456750, at *9–14. Plaintiff urges us to reverse the district court's grant of summary judgment to Defendants and, in doing so, makes two arguments. First, Plaintiff argues that the district court erred in granting summary judgment when there remained a factual dispute regarding whether Plaintiff could meet his burden under the *McVey* test's second prong. Specifically,

Plaintiff maintains that his January 20 speech did not disrupt the Department or cause a reasonable apprehension of disruption, such that the Department's interest in maintaining an efficient workplace outweighed Plaintiff's interest in speaking. Second, Plaintiff argues that the district court erred in finding that his January 23 and February 17 posts and "like" were not on a matter of public concern and, therefore, failed *McVey's* first prong. For the reasons below, we hold that the district court properly granted summary judgment to Defendants.

1.

[10] [11] We first address whether Plaintiff's Facebook posts and "likes" addressed ***343** matters of public concern. In determining whether speech addresses matters of public concern, "we examine the content, context, and form of the speech at issue in light of the entire record." *Urofsky v. Gilmore*, 216 F.3d 401, 406 (4th Cir. 2000) (en banc). "Speech involves a matter of public concern when it involves an issue of social, political, or other interest to a community." *Id.* This "public-concern inquiry centers on whether 'the public or the community is likely to be truly concerned with or interested in the particular expression.'" *Kirby v. City of Elizabeth City*, 388 F.3d 440, 446 (4th Cir. 2004) (quoting *Arvinger v. Mayor of Baltimore*, 862 F.2d 75, 79 (4th Cir. 1988)); *see also Goldstein v. Chestnut Ridge Volunteer Fire Co.*, 218 F.3d 337, 352–53 (4th Cir. 2000) ("This is a subtle, qualitative inquiry; we use the content, form, and context as guideposts in the exercise of common sense, asking throughout: would a member of the community be truly concerned with the employee's speech?").

[12] [13] [14] Conversely, "[i]n the absence of unusual circumstances, a public employee's speech 'upon matters only of personal interest' is not afforded constitutional protection." *Seemuller v. Fairfax Cty. Sch. Bd.*, 878 F.2d 1578, 1581 (4th Cir. 1989) (quoting *Connick*, 461 U.S. at 147, 103 S.Ct. 1684); *see also Jurgensen v. Fairfax County*, 745 F.2d 868, 879 (4th Cir. 1984) ("If the speech relates primarily to a matter of 'limited public interest' and ... center[s] instead on matters primarily, if not exclusively 'of personal interest' to the employee ... that fact must be weighed in determining whether a matter of true public concern is involved ...."). To that end, "[t]he Supreme Court has warned us to guard against 'attempt[s] to constitutionalize the employee grievance.'" *Brooks v. Arthur*, 685 F.3d 367, 373 (4th Cir. 2012) (second alteration in original) (quoting *Connick*, 461 U.S. at 154, 103 S.Ct. 1684). Accordingly, "[p]ersonal grievances[and] complaints about conditions of employment ... do not constitute speech about matters of public concern." *Campbell v. Galloway*, 483 F.3d 258, 267 (4th Cir. 2007) (internal quotation marks omitted) (quoting *Stroman v. Colleton Cty. Sch. Dist.*, 981 F.2d 152, 156 (4th Cir. 1992)). Likewise, we must also "ensure that matters of internal policy, including mere allegations of favoritism, employment rumors, and other complaints of interpersonal discord, are not treated as matters of public policy." *Goldstein*, 218 F.3d at 352.

[15] Set against this backdrop, at least some of Plaintiff's Facebook activity referenced in the Department's charging document touched on issues of public concern. In particular, Plaintiff's and Grutzmacher's January 20, 2013, discussion about

"liberal[s]" and "assault liberal[s]" was, according to an expert report submitted by Plaintiff, a commentary on gun control legislation using "a lexicon that is extremely common in contemporary American gun culture." J.A. 566–71. The report maintains that Plaintiff's and Grutzmacher's exchange reflects a "well-known meta-narrative" under which "'liberal' ... is a collectivist ideologue, a statist, who believes in the absolute power of government even at the expense of individual autonomy and rights, including an individual's right to own, carry and use firearms." J.A. 567–68. Courts have long recognized that "[t]he debate over the propriety of gun control legislation is ... a matter of public concern." *Thomas v. Whalen*, 51 F.3d 1285, 1290 (6th Cir. 1995). Consequently, the "liberal" and "assault liberal" post and comment implicated a matter of public concern.

[16] Likewise, Plaintiff's January 23, 2013, post describing the Department's Social Media Guidelines and expressing concern ***344** that those guidelines infringed on Plaintiff's First Amendment rights also addressed a matter of public concern. As explained above, the public employee speech doctrine recognizes the unique role government employees—individuals who "are often in the best position to know what ails the agencies for which they work"—play in keeping the electorate informed about the operations of public employers. *See Liverman v. City of Petersburg*, 844 F.3d 400, 408 (4th Cir. 2016) (internal quotation marks omitted) (quoting *Waters v. Churchill*, 511 U.S. 661, 674, 114 S.Ct. 1878, 128 L.Ed.2d 686 (1994) (plurality opinion)). To that end, the interest advanced by the public employee speech doctrine "is as much *the public's interest in receiving informed opinion* as it is the employee's own right to disseminate it." *Roe*, 543 U.S. at 82, 125 S.Ct. 521 (emphasis added); *see also Garcetti*, 547 U.S. at 419, 126 S.Ct. 1951 ("The Court has acknowledged the importance of promoting the public's interest in receiving the well-informed views of government employees engaging in civic discussion."); *United States v. Nat'l Treasury Emps. Union*, 513 U.S. 454, 470, 115 S.Ct. 1003, 130 L.Ed.2d 964 (1995) ("The large-scale disincentive to Government employees' expression also imposes a significant burden on the public's right to read and hear what the employees would otherwise have written and said."). Because the public has an interest in receiving the "informed" opinions of public employees, it necessarily also has an interest in information about policies that circumscribe public employees' speech and public employees' opinions of such policies.

However, we also acknowledge that some of the Facebook activity prompting Plaintiff's termination did not implicate matters of public concern. For instance, Plaintiff's "like" of the image depicting an elderly woman raising her middle finger and entitled "for you Chief"—on the heels of the Department's investigation into Plaintiff's January 20 and 23 Facebook activity—"amounted to no more than an employee grievance not protected by the First Amendment." *Stroman*, 981 F.2d at 157.

When "a single expression of speech" encompasses both matters of public concern and matters of purely personal interest, "the proper approach is to consider [the speech] ... in its entirety." *Id.* Whether a series of related posts and "likes" over a several-week period to a dynamic social networking platform—like the posts and "likes" that prompted Plaintiff's termination—constitute "a single expression of speech" is an open question. Rather than resolve that unsettled question—and because at least some of Plaintiff's speech addressed matters of public concern—we will "weigh whatever public interest commentary may be contained in [Plaintiff's Facebook activity] against the [Department's] dual interest as a provider of public service and employer of persons hired to provide that service." *Id.* at 158 (citing *Pickering*, 391 U.S. at 568, 88 S.Ct. 1731). We note that this approach accords with the Department's decision to terminate Plaintiff, which was based on the "public statements [Plaintiff] made over a number of days (not simply one incident—one day)" and "the totality of the circumstances [of] his violations." J.A. 119, 242.

2.

[17] [18] [19] Having concluded that at least some of the Facebook activity prompting Plaintiff's termination implicated matters of public concern, we now must determine "whether [Plaintiff's] interest in speaking upon the matter[s] of public concern outweighed the [Department's] interest in providing effective and efficient services to *345 the public." *McVey*, 157 F.3d at 277.* "Whether [an] employee's interest in speaking outweighs the government's interest is a question of law for the court." *Smith*, 749 F.3d at 309. In balancing these interests, we must "consider the context in which the speech was made, including the employee's role and the extent to which the speech impairs the efficiency of the workplace." *Id.* (citing *Rankin v. McPherson*, 483 U.S. 378, 388, 107 S.Ct. 2891, 97 L.Ed.2d 315 (1987)).

> Factors relevant to this inquiry include whether a public employee's speech (1) impaired the maintenance of discipline by supervisors; (2) impaired harmony among coworkers; (3) damaged close personal relationships; (4) impeded the performance of the public employee's duties; (5) interfered with the operation of the institution; (6) undermined the mission of the institution; (7) was communicated to the public or to coworkers in private; (8) conflicted with the responsibilities of the employee within the institution; and (9) abused the authority and public accountability that the employee's role entailed.

Ridpath, 447 F.3d at 317 (citing *McVey*, 157 F.3d at 278).

[20] To demonstrate that an employee's speech impaired efficiency, a government employer need not "prove that the employee's speech actually disrupted efficiency, but only that an adverse effect was 'reasonably to be apprehended.'" *Maciariello v. Sumner*, 973 F.2d 295, 300 (4th Cir. 1992) (quoting *Jurgensen*, 745 F.2d at 879);

---

* Although the district court concluded that Plaintiff's January 23 and February 17 Facebook activity did not address matters of public concern, *Buker*, 2015 WL 3456750, at *13–14, "[o]ur review is not limited to the grounds the district court relied upon, and we may affirm 'on any basis fairly supported by the record,'" *Lawson*, 828 F.3d at 247 (quoting *Eisenberg v. Wachovia Bank, N.A.*, 301 F.3d 220, 222 (4th Cir. 2002)).

see also Durham v. Jones, 737 F.3d 291, 302 (4th Cir. 2013) ("While [it] is correct that 'concrete evidence' of an actual disruption is not required, there must still be a reasonable apprehension of such a disruption."). Additionally, this Court has previously recognized that "[a] social media platform amplifies the distribution of the speaker's message—which favors the employee's free speech interests—but also increases the potential, in some cases exponentially, for departmental disruption, thereby favoring the employer's interest in efficiency." *Liverman*, 844 F.3d at 407.

[21] For several reasons, we conclude that the Department's interest in efficiency and preventing disruption outweighed Plaintiff's interest in speaking in the manner he did regarding gun control and the Department's social media policy. First, Plaintiff's Facebook activity interfered with and impaired Department operations and discipline as well as working relationships within the Department. "[F]ire companies have a strong interest in the promotion of camaraderie and efficiency" as well as "internal harmony [and] trust," and therefore we accord "substantial weight" to a fire department's interest in limiting dissension and discord. *Goldstein*, 218 F.3d at 355; *see also Janusaitis v. Middlebury Volunteer Fire Dep't*, 607 F.2d 17, 26 (2d Cir. 1979) ("When lives may be at stake in a fire, an *esprit de corps* is essential to the success of the joint endeavor. Carping criticism and abrasive conduct have no place in a small organization that depends upon common loyalty—'harmony among coworkers.'" (quoting *Pickering*, 391 U.S. at 570, 88 S.Ct. 1731)).

Here, Plaintiff's Facebook activity led to "dissension in the [D]epartment" and resulted in "[n]umerous" conversations between at least one battalion chief and lower-level ***346** employees in which the battalion chief "had to[,] ... as a supervisor[,] justify[ ] that it's okay for anybody to say or do anything against the policy." J.A. 550. Additionally, at least one lieutenant perceived Grutzmacher's comment regarding "picking a black one," which Plaintiff "liked," as "referr[ing] to a black person." J.A. 337. Three African-American employees within the Department approached the president of the Phoenix Sentinels—the Howard County affiliate of the International Association of Black Professional Firefighters, a constituent group representing African-American and other minority firefighters—about the posts, with one member stating, "I don't want to work for [Plaintiff] anymore. I don't trust him."* J.A. 240. Accordingly, we accord "substantial weight" to Defendants' interest in preventing Plaintiff from causing further dissension and disharmony.

---

* Although Plaintiff maintains that this testimony is inadmissible hearsay and that the district court should not have considered it, the district court did not rely on the statement for the truth of the matter asserted, but relied on it to illustrate the disruptive effect of Plaintiff's speech. *See United States v. Pratt*, 239 F.3d 640, 644 (4th Cir. 2001) (finding that an out-of-court statement not intended to prove the truth of the matter asserted is not hearsay and, thus, is not excluded by the hearsay rule). Thus, the district court did not err in considering this testimony.

Second, Plaintiff's Facebook activity significantly conflicted with Plaintiff's responsibilities as a battalion chief. Courts have long recognized that "[t]he expressive activities of a highly placed supervisory ... employee will be more disruptive to the operation of the workplace than similar activity by a low-level employee with little authority or discretion." *McEvoy v. Spencer*, 124 F.3d 92, 103 (2d Cir. 1997) (citing authorities); *see also Brown v. Dep't of Transp.*, 735 F.2d 543, 547 (Fed. Cir. 1984) ("[Plaintiff's] position as a supervisor ... weighs heavily on the agency's side."). As a leader within the Department, Plaintiff was responsible for acting as an impartial decisionmaker and "enforcing Departmental policies and taking appropriate action for violations of those policies." J.A. 105. The record demonstrates that Plaintiff's actions led to concerns regarding Plaintiff's fitness as a supervisor and role model, and concerns that Plaintiff's subordinates would not take him seriously if Plaintiff tried to discipline them in the future. By flouting Department policies he was expected to enforce, Plaintiff "violated the trust [his inferiors] have in him to be in his administrative role as a battalion chief, because people count on him to be fair." J.A. 226–27. Accordingly, Plaintiff's managerial position also weighs in the Department's favor.

[22] Third, Plaintiff's speech frustrated the Department's public safety mission and threatened "community trust" in the Department, which is "vitally important" to its function. J.A. 284–85. "[T]he more the employee's job requires ... public contact, the greater the state's interest in firing her for expression that offends her employer." *McEvoy*, 124 F.3d at 103 (alteration in original) (internal quotation marks omitted) (quoting Craig D. Singer, Comment, Conduct and Belief: Public Employees' First Amendment Rights to Free Expression and Political Affiliation, 59 U. *Chi. L. Rev.* 897, 901 (1992)). "[F]irefighters ... are quintessentially public servants. As such, part of their job is to safeguard the public's opinion of them, particularly with regard to a community's view of the respect that ... firefighters accord the members of that community." *Locurto v. Giuliani*, 447 F.3d 159, 178–79 (2d Cir. 2006).

*347 Here, Plaintiff's January 20 post, made while he was on-duty and in his office, "advocat[ed] violence to certain classes of people" and "advocated using violence to [e]ffect a political agenda." J.A. 183, 646. Additionally, the Department reasonably was concerned that Plaintiff's Facebook activity—particularly his "like" of Grutzmacher's comment regarding "black one[s]"—could be interpreted as supporting "racism" or "bias," J.A. 283, and thereby "interfere with the public trust of [Plaintiff] being able to make fair decisions for everybody," J.A. 231; *see also Locurto*, 447 F.3d at 182–83 ("[E]ffective police and fire service presupposes respect for the members of [African-American and other minority] communities, and the defendants were permitted to account for this fact in disciplining the plaintiffs."). The potential for Plaintiff's statements to diminish the Department's standing with the public further weighs in favor of the Department.

Fourth, Plaintiff's speech—particularly his "like" of the image depicting a woman raising her middle finger—"expressly disrespect[ed] [his] superiors." *LeFande v. District of Columbia*, 841 F.3d 485, 495 (D.C. Cir. 2016). A public employee's interest in speaking on matters of public concern "does not require that [a public] employer[ ] tolerate associated behavior that [it] reasonably believed was disruptive and insubordinate." *Dwyer v. Smith*, 867 F.2d 184, 194 (4th Cir. 1989); *see also Connick*, 461 U.S. at 154, 103 S.Ct. 1684 ("The limited First Amendment interest involved here does not require that Connick tolerate action which he reasonably believed would disrupt the office, undermine his authority, and destroy close working relationships."). Here, Plaintiff's "continued unrestrained conduct" after already being reprimanded "'smack[ed] of insubordination.'" *See Graziosi v. City of Greenville*, 775 F.3d 731, 740 (5th Cir. 2015) (quoting *Nixon v. City of Houston*, 511 F.3d 494, 499 (5th Cir. 2007)). Employees within the Department viewed Plaintiff's "like" of Donnelly's Facebook picture of an older woman with her middle finger raised as a "sparring match between the battalion chief and an assistant chief [that publicly] escalated to the level of telling the fire chief to fuck off." J.A. 297–98. Therefore, the disrespectful and insubordinate tone of Plaintiff's relevant Facebook activity also weighs in the Department's favor.

Lastly, we observe that the record is rife with observations of how Plaintiff's Facebook activity, subsequent to Assistant Chief Jerome's request that Plaintiff remove any offending posts, disregarded and upset the chain of command upon which the Department relies. Fire departments operate as "paramilitary" organizations in which "discipline is demanded, and freedom must be correspondingly denied." *Maciariello*, 973 F.2d at 300. Accordingly, we afford fire departments "greater latitude ... in dealing with dissension in their ranks." *Id.* Although the Department's status as a paramilitary organization is not dispositive of the *Pickering* analysis, *see Liverman*, 844 F.3d at 408, it does further tip the scale in the Department's favor.

By contrast, though we recognize that at least some of Plaintiff's speech addressed matters of public concern—gun control and the Department's Social Media Guidelines—the public's interest in Plaintiff speaking on those matters of public concern does not outweigh the significant governmental interests set forth above. In particular, we have recognized that a public safety official's interest in speaking on matters of public concern is sufficient to outweigh the compelling government interests set forth above when, for example, the official's speech is "grounded ... in specialized knowledge [or] expresse[s] a general 'concern about the inability of the [Department] to carry out its vital public ***348** mission effectively.'"* *Liverman*, 844 F.3d at 410 (third alteration in original) (quoting *Cromer v. Brown*, 88 F.3d 1315, 1325–26 (4th Cir. 1996)). For instance, in *Liverman*, we found statements by veteran police officers raising "[s]erious concerns regarding officer training and supervision" were sufficient to overcome the government's interest in preventing workplace disruption. *Id.* at 411; *see also Durham*, 737 F.3d at 302 ("Serious, to say nothing of corrupt, law enforcement misconduct is a substantial concern that must

---

* By identifying speech grounded in a public employee's specialized knowledge or raising questions about public safety as *examples* of public employee speech warranting the highest level of First Amendment protection, we do not suggest that those are the *only* two categories of public employee speech warranting such protection.

be met with a similarly substantial disruption in the calibration of the controlling balancing test."); *Goldstein*, 218 F.3d at 355 ("[T]he substance of the public concern included allegations that some emergency personnel lacked required training and certifications; that the leadership of the company was overlooking violations of safety regulations; and that the conduct of crewmembers was jeopardizing the safety of the crew and of the public. These allegations were a matter of the highest public concern, and as such, they were entitled to the highest level of First Amendment protection." (footnote omitted)). Plaintiff's Facebook activity is not of the same ilk as the speech at issue in *Liverman*, *Durham*, and *Goldstein*, which this Court found sufficient to outweigh the types of significant governmental interests at issue here.

In sum, we conclude the Department's interest in workplace efficiency and preventing disruption outweighed the public interest commentary contained in Plaintiff's Facebook activity. In reaching this conclusion, we emphasize that this balancing test is a "particularized" inquiry. *Goldstein*, 218 F.3d at 356. Therefore, although we resolve the balancing test in favor of the Department, we expressly caution that a fire department's interest in maintaining efficiency will not always outweigh the interests of an employee in speaking on matters of public concern. *See id.*

Because the Department's interest in managing its internal affairs outweighs the public interest in Plaintiff's speech, we need not reach the third prong of the *McVey* test. As such, we conclude that the district court properly granted summary judgment in favor of Defendants on Plaintiff's First Amendment retaliation claim.

B.

Plaintiff also contends that the district court improperly dismissed his facial challenge to the Department's Social Media Guidelines and Code of Conduct as moot. When a plaintiff challenges a government policy "for vagueness or overbreadth, the Supreme Court has concluded that [he] ha[s] standing to assert the rights of third parties whose protected speech may have been impermissibly curtailed by the challenged prohibition, even though as applied to the plaintiff[ ], the [policy] only curtailed unprotected expression." *Brandywine, Inc. v. City of Richmond*, 359 F.3d 830, 835 (6th Cir. 2004) (citing *Young v. Am. Mini Theatres, Inc.*, 427 U.S. 50, 59 n.17, 96 S.Ct. 2440, 49 L.Ed.2d 310 (1976)). Because we find the district court properly granted Defendants' motion for summary judgment against Plaintiff, we decline to review Plaintiff's as-applied facial challenge and review only the district court's determination regarding Plaintiff's third-party facial challenge.

[23] [24] "We review the district court's mootness determination de novo." ***349** S.C. Coastal Conservation League v. U.S. Army Corps of Eng'rs*, 789 F.3d 475, 482 (4th Cir. 2015). A claim becomes moot "when the issues presented are no longer 'live' or the parties lack a legally cognizable interest in the outcome." *County of Los Angeles v. Davis*, 440 U.S. 625, 631, 99 S.Ct. 1379, 59 L.Ed.2d 642 (1979) (internal quotation marks omitted) (quoting *Powell v. McCormack*, 395 U.S. 486, 496, 89 S.Ct. 1944, 23 L.Ed.2d 491 (1969)).

[25] On appeal, Defendants contend that the district court's mootness finding was proper because the Department has repealed the previous Social Media Guidelines and Code of Conduct in operation at the time of Plaintiff's termination; the revised policies did not include any of the provisions Plaintiff challenged in the prior iterations of the policies; and the Department "did not intend to readopt or enforce the challenged prior versions of either policy." Appellees' Br. at 17. Conversely, Plaintiff argues that the Department's subsequent actions have not mooted his facial challenge, as the Department is free to "re-enact the unconstitutional provisions of the old policies." Appellant's Br. at 36. We reject Plaintiff's contention.

[26] "It is well established that a defendant's 'voluntary cessation of a challenged practice' moots an action only if 'subsequent events made it absolutely clear that the allegedly wrongful behavior could not reasonably be expected to recur.'" *Wall v. Wade*, 741 F.3d 492, 497 (4th Cir. 2014) (quoting *Friends of the Earth, Inc. v. Laidlaw Envtl. Servs., Inc.*, 528 U.S. 167, 189, 120 S.Ct. 693, 145 L.Ed.2d 610 (2000)). Here, in addition to adopting a new Social Media Policy and revised Code of Conduct, current Fire Chief Butler submitted a sworn affidavit that, "[a]s head of the Fire Department, [he] fully intend[s] to operate under the newly issued [policies] and do[es] not intend to re-issue the original versions." J.A. 924. Additionally, Defendants' counsel declared at oral argument that the Department has no intent to reenact the offending policies. And from the record, we discern "no hint" that the Department has any intention of reinstituting the prior policies. *See Troiano v. Supervisor of Elections*, 382 F.3d 1276, 1284–85 (11th Cir. 2004). Based on these formal assurances and the absence of any evidence to the contrary, Defendants have met their "heavy burden of persuad[ing]" this Court that they will not revert to the challenged policies. *Wall*, 741 F.3d at 497 (alteration in original) (internal quotation marks omitted) (quoting *Laidlaw*, 528 U.S. at 189, 120 S.Ct. 693); *see Winsness v. Yocom*, 433 F.3d 727, 736 (10th Cir. 2006) (finding that public officials' alteration of challenged policy, coupled with sworn affirmation that they would not revert to policy previously in effect, rendered plaintiff's challenge moot). Thus, the district court properly dismissed Plaintiff's third-party facial challenge as moot.

III.

For these reasons, the judgment of the district court is

AFFIRMED.

ALL CITATIONS

851 F.3d 332, 101 Empl. Prac. Dec. P 45,764, 41 IER Cases 1738

16 Retirement and Welfare Laws

******

Chapter Objectives:

1. Acquire an understanding of the federal retirement laws.
2. Acquire an understanding of the unemployment laws.
3. Identify the impact of these laws with the safety and health function.
4. Acquire an understanding of the Employee Retirement Income Security Act (ERISA) and Social Security.

******

With the "greying" of the American workforce, safety and health professionals are often involved in the correlated human resource areas of retirements, pensions, social security, and unemployment insurance. Although safety and health professionals are definitely not experts in these unique areas of the law, it is important for safety and health professionals to, at a minimum, be able to recognize the laws and be able to understand the potential impact of the specific law on the safety and health function. Safety and health professionals should be aware that the federal retirement and welfare laws may also have correlating state laws which impact this area of the law.

The Employee Retirement Income Security Act (known as ERISA)* is a federal law enacted originally in 1974, and amended by the Multi-Employer Pension Plan amendment in 1980.[†] In 1984, the original statute was amended by the Retirement Equity Act. ERISA established minimum standards for most voluntarily established pension plans as well as health plans by private sector employers. In general, ERISA required pension and health plans to provide participating employees with plan information, including plan features and plan funding, established minimum standards for participation in the plan, establishes vesting periods, benefit accruals, and funding. Additionally, ERISA created fiduciary responsibilities for the entities managing and controlling plan assets, requires grievance and appeal processes, provides a right to sue for benefits and breach of fiduciary duty, and, in the event of a defined benefit plan being terminated, guarantees payment of certain benefits through the Pension Benefit Guaranty Corporation (known as the "PBGC").

---

* 29 U.S.C.A. § 1001 *et seq.*
[†] Public Law, 96-384, Sept. 26, 1980, 94 Stat. 1207.

ERISA is divided into four titles: Title I is administered by the U.S. Department of Labor and addresses employee rights;* Title II is administered by the Internal Revenue Service and is addressed in the IRS Code;† Title III addresses jurisdiction, administration, and enforcement of ERISA between the Department of Labor, IRS, and PDGC;‡ and Title IV creates the PBGC and establishes a protocol for employee plan termination insurance.§ ERISA applies to two types of employee benefit plans, namely, pension plans and welfare plans. Pension plans are designed to provide retirement or deferral income to employees and can be defined benefits plans (retirement amount determined in advance, but the amount contributed may vary) and defined contribution plans (contributions to the fund are fixed, but the retirement benefits are unknown until retirement). Welfare plans include programs providing benefits in the event of employee sickness, accident, hospitalization, death, disability, unemployment, vacation, scholarship funds, daycare services, or other related benefits described in the Labor Management Relations Act.¶

ERISA is a very extensive law with recordkeeping, reporting, and disclosure requirements, as well as specific transaction prohibitions and fiduciary responsibilities. Safety and health professionals should be aware that ERISA provides the Secretary of Labor with broad powers to conduct investigations, subpoena information, and pursue criminal or civil penalties for violation of a subpoena under Section 9 of the Federal Trade Commission Act.** Not unlike OSHA, safety and health professionals should be aware that making false statement or misrepresentation of fact, as well as interference with employee rights and theft or embezzlement can result in fines and/or imprisonment upon conviction.††

Although normally not impacting the safety and health function, the Social Security Act definitely impacts the safety and health professional.‡‡ The Social Security Act provides protection to workers and their families against the loss of income due to retirement, age, disability, and death. Safety and health professionals should be aware that in order to be eligible for social security benefits in the areas of disability, old age, and survivor insurance benefits, insured individuals must fulfill certain work history and age requirements as provided in the statute. The Social Security Administration determines whether an individual has fulfilled the work history requirement by applying a quarter of coverage measurement,§§ and calculating the amount an individual is entitled to involves a very detailed and complicated process. The amount identified after this computation is known as the primary insurance amount, and when an individual meets the requirements to be fully insured, the individual is entitled to the primary insurance amount. Disability benefits are provided to individuals under the age of 65 who are unable to engage in any substantial gainful activity due to any medically determined physical or

---

* 29 U.S.C.A. § 1001–1145.
† 29 U.S.C.A. § 401 *et. seq.*
‡ 29 U.S.C.A. § 1201–1242.
§ 29 U.S.C.A. § 1301–1461.
¶ 29 U.S.C.A. § 186.
** 15 U.S.C.A. § 50.
†† 18 U.S.C.A. § 1027.
‡‡ 42 U.S.C.A. § 415.
§§ 42 U.S.C.A. § 414.

mental impairment that can result in death or can last beyond a minimum of one year.*

Of particular importance for safety and health professionals is the fact that social security benefits are not automatic. Social security benefits are not payable to the insured individual unless an application is submitted to the Social Security Administration.† Additionally, if the claim is denied, the Social Security Administration has an established internal hearing and appeals process which includes a non-adversarial hearing before a hearing officer, an appeal to the Appeals Council of the Bureau of Hearings and Appeals, and a civil action in the appropriate U.S. District Court.

Although most states possess unemployment insurance statutes, safety and health professionals should be aware that the Federal Social Security Act also established a federal-state system of insurance to protect workers and their families against loss of income due to unemployment.‡ Federal employees and veterans of the armed forces are eligible for unemployment compensation benefits if qualified. The individual state unemployment agencies serve as the agents of the Secretary of Labor in processing claims and paying benefits to eligible individuals. In general, the application for benefits processed in the state unemployment insurance office are processed in the same manner and to the same extent as state claims and determined under the individual state unemployment statute.§ Veterans are usually ineligible for unemployment benefits for a period of time after subsistence or educational allowances have ended.¶

Safety and health professionals should be aware that there are other federal unemployment benefits specifically for qualified railroad workers (Railroad Unemployment Insurance Act), as well as workers' compensation-type coverage for injury or death for maritime employment (Longshoremen's and Harbor Workers' Compensation Act), seamen aboard ships on U.S. navigable waters (Jones Act), and federal civilian employees (Federal Employees' Compensation Act).

RETIREMENT PLANS AND ERISA FAQS

WHAT IS ERISA?

The Employee Retirement Income Security Act of 1974, or ERISA, protects the assets of millions of Americans so that funds placed in retirement plans during their working lives will be there when they retire.

ERISA is a federal law that sets minimum standards for retirement plans in private industry. For example, if your employer maintains a retirement plan, ERISA specifies when you must be allowed to become a participant, how long you have to

---

* 42 U.S.C.A. § 423(a)(1) and (d)(1)(a).
† 20 C.F.R. § 404.701.
‡ 72 U.S.C.A. 301 *et seq.*
§ 5 U.S.C.A. § 85002(d).
¶ 5 U.S.C.A. § 8525(b).

work before you have a non-forfeitable interest in your benefit, how long you can be away from your job before it might affect your benefit, and whether your spouse has a right to part of your benefit in the event of your death. Most of the provisions of ERISA are effective for plan years beginning on or after January 1, 1975.

ERISA does not require any employer to establish a retirement plan. It only requires that those who establish plans must meet certain minimum standards. The law generally does not specify how much money a participant must be paid as a benefit.

ERISA does the following:

- Requires plans to provide participants with information about the plan including important information about plan features and funding. The plan must furnish some information regularly and automatically. Some is available free of charge, some is not.
- Sets minimum standards for participation, vesting, benefit accrual and funding. The law defines how long a person may be required to work before becoming eligible to participate in a plan, to accumulate benefits, and to have a non-forfeitable right to those benefits. The law also establishes detailed funding rules that require plan sponsors to provide adequate funding for your plan.
- Requires accountability of plan fiduciaries. ERISA generally defines a fiduciary as anyone who exercises discretionary authority or control over a plan's management or assets, including anyone who provides investment advice to the plan. Fiduciaries who do not follow the principles of conduct may be held responsible for restoring losses to the plan.
- Gives participants the right to sue for benefits and breaches of fiduciary duty.
- Guarantees payment of certain benefits if a defined plan is terminated, through a federally chartered corporation, known as the Pension Benefit Guaranty Corporation.

What is a defined benefit plan?

A defined benefit plan, funded by the employer, promises you a specific monthly benefit at retirement. The plan may state this promised benefit as an exact dollar amount, such as $100 per month at retirement. Or, more often, it may calculate your benefit through a formula that includes factors such as your salary, your age, and the number of years you worked at the company. For example, your pension benefit might be equal to 1 percent of your average salary for the last 5 years of employment times your total years of service.

What is a defined contribution plan?

A defined contribution plan, on the other hand, does not promise you a specific benefit amount at retirement. Instead, you and/or your employer contribute money to your individual account in the plan. In many cases, you are responsible for choosing how these contributions are invested, and deciding how much to contribute from your paycheck through pretax deductions. Your employer may add to your account, in some cases by matching a certain percentage of your contributions. The value of your account depends on how much is contributed and how well the investments perform. At retirement, you receive the balance in your account, reflecting the contributions, investment gains or losses, and any fees charged against your account. The 401(k) plan is a popular type of defined contribution plan. There are four types of 401(k) plans: traditional 401(k), safe harbor 401(k), SIMPLE 401(k), and automatic enrollment 401(k) plans. The SIMPLE IRA plan, SEP, employee stock ownership plan (ESOP), and profit sharing plan are other examples of defined contribution plans.

WHAT ARE SIMPLIFIED EMPLOYEE RETIREMENT PLANS (SEPs)?

Simplified Employee Pension Plan (SEP) – A plan in which the employer makes contributions on a tax-favored basis to individual retirement accounts (IRAs) owned by the employees. If certain conditions are met, the employer is not subject to the reporting and disclosure requirements of most retirement plans. Under a SEP, an IRA is set up by or for an employee to accept the employer's contributions.

WHAT ARE 401(K) PLANS?

401(k) Plan – In this type of defined contribution plan, the employee can make contributions from his or her paycheck before taxes are taken out. The contributions go into a 401(k) account, with the employee often choosing the investments based on options provided under the plan. In some plans, the employer also makes contributions, matching the employee's contributions up to a certain percentage. SIMPLE and safe harbor 401(k) plans have additional employer contribution and vesting requirements.

WHAT ARE PROFIT SHARING PLANS OR STOCK BONUS PLANS?

Profit Sharing Plan – A profit sharing plan allows the employer each year to determine how much to contribute to the plan (out of profits or otherwise) in cash or employer stock. The plan contains a formula for allocating the annual contribution among the participants.

WHAT ARE EMPLOYEE STOCK OWNERSHIP PLANS (ESOPs)?

Employee Stock Ownership Plan (ESOP) – A type of defined contribution plan that is invested primarily in employer stock.

WHO CAN PARTICIPATE IN YOUR EMPLOYER'S RETIREMENT PLAN?

Once you have learned what type of retirement plan your employer offers, you need to find out when you can participate in the plan and begin to earn benefits. Plan rules can vary as long as they meet the requirements under Federal law. You need to check with your plan or review the plan booklet (called the Summary Plan Description) to learn your plan's rules and requirements. Your plan may require you to work for the company for a period of time before you may participate in the plan. In addition, there typically is a time frame for when you begin to accumulate benefits and earn the right to them (sometimes referred to as "vesting").

Find out if you are within the group of employees covered by your employer's retirement plan. Federal law allows employers to include certain groups of employees and exclude others from a retirement plan. For example, your employer may sponsor one plan for salaried employees and another for union employees. Part-time employees may be eligible if they work at least 1,000 hours per year, which is about 20 hours per week. So if you work part-time, find out if you are covered.

WHEN CAN YOUR PARTICIPATION BEGIN?

Once you know you are covered, you need to find out when you can begin to participate in the plan. You can find this information in your plan's Summary Plan Description. Federal law sets minimum requirements, but a plan may be more generous. Generally, a plan may require an employee to be at least 21 years old and to have a year of service with the company before the employee can participate in a plan. However, plans may allow employees to begin participation before reaching age 21 or completing one year of service. For administrative reasons, your participation may be delayed up to 6 months after you meet these age and service criteria, or until the start of the next plan year, whichever is sooner. The plan year is the calendar year, or an alternative 12-month period, that a retirement plan uses for plan administration. Because the rules can vary, it is important that you learn the rules for your plan.

Employers have some flexibility to require additional years of service in some circumstances. For example, if your plan allows you to vest (discussed in detail later) immediately upon participating in the plan, it may require that you work for the company for two years before you may participate in the plan.

Federal law also imposes other participation rules for certain circumstances. For example, if you were an older worker when you were hired, you cannot be excluded from participating in the plan just because you are close to retirement age.

Some 401(k) and SIMPLE IRA plans enroll employees automatically. This means that you will automatically become a participant in the plan unless you choose to opt out. The plan will deduct a set contribution level from your paycheck and put it into a predetermined investment. If your employer has an automatic enrollment plan, you should receive a notice describing the automatic contribution process, when your participation begins, your opportunity to opt out of the plan or change your contribution level, and where your automatic contributions are invested. If you are in a 401(k), the notice will also describe your right to change investments, or if you are in a SIMPLE IRA plan, your right to change the financial institution where your contributions are invested.

WHEN DO YOU BEGIN TO ACCUMULATE BENEFITS?

Once you begin to participate in a retirement plan, you need to understand how you accrue or earn benefits. Your accrued benefit is the amount of retirement benefits that you have accumulated or that have been allocated to you under the plan at any particular point in time.

Defined benefit plans often count your years of service in order to determine whether you have earned a benefit and also to calculate how much you will receive in benefits at retirement. Employees in the plan who work part-time, but who work 1,000 hours or more each year, must be credited with a portion of the benefit in proportion to what they would have earned if they were employed full time. In a defined contribution plan, your benefit accrual is the amount of contributions and earnings that have accumulated in your 401(k) or other retirement plan account, minus any fees charged to your account by your plan.

Special rules for when you begin to accumulate benefits may apply to certain types of retirement plans. For example, in a Simplified Employee Pension Plan (SEP), all participants who earn at least $550 a year from their employers are entitled to receive a contribution.

CAN A PLAN REDUCE PROMISED BENEFITS?

Defined benefit plans may change the rate at which you earn future benefits, but cannot reduce the amount of benefits you have already accumulated. For example, a plan that accrues benefits at the rate of $5 a month for years of service through 2010 may be amended to provide that for years of service beginning in 2011 benefits will be credited at the rate of $4 per month. Plans that make a significant reduction in the rate at which benefits accumulate must provide you with written notice generally at least 45 days before the change goes into effect.

Also, in most situations, if a company terminates a defined benefit plan that does not have enough funding to pay all of the promised benefits, the Pension Benefit Guaranty Corporation (PBGC) will pay plan participants and beneficiaries some retirement benefits, but possibly less than the level of benefits promised. (For more information, see the PBGC's Website.)

In a defined contribution plan, the employer may change the amount of employer contributions in the future. Depending on the plan terms, the employer may also be able to stop making contributions for a few years or indefinitely.

An employer may terminate a defined benefit or a defined contribution plan, but may not reduce the benefit you have already accrued in the plan.

HOW SOON DO YOU HAVE A RIGHT TO YOUR ACCUMULATED BENEFITS?

You immediately vest in your own contributions and the earnings on them. This means you have earned the right to these amounts without the risk of forfeiting them. But note – there are restrictions on actually taking them out of the plan.

However, you do not necessarily have an immediate right to any contributions made by your employer. Federal law provides a maximum number of years a company may require employees to work to earn the vested right to all or some of these benefits.

In a defined benefit plan, an employer can require that employees have 5 years of service in order to become 100 percent vested in the employer funded benefits (called cliff vesting). Employers also can choose a graduated vesting schedule, which requires an employee to work 7 years in order to be 100 percent vested, but provides at least 20 percent vesting after 3 years, 40 percent after 4 years, 60 percent after 5 years, and 80 percent after 6 years of service. Plans may provide a different schedule as long as it is more generous than these vesting schedules. (Unlike most defined benefit plans, in a cash balance plan, employees vest in employer contributions after 3 years.)

In a defined contribution plan such as a 401(k) plan, you are always 100 percent vested in your own contributions to a plan, and in any subsequent earnings from your contributions. However, in most defined contribution plans you may have to work several years before you are vested in the employer's matching contributions. (There are exceptions, such as the SIMPLE 401(k) and the safe harbor 401(k), in which you are immediately vested in all required employer contributions. You also vest immediately in the SIMPLE IRA and the SEP.)

Currently, employers have a choice of two different vesting schedules for employer matching 401(k) contributions. Your employer may use a schedule in which employees are 100 percent vested in employer contributions after 3 years of service (cliff vesting). Under graduated vesting, an employee must be at least 20 percent vested after 2 years, 40 percent after 3 years, 60 percent after 4 years, 80 percent after 5 years, and 100 percent after 6 years. If your automatic enrollment 401(k) plan requires employer contributions, you vest in those contributions after 2 years. Automatic enrollment 401(k) plans with optional matching contributions follow one of the vesting schedules noted above.

Employers making other contributions to defined contribution plans, such as a 401(k) plan, also can choose between two vesting schedules. For those contributions made since 2007, they can choose between the graduated and cliff vesting schedules. For contributions made prior to 2007, they can choose between schedules.

You may lose some of the employer-provided benefits you have earned if you leave your job before you have worked long enough to be vested. However, once vested, you have the right to receive the vested portion of your benefits even if you leave your job before retirement. But even though you have the right to certain benefits, your defined contribution plan account value could decrease after you leave your job as a result of investment performance.

Note

If you leave your company and return, you may be able to count your earlier period of employment towards the years of service needed for vesting in the employer-provided benefits. Unless your break in service with the company was 5 years or a time equal to the length of your pre-break employment, whichever is greater, you likely can count that time prior to your break. Because these rules are very specific, you should read your plan document carefully if you are contemplating a short-term break from your employer, and then discuss it with your plan administrator. If you left employment prior to January 1, 1985, different rules apply. For more information, contact the Department of Labor toll free at 1.866.444.3272.

For Reserve and National Guard units called to active duty, the Uniformed Services Employment and Reemployment Rights Act (USERRA) requires that the period of military duty be counted as covered service with the employer for eligibility, vesting, and benefit accrual purposes. Returning service members are treated as if they had been continuously employed regardless of the type of retirement plan the employer has adopted. However, a person who is reemployed is entitled to accrued benefits resulting from employee contributions only to the extent that he or she actually makes the contributions to the plan.

INFORMATION PROVIDED BY THE RETIREMENT PLAN

Each retirement plan is required to have a formal, written plan document that details how it operates and its requirements. As noted previously, there is also a booklet that describes the key plan rules, called the Summary Plan Description (SPD), which should be much easier to read and understand. The SPD also should include a summary of any material changes to the plan or to the information required to be in the SPD. In many cases, you can start with the SPD and then look at the plan document if you still have questions.

In addition, plans must provide you with a number of notices.

For example, defined contribution plans, such as 401(k) plans, generally are required to provide advance notice to employees when a "blackout period" occurs. A blackout period is when a participant's right to direct investments, take loans, or obtain distributions is suspended for a period of at least three consecutive business days. Blackout periods can often occur when plans change recordkeepers or investment options.

Some plan information, such as the Summary Plan Description, must be provided to you automatically and without charge at the time periods indicated below. You may request a Summary Plan Description at other times, but your employer might charge you a copying fee. You must ask the plan if you want other information, such as a copy of the written plan document or the plan's Form 5500 annual financial report, and you may have to pay a copying fee. Many employers provide benefit information on a Website.

In some cases, plans provide information more frequently than required by Federal law. For instance, some plans allow participants to check their statements online or by telephone.

The plan's annual financial report (Form 5500) is also available (there is a copying fee if over 100 pages) by contacting the U.S. Department of Labor, EBSA Public Disclosure Facility, Room N-1513, 200 Constitution Avenue, NW, Washington, D.C. 20210, Tel: 202.693.8673. For annual reports for 2009 and later years, you can also find the report online at www.efast.dol.gov. In addition, if your plan administrator does not provide you, as a participant covered under the plan, with a copy of the Summary Plan Description automatically or after you request it, you may contact the Department of Labor toll free at 1.866.444.3272 for help.

WHAT PLAN INFORMATION SHOULD YOU REVIEW REGULARLY?

If you are in a defined benefit plan, you will receive an individual benefit statement once every 3 years. Review its description of the total benefits you have earned and whether you are vested in those benefits. Also check to make sure your date of birth, date of hire, and the other information included is correct. You will also receive an annual notice of the plan's funding status.

Defined contribution plans, including 401(k) plans, also must send participants individual benefit statements either quarterly, if participants direct investments of their accounts, or annually, if they do not. When you receive a statement, check it to make sure all of the information is accurate. This information may include:

- Salary level
- Amounts that you and your employer have contributed
- Years of service with the employer
- Home address
- Social Security number
- Beneficiary designation
- Marital status
- The performance of your investments (defined contribution plan participants)
- Fees paid by the plan and/or charged to participants. (For more information, see the Department of Labor brochure A Look at 401(k) Plan Fees or call the Department of Labor toll free at 1.866.444.3272.) Check with your plan to see if this information is included in materials on your investment options, the benefit statement, the Summary Plan Description or the plan's Annual Report (Form 5500).

When can you begin to receive retirement benefits?

There are several points to keep in mind in determining when you can receive benefits:

1. Federal law provides guidelines for when plans must start paying retirement benefits. Under Federal law, your plan must allow you to begin receiving benefits
   - the later of - Reaching age 65 or the age your plan considers to be normal retirement age (if earlier)
   - Or - 10 years of service
   - Or - Terminating your service with the employer
2. Plans can choose to start paying benefits sooner. The plan documents will state when you may begin receiving payments from your plan.
3. You must file a claim for benefits for your payments to begin. This takes some time for administrative reasons.

For administrative reasons, benefits do not begin immediately after meeting these conditions. At a minimum, your plan must provide that you will start receiving benefits within 60 days after the end of the plan year in which you satisfy the conditions. Also, you need to file a claim under your plan's procedures.

Under certain circumstances, your benefit payments may be suspended if you continue to work beyond normal retirement age. The plan must notify you of the suspension during the first calendar month or payroll period in which payments are withheld. This information should also be included in the Summary Plan Description. A plan also must advise you of its procedures for requesting an advance determination of whether a particular type of reemployment would result in a suspension of benefit payments. If you are a retiree and are considering taking a job, you may wish to write to your plan administrator and ask if your benefits would be suspended.

Listed below are some permitted variations:

- Although defined benefit plans and money purchase plans generally allow you to receive benefits only when you reach the plan's retirement age, some have provisions for early retirement.
- 401(k) plans often allow you to receive your account balance when you leave your job.
- 401(k) plans may allow for distributions while still employed if you have reached age 59½ or if you suffer a hardship.
- Profit sharing plans may permit you to receive your vested benefit after a specific number of years or whenever you leave your job.
- A phased retirement option allows employees at or near retirement age to reduce their work hours to part time, receive benefits, and continue to earn additional funds.
- ESOPs do not have to pay out any benefits until 1 year after the plan year in which you retire, or as many as 6 years if you leave for reasons other than retirement, death, or disability.

Warning

1. You may owe current income taxes – and possibly tax penalties – on your distribution if you take money out before age 59½, unless you transfer it to an IRA or another tax-qualified retirement plan.
2. Taking all or a portion of your funds out of your account before retirement age will mean you have less in retirement benefits.

When is the latest you may begin to take payment of your benefits?

Federal law sets a mandatory date by which you must start receiving your retirement benefits, even if you would like to wait longer. This mandatory start date generally is set to begin on April 1 following the calendar year in which you turn 70½ or, if later, when you retire. However, your plan may require you to begin receiving distributions even if you have not retired by age 70½.

In what form will your benefits be paid?

If you are in a defined benefit or money purchase plan, the plan must offer you a benefit in the form of a life annuity, which means that you will receive equal, periodic payments, often as a monthly benefit, which will continue for the rest of your life. Defined benefit and money purchase plans may also offer other payment options, so check with the plan. If you are in a defined contribution plan (other than a money purchase plan), the plan may pay your benefits in a single lump sum payment as well as offer other options, including payments over a set period of time (such as 5 or 10 years) or an annuity with monthly lifetime payments.

CAN A BENEFIT CONTINUE FOR YOUR SPOUSE SHOULD YOU DIE FIRST?

In a defined benefit or money purchase plan, unless you and your spouse choose otherwise, the form of payment will include a survivor's benefit. This survivor's benefit, called a qualified joint and survivor annuity (QJSA), will provide payments over your lifetime and your spouse's lifetime. The benefit payment that your surviving spouse receives must be at least half of the benefit payment you received during your joint lives. If you choose not to receive the survivor's benefit, both you and your spouse must receive a written explanation of the QJSA and, within certain time limits, you must make a written waiver and your spouse must sign a written consent to the alternative payment form without a survivor's benefit. Your spouse's signature must be witnessed by a notary or plan representative.

In most 401(k) plans and other defined contribution plans, the plan is written so different protections apply for surviving spouses. In general, in most defined contribution plans, if you should die before you receive your benefits, your surviving spouse will automatically receive them. If you wish to select a different beneficiary, your spouse must consent by signing a waiver, witnessed by a notary or plan representative.

If you were single when you enrolled in the plan and subsequently married, it is important that you notify your employer and/or plan administrator and change your status under the plan. If you do not have a spouse, it is important to name a beneficiary.

If you or your spouse left employment prior to January 1, 1985, different rules apply. For more information on these rules, contact the Department of Labor toll free at 1.866.444.3272.

CAN YOU BORROW FROM YOUR 401(K) PLAN ACCOUNT?

401(k) plans are permitted to – but not required to – offer loans to participants. The loans must charge a reasonable rate of interest and be adequately secured. The plan must include a procedure for applying for the loans and the plan's policy for granting them. Loan amounts are limited to the lesser of 50 percent of your account balance or $50,000 and must be repaid within 5 years (unless the loan is used to purchase a principal residence).

CAN YOU GET A DISTRIBUTION FROM YOUR PLAN IF YOU ARE NOT YET 65 OR YOUR PLAN'S NORMAL RETIREMENT AGE BUT ARE FACING A SIGNIFICANT FINANCIAL HARDSHIP?

Again, defined contribution plans are permitted to – but not required to – provide distributions in case of hardship. Check your plan booklet to see if it does permit them and what circumstances are included as hardships.

If you are in a defined benefit plan (other than a cash balance plan), you most likely will be required to leave the benefits with the retirement plan until you become eligible to receive them. As a result, it is very important that you update your personal information with the plan administrator regularly and keep current on any changes in your former employer's ownership or address.

If you are in a cash balance plan, you probably will have the option of transferring at least a portion of your account balance to an individual retirement account or to a new employer's plan.

If you leave your employer before retirement age and you are in a defined contribution plan (such as a 401(k) plan), in most cases you will be able to transfer your account balance out of your employer's plan.

WHAT CHOICES DO YOU HAVE FOR TAKING YOUR DEFINED CONTRIBUTION BENEFITS?

- A lump sum – you can choose to receive your benefits as a single payment from your plan, effectively cashing out your account. You may need to pay income taxes on the amount you receive, and possibly a penalty.
- A rollover to another retirement plan – you can ask your employer to transfer your account balance directly to your new employer's plan if it accepts such transfers.
- A rollover to an IRA – you can ask your employer to transfer your account balance directly to an individual retirement account (IRA).
- If your account balance is less than $5,000 when you leave the employer, the plan can make an immediate distribution without your consent. If this distribution is more than $1,000, the plan must automatically roll the funds into an IRA it selects, unless you elect to receive a lump sum payment or to roll it over into an IRA you choose. The plan must first send you a notice allowing you to make other arrangements, and it must follow rules regarding what type of IRA can be used (i.e., it cannot combine the distribution with savings you have deposited directly in an IRA). Rollovers must be made to an entity that is qualified to offer individual retirement plans. Also, the rollover IRA must have investments designed to preserve principal. The IRA provider may not charge more in fees and expenses for such plans than it would to its other individual retirement plan customers.

Please note: If you elect a lump sum payment and do not transfer the money to another retirement account (employer plan or IRA other than a Roth IRA), you will owe a tax penalty if you are under age 59½ and do not meet certain exceptions. In addition, you may have less to live on during your retirement. Transferring your retirement plan account balance to another plan or an IRA when you leave your job will protect the tax advantages of your account and preserve the benefits for retirement.

WHAT HAPPENS IF YOU LEAVE A JOB AND LATER RETURN?

If you leave an employer for whom you have worked for several years and later return, you may be able to count those earlier years toward vesting. Generally, a plan must preserve the service credit you have accumulated if you leave your employer and then return within five years. Service credit refers to the years of service that count towards vesting. Because these rules are very specific, you should read your plan document carefully if you are contemplating a short-term break from your employer, and then discuss it with your plan administrator. If you left employment prior to January 1, 1985, different rules apply.

If you retire and later go back to work for a former employer, you must be allowed to continue to accrue additional benefits, subject to a plan limit on the total years of service credited under the plan.

How do you make a claim for benefits?

Federal retirement law requires all plans to have a reasonable written procedure for processing your benefits claim and appeal if your claim is denied. The Summary Plan Description (SPD) should include your plan's claims procedures. Usually, you fill out the required paperwork and submit it to the plan administrator, who then can tell you what your benefits will be and when they will start.

If there is a problem or a dispute about whether you qualify for benefits or what amount you should receive, check your plan's claims procedure. Federal law outlines the following claims procedures requirements:

- Once your claim is filed, the plan can take up to 90 days to reach a decision, or 180 days if it notifies you that it needs an extension.
- If your claim is denied, you must receive a written notice, including specific information about why your claim was denied and how to file an appeal.
- You have 60 days to request a full and fair review of your denied claim, using your plan's appeals procedure.
- The plan can take up to 60 days to review your appeal, as well as an additional 60 days if it notifies you of the need for an extension. The plan must then send a written notice telling you whether the appeal was granted or denied.
- If the appeal is denied, the written notice must tell you the reason, describe any additional appeal levels, and give you a statement regarding your rights to seek judicial review of the plan's decision.

If you believe the plan failed to follow ERISA's requirements, you may decide to seek legal advice if the plan denies your appeal. You also can contact the Department of Labor concerning your rights under ERISA by calling toll free 1.866.444.3272.

For more information on claims procedures, see the Department of Labor publication Filing a Claim for Your Retirement Benefits or call toll free 1.866.444.3272.

Does your plan have to identify those responsible for operating the plan?

In every retirement plan, there are individuals or groups of people who use their own judgment or discretion in administering and managing the plan or who have the power to or actually control the plan's assets. These individuals or groups are called plan fiduciaries. Fiduciary status is based on the functions that the person performs for the plan, not just the person's title.

A plan must name at least one fiduciary in the written plan document, or through a process described in the plan, as having control over the plan's operations. This fiduciary can be identified by office or by name. For some plans, it may be an administrative committee or the company's board of directors. Usually, a plan's fiduciaries will include the trustee, investment managers, and the plan administrator. The plan administrator is usually the best starting point for questions you might have about the plan.

WHAT ARE THE RESPONSIBILITIES OF PLAN FIDUCIARIES?

Fiduciaries have important responsibilities and are subject to certain standards of conduct because they act on behalf of the participants in the plan. These responsibilities include:

- Acting solely in the interest of plan participants and their beneficiaries, with the exclusive purpose of providing benefits to them;
- Carrying out their duties with skill, prudence, and diligence;
- Following the plan documents (unless inconsistent with ERISA);
- Diversifying plan investments;
- Paying only reasonable expenses of administering the plan and investing its assets; and
- Avoiding conflicts of interest.

The fiduciary also is responsible for selecting the investment providers and the investment options, and for monitoring their performance. Some plans, such as most 401(k) or profit sharing plans, can be set up to permit participants to choose the investments in their accounts (within certain investment options provided by the plan). If the plan is properly set up to give participants control over their investments, then the fiduciary is not liable for losses resulting from the participant's investment decisions. Department of Labor rules provide guidance designed to make sure participants have sufficient information on the specifics of their investment options so they can make informed decisions. This information includes:

- A description of each investment option, including the investment goals, risk, and return characteristics;
- Information about any designated investment managers;
- An explanation of when and how to request changes in investments, plus any restrictions on when you can change investments;
- A statement of the fees that may be charged to your account when you change investment options or buy and sell investments; and
- The name, address, and telephone number of the plan fiduciary or other person designated to provide certain additional information on request.

A statement that the plan is intended to follow the Department of Labor rules and that the fiduciaries may be relieved of liability for losses that are the direct and necessary result of a participant's investment instructions also must be included.

For an automatic enrollment plan, such as a 401(k), the plan fiduciary selects the investments for employees' automatic contributions if the employees do not provide direction. If the plan is properly set up, using certain default investments that generally minimize the risk of large losses and provide long term growth, and providing notice of the plan's automatic enrollment process, then the fiduciary may be relieved of liability for losses resulting from investing in these default alternatives for participants. The plan also must provide a broad range of investments for participants to choose from and information on the plan's investments so participants can make informed decisions. Department of Labor rules provide guidance on the default investment alternatives that can be used and the notice and information to be provided to participants.

WHAT IF A PLAN FIDUCIARY FAILS TO CARRY OUT ITS RESPONSIBILITIES?

Fiduciaries that do not follow the required standards of conduct may be personally liable. If the plan lost money because of a breach of their duties, fiduciaries would have to restore those losses, or any profits received through their improper actions. For example, if an employer did not forward participants' 401(k) contributions to the plan, they would have to pay back the contributions to the plan as well as any lost earnings, and return any profits they improperly received. Fiduciaries also can be removed from their positions as fiduciaries if they fail to follow the standards of conduct.

WHEN DOES THE EMPLOYER NEED TO DEPOSIT EMPLOYEE CONTRIBUTIONS IN THE PLAN?

If you contribute to your retirement plan through deductions from your paycheck, then the employer must follow certain rules to make sure that it deposits the contributions in a timely manner. The law says that the employer must deposit participant contributions as soon as it is reasonably possible to separate them from the company's assets, but no later than the 15th business day of the month following the payday. For small plans (those with fewer than 100 participants), salary reduction contributions deposited with the plan no later than the 7th business day following withholding by the employer will be considered contributed in compliance with the law. In the Annual Report (Form 5500), the plan administrator is required to include information on whether deposits of contributions were made on a timely basis. For more information, see the Department of Labor's Ten Warning Signs That Your 401(k) Contributions Are Being Misused for indicators of possible delays in depositing contributions.

WHAT ARE THE PLAN FIDUCIARIES' OBLIGATIONS REGARDING THE FEES AND EXPENSES PAID BY THE PLAN? CAN THE PLAN CHARGE MY DEFINED CONTRIBUTION PLAN ACCOUNT FOR FEES?

Plan fiduciaries have a specific obligation to consider the fees and expenses paid by your plan for its operations. ERISA's fiduciary standards, discussed above, mean that fiduciaries must establish a prudent process for selecting investment alternatives and service providers to the plan; ensure that fees paid to service providers and other expenses of the plan are reasonable in light of the level and quality of services provided; select investment alternatives that are prudent and adequately diversified; and monitor investment alternatives and service providers once selected to see that they continue to be appropriate choices.

The plan may deduct fees from your defined contribution plan account. Plan administration fees and investment fees can be deducted from your account either as a direct charge or indirectly as a reduction of your account's investment returns. Fees for individual services, such as for processing a loan from the plan or a Qualified Domestic Relations Order, also may be charged to your account.

For more information, see the Department of Labor brochure A Look at 401(k) Plan Fees or call the Department of Labor toll free at 1.866.444.3272.

WHAT HAPPENS WHEN A PLAN IS TERMINATED?

Federal law provides some measures to protect employees who participated in plans that are terminated, both defined benefit and defined contribution. When a plan is terminated, the current employees must become 100 percent vested in their accrued benefits. This means you have a right to all the benefits that you have earned at the time of the plan termination, even benefits in which you were not vested and would have lost if you had left the employer. If there is a partial termination of a plan (for example, if your employer closes a particular plant or division that results in the end of employment of a substantial percentage of plan participants), the affected employees must be immediately 100 percent vested to the extent the plan is funded.

WHAT IF YOUR TERMINATED DEFINED BENEFIT PLAN DOES NOT HAVE ENOUGH MONEY TO PAY THE BENEFITS?

The Federal government, through the Pension Benefit Guaranty Corporation (PBGC), insures most private defined benefit plans. For terminated defined benefit plans with insufficient money to pay all of the benefits, the PBGC will guarantee the payment of your vested pension benefits up to the limits set by law. For further information on plan termination guarantees, contact the Pension Benefit Guaranty Corporation toll free at 1.800.400.7242, or visit the Website.

WHAT HAPPENS IF A DEFINED CONTRIBUTION PLAN IS TERMINATED?

The PBGC does not guarantee benefits for defined contribution plans. If you are in a defined contribution plan that is in the process of terminating, the plan fiduciaries and trustees should take actions to maintain the plan until they terminate it and pay out the assets.

Is your accrued benefit protected if your plan merges with another plan?

Your plan rules and investment choices are likely to change if your company merges with another. Your employer may choose to merge your plan with another plan. If your plan is terminated as a result of the merger, the benefits that you have accrued cannot be reduced. You must receive a benefit that is at least equal to the benefit you were entitled to before the merger. In a defined contribution plan, the value of your account may still fluctuate after the merger based on the performance of the investments.

Special rules apply to mergers of multiemployer defined benefit plans, which generally are under the jurisdiction of the PBGC. Contact the PBGC for further information.

What if your employer goes bankrupt?

Generally, your retirement assets should not be at risk if your employer declares bankruptcy. Federal law requires that retirement plans fund promised benefits adequately and keep plan assets separate from the employer's business assets. The funds must be held in trust or invested in an insurance contract. The employers' creditors cannot make a claim on retirement plan funds. However, it is a good idea to confirm that any contributions your employer deducts from your paycheck are forwarded to the plan's trust or insurance contract in a timely manner.

Significant business events such as bankruptcies, mergers, and acquisitions can result in employers abandoning their individual account plans (e.g., 401(k) plans), leaving no plan fiduciary to manage it. In this situation, participants often have great difficulty in accessing the benefits they have earned and have no one to contact with questions. Custodians such as banks, insurers, and mutual fund companies are left holding the assets of these plans but do not have the authority to terminate the plans and distribute the assets. In response, the Department of Labor issued rules to create a voluntary process for the custodian to wind up the plan's business so that benefit distributions can be made and the plan terminated. Information about this program can be found on the Department of Labor's Website.

Can other people make claims against your benefit (divorce)?

In general, your retirement plan is safe from claims by other people. Creditors to whom you owe money cannot make a claim against funds that you have in a retirement plan. For example, if you leave your employer and transfer your 401(k) account into an individual retirement account (IRA), creditors generally cannot get access to those IRA funds even if you declare bankruptcy.

Federal law does make an exception for family support and the division of property at divorce. A state court can award part or all of a participant's retirement benefit to the spouse, former spouse, child, or other dependent. The recipient named in the order is called the alternate payee. The court issues a specific court order, called a domestic relations order, which can be in the form of a state court judgment, decree or order, or court approval of a property settlement agreement. The order must relate to child support, alimony, or marital property rights, and must be made under state domestic relations law. The plan administrator determines if the order is a qualified domestic relations order (QDRO) under the plan's procedures and then notifies the participant and the alternate payee. If the participant is still employed, a QDRO can require payment to the alternate payee to begin on or after the participant's earliest possible retirement age available under the plan. These rules apply to both defined benefit and defined contribution plans. For additional information, see EBSA's publication, QDROs – The Division of Retirement Benefits Through Qualified Domestic Relations Orders, available by calling toll free 1.866.444.3272 or on the Website.

If you are involved in a divorce, you should discuss these issues with your plan administrator and your attorney.

WHAT DO YOU DO IF YOU HAVE A PROBLEM?

Sometimes, retirement plan administrators, managers, and others involved with the plan make mistakes. Some examples include:

- Your 401(k) or individual account statement is consistently late or comes at irregular intervals;
- Your account balance does not appear to be accurate;
- Your employer fails to transmit your contribution to the plan on a timely basis;
- Your plan administrator does not give or send you a copy of the Summary Plan Description; or
- Your benefit is calculated incorrectly.

It is important for you to know that you can follow up on any possible mistakes without fear of retribution. Employers are prohibited by law from firing or disciplining employees to avoid paying a benefit, as a reprisal for exercising any of the rights provided under a plan or Federal retirement law (ERISA), or for giving information or testimony in any inquiry or proceeding related to ERISA.

START WITH YOUR EMPLOYER AND/OR PLAN ADMINISTRATOR

If you find an error or have a question, in most cases, you can start by looking for information in your Summary Plan Description. In addition, you can contact your employer and/or the plan administrator and ask them to explain what has happened and/or make a correction.

IS IT POSSIBLE TO SUE UNDER ERISA?

Yes, you have a right to sue your plan and its fiduciaries to enforce or clarify your rights under ERISA and your plan in the following situations:

- To appeal a denied claim for benefits after exhausting your plan's claims review process;
- To recover benefits due you;
- To clarify your right to future benefits;

- To obtain plan documents that you previously requested in writing but did not receive;
- To address a breach of a plan fiduciary's duties; or
- To stop the plan from continuing any act or practice that violates the terms of the plan or ERISA.

WHAT IS THE ROLE OF THE LABOR DEPARTMENT?

The U.S. Department of Labor's Employee Benefits Security Administration (EBSA) is the agency responsible for enforcing the provisions of ERISA that govern the conduct of plan fiduciaries, the investment and protection of plan assets, the reporting and disclosure of plan information, and participants' benefit rights and responsibilities.

However, not all retirement plans are covered by ERISA. For example, Federal, state, or local government plans and some church plans are not covered.

The Department of Labor enforces the law by informally resolving benefit disputes, conducting investigations, and seeking correction of violations of the law, including bringing lawsuits when necessary.

The Department has benefits advisors committed to providing individual assistance to participants and beneficiaries. Participants will receive information on their rights and responsibilities under the law and help in obtaining benefits to which they are entitled.

Contact a benefits advisor by calling toll free at 1.866.444.3272 or electronically at https://www.askebsa.dol.gov/.

WHAT OTHER FEDERAL AGENCIES CAN ASSIST PARTICIPANTS AND BENEFICIARIES?

The Pension Benefit Guaranty Corporation (PBGC) is a Federally created corporation that guarantees payment of certain pension benefits under most private defined benefit plans when they are terminated with insufficient money to pay benefits.

You may contact the PBGC at:

Pension Benefit Guaranty Corporation
1200 K Street, NW
Washington, DC 20005-4026
Tel: 202.326.4000
Toll free: 1.800.400.PBGC (7242)

The Treasury Department's Internal Revenue Service is responsible for the rules that allow tax benefits for both employees and employers related to retirement plans, including vesting and distribution requirements. The IRS maintains a taxpayer assistance line for retirement plans at: 1.877.829.5500 (toll-free number). The call center is open Monday through Friday.*

---

* www.dol.gov.

17 Emerging Issues
Guns, Grass, Drones, and the Safety Profession

Chapter Objectives:

1. Acquire a knowledge of the issues involving the legalization of marijuana.
2. Understand the issues involved in the 4th Amendment and gun issues in the workplace.
3. Acquire an understanding of the issues involved in the use of drones in the workplace.
4. Acquire an understanding of the issues regarding the safety "profession."

SAFETY AND HEALTH PROFESSION

Is safety and health truly a "profession," or is it simply an operational position within a corporation or a function required to meet regulatory needs? In comparing safety and health to other identifiable "professions," such as medicine or law, arguably safety and health appears to be deficient. According to the Cambridge English dictionary, a "profession" is defined as "any type of work, especially one that needs a high level of education or a particular skill."* In analyzing the safety and health profession, there are no specific educational requirements prescribed by law to be a safety and health professional. There are no specific qualifications necessary to be a safety and health professional. And there is no competency testing required to manage the safety and health function.

In analyzing the medical and legal professions, each of these professions required substantial and high levels of education, both possess licensure requirements established by the state, possess competency testing, require continuing education, and have a mandatory code of professional conduct which is enforceable. When analyzing the safety and health profession, there is no educational requirement, no licensure requirements, no competency testing, no required continuing education requirements, and no mandatory and enforceable code of professional conduct. The safety and health profession does have several safety and health focused organizations, such as the American Society of Safety Professionals and the National Safety Council, which offer voluntary competency education and specified levels, and they do possess a code of professional conduct; however, this code is largely unenforceable outside of the membership.

When comparing the requirements of the legal or medical professions with that of the safety and health profession, is the safety and health profession truly a profession?

---

* See www.dictionary.cambridge.org.

Arguably, the safety and health profession, a function which safeguards the lives and welfare of large number of workers throughout the United States and beyond, does not meet the criteria to be called a "profession." In its current status, individual companies or organizations can determine the criteria from which they select the individual(s) who will be responsible for the safety and health of their employees. The individual company or organization establishes the educational requirements and enforces the individual's conduct through internal company policy. Arguably, a company or organization could establish the educational level at high school level, and hire any individual achieving a high school diploma and designate this individual as their safety and health professional.

For the individuals reading this text who are or who plan to work within the safety and health function, please ask yourself the question whether or not safety and health is truly a profession. For Occupational Safety and Health Administration (OSHA) and other governmental entities, please ask what qualifications and level of education should the individual(s) responsible for the safety and health of the American workforce possess to be effective in reducing or eliminating risks and thus injuries and illnesses in the workplace.

Has the risk of injury or illness in the workplace reduced or increased through permitting unqualified individuals to serve in the important safety and health function in the workplace?

Arguably, for safety and health to truly be consider a profession along the lines of the medical and legal professions, federal and/or state regulatory bodies should consider the establishment of a licensure requirement for individuals working within the safety and health function with minimum continuing education requirements. Furthermore, the federal and/or state regulatory bodies should establish a mandatory code of professional conduct which is enforceable through suspension or removal of the license, as well as civil and criminal penalties. To enter the safety and health profession, the federal and/or state regulatory bodies should establish minimum educational levels and appropriate competency testing should be established to ensure the skill and ability levels of those entering the safety and health profession.

Who do you want to be responsible for your safety and health in the workplace? Is there a difference between a high school graduate being placed in a safety and health job and a graduate from an undergraduate occupational safety and health program from a university who has successfully completed a competency examination, is required to continue his/her education throughout his/her career, is licensed, and is required to follow a specified conde of professional conduct? The answer should be crystal clear. If we are truly to be recognized as a profession paralleling the medical, legal, and teaching professions, safety and health professionals must bolster the requirements to enter and working within the function for the safety and health profession to truly be called a "profession."

GUNS AND THE WORKPLACE

With the 4th Amendment of the United States Constitution providing citizens with the rights to own and possess firearms and many states having concealed carry laws permitting qualified individuals to acquire a license to carry a concealed weapon, many individuals believe that they can rightfully and legally carry a concealed weapon in all locations. However, with the prevalence of workplace shootings, school shootings, and other gun-related injuries and deaths, many companies and organizations have established policies and protocols prohibiting employees and others from carrying a gun or other weapon on company property.

For most safety and health professionals, as well as companies and organizations, the risk of harm which could result from employees carrying concealed weapons usually far exceeds the individual employee's right to carry a concealed weapon in the workplace. Generally, safety and health professionals should be aware of the potential legal liabilities and risks involved, including, but not limited to, workers' compensation claims, OSH Act violations (under the General Duty clause), negligence actions, and actions under state tort laws for injuries resulting from the discharge of a weapon on company property. In order to limit potential legal liabilities, many companies and organizations have established policies in which firearms, concealed or open, as well as all items of weaponry, are prohibited in the workplace or on company-owned or controlled property. Although state laws vary widely, companies and organizations have generally been permitted to implement and enforce policies prohibiting firearms, as well as other identified weapons from the workplace.

Safety and health professionals should become familiar with not only the federal laws, but also individual state laws regarding concealed carry, due to the fact that both may impact whether an employee is permitted to carry a concealed weapon in the workplace. Additionally, safety and health professionals should be aware that in certain states, there can be a legal risk associated with prohibiting employees from carrying a concealed weapon. Under several state concealed carry laws, employees and others are permitted to keep their firearm in their personal vehicle, even if the vehicle is on company property. Most states require a posted notice to employees when implementing a policy and protocol to ban firearms and weapons on company property.

In developing a policy and protocol for a firearm and weapon-free workplace, safety and health professionals should be very specific in the wording utilized in the policy to avoid potential conflicts or misunderstandings. Additionally, safety and health professionals should identify when the policy goes into effect and note that the policy will be strictly enforced. Posting of the policy and protocol, as well as appropriate signage, is usually required. Additionally, to ensure complete clarity, terminology, such as firearm and weapon, should be clearly defined, as well as the scope and boundaries in which the policy is in effect. Careful consideration should be provided to the requirements of the individual state laws with regards to parking lots, company vehicles, and job sites off company property.

As with most human resource policies, safety and health professionals should address the prohibited conduct (e.g., bringing a concealed handgun to work), as well as the prescribed disciplinary action (disciplinary action up to and including discharges). Depending on the company's operations, safety and health professionals should consider the inclusion of temporary workers, subcontractors, and other nonemployees within the scope of the policy if working on company property. Additionally, safety and health professionals should consider incorporating specific information in the policy regarding the reasonable search of employee lockers, employee vehicles, and company property, as well as desks, boxes, lunch boxes, and clothing bags. Within the policy should be specific language identifying what would happen if an employee refused to permit the search (e.g., voluntary termination) as well as nonemployees, such as vendors, who refuse to permit a search (e.g., prohibited from entry).

Safety and health professionals should be aware that the issue of concealed carry could also impact your emergency and disaster program and workplace violence program, as well as such programs as active shooter protocols. Prudent safety and health professionals should work with your human resource department and legal counsel to ensure that all federal and state laws are addressed when developing a firearms and weapon-free workplace policy. Safety and health professionals should be aware that this controversial issue will continue within our society; however, it is your job to safeguard all employees and others within your workplace. Remember, the Occupational Safety and Health Act requires companies and organizations to create and maintain a workplace free from recognized hazards.

Marijuana Laws

With a substantial number of states legalizing the use of medical marijuana, as well as for recreational use, safety and health professionals and employers are challenged with issues involving controlled substance testing and marijuana use on-the-job while maintaining a safe and healthful environment. Adding to the myriad of issues involved in marijuana use is the current economic conditions where the unemployment rate in the United States has dropped below 4% and the companies and organizations cannot acquire qualified employees (and/or applicants who can successfully pass the controlled substance testing). Although most safety and health professionals rely on company or organization-mandated controlled substance testing to ensure that employees are not working under the influence, many companies and organizations are reducing marijuana from their drug testing panels in order to increase the applicant pools, as well as potentially reduce costs in testing.

For safety and health professionals, the increased use of marijuana posed the potential of increased workplace injuries. "Marijuana use by adults in the United States almost doubled between 1984 and 2015, according to a 2017 study."* Although marijuana continues to be illegal on the federal level, the legalization at the state level, whether medical or recreational use, has steadily gained ground. With the legalization combined with increased productivity due to a growing economy, a greying workforce, a smaller job applicant pool, and other factors, many companies have evaluated their previous zero-tolerance controlled substance programs and have removed THC from their testing panels. With the elimination of this testing, safety and health professionals should be cognizant of

---

* www.newsweek.com/marijuana-use-us-has-increased-2005-not-because-legislation-study-says-663826).

the potential of increased work-related accidents resulting from impairment from marijuana use.

Safety and health professionals should be aware that the trend toward legalization appears likely to continue, and a recent poll identified that 60% of people in the United States favor the legalization of marijuana.* With companies removing THC from their testing panels, the risk will shift to safety and health professionals to train their supervisors to identify employees who may be working in an impaired condition, provide safeguards for safety-sensitive positions, ensure that disciplinary policies and related policies are up-to-date and in line with current law, and design programs which will create and maintain a safe and healthful environment within the parameters of the potential risk probability.

Drones

With the use of drones of various levels and with various functions, safety and health professionals should be aware of laws and regulations created within this emerging area. The use of drones is a relatively new area and has steadily increased, primarily due to the improved technology and the lower price. There is a wide range of drones used for varying uses from military to recreational. Many safety and health professionals have adopted drone technology to assist in assessments, observe and analyze equipment, carry out aerial assessment of large areas, and numerous other uses. Companies, as well as individuals, have adopted this technology to provide an aerial assessment in a cost-effective manner.

However, safety and health professionals should be aware that government agency possessing jurisdiction over the airways of the United States is the Federal Aviation Administration (known as the "FAA") and the FAA has established new rules and regulations regarding the operation of drones. Safety and health professionals should also be aware that if a state or local law addressing the use of drones is enacted, the FAA rules and regulations would be preempted by the FAA rules and regulations, thus invalidating the state or local law or rule. However, state and local laws or ordinances within the specific policing power of the state or local government, such as land use or local police power, generally are outside of the FAA regulations.

There can be many positive aspects to the use of drones by safety and health professionals within the scope of work and within operational boundaries. However, safety and health professionals should be aware that drones can also create new potential risks when used by others to impact the workplace. Issues involving privacy have been addressed within the FAA regulations as well as by many states. Additionally, issues regarding the use by law enforcement for surveillance has been addressed by the FAA, requiring law enforcement to secure a warrant similar to other types of surveillance. In a number of states, specific laws have been passed addressing surveillance by drones by private citizens. Drones can be used for a multitude of purposes, including terrorism. The FAA has established rules regarding

---

* www.nbcnews.com/politics/first-read/nbc-ws).

the airspace and use of drones around prisons, stadiums, airports, and other critical infrastructure.

Safety and health professionals should be aware that the FAA has adopted the term "unmanned aircraft" (UA) to describe drones. Although there are many different aircraft in the category of drones, including but not limited to, UAV (unmanned aerial vehicle), RPV (remoted piloted vehicle), and ROA (remotely operated aircraft), the category has been primarily divided by recreational and non-recreational use. Recreational use vehicles have limited size and are flown solely for recreational or sport purposes. These types of drones are governed by the voluntary safety standards adopted by the Academy of Model Aeronautics. Drones operated for non-recreational purposes in the United States and governed by the FAA are required to obtain a Certificate of Authorization (COA) to operate in national airspace. In 2015, the FAA, under the FAA Modernization and Reform Act, required all drones weighing more than 250 grams flown for any purpose to be registered with the FAA.* Failure to register can result in civil penalties up to $27,500 and criminal penalties of up to $250,000.00 and/or up to three years in prison.[†]

Safety and health professionals who are using or plan to use a drone within their operations should be cognizant of the FAA regulations, as well as state and local laws, regulations, and ordinances. Additionally, safety and health professionals should verify and discuss surveillance and privacy issues as well as licensure issues with their legal and human resource departments. Conversely, safety and health professionals should address the potential use and risks which can be created by others against your operations or structures. As with most tools, a drone can be very beneficial in providing an aerial view of equipment, disaster situations, and non-traditional workplaces; however, drones also possess the potential for use in terrorism events, invasion of privacy, improper surveillance, and other uses.

Case Study: The following case may have been modified for the purposes of this text.

IN THE UNITED STATES DISTRICT COURT DISTRICT OF KANSAS

Lee Lain, Plaintiff, v. BNSF Railway Company, Defendant.

Case No. 13-CV-2201

MEMORANDUM & ORDER Plaintiff Lee Lain filed this negligence action under the Federal Employers' Liability Act against BNSF Railway Company, his employer, for injuries he sustained after he slipped and fell due to an "unnatural accumulation of ice" on a pedestrian pathway at the BNSF Technical Training Center on the campus of Johnson County Community College. This matter is presently before the court on BNSF's motion to exclude the opinions of Frank Burg, a liability expert retained by plaintiff, under Federal Rules of Evidence 702–705 and the rule set forth in *Daubert v. Merrell Dow Pharmaceuticals, Inc.*, 509 U.S. 579 (1993) and *Kumho*

---

* PL 112-95.
[†] *Id.* Also see, *National Law Review*, Neal, Gerber and Eisenberg LLP, Dec. 17, 2015.

Tire Co. v. Carmichael, 526 U.S. 137 (1999). As will be explained, the motion is granted in part and denied in part.

1. Governing Standards

In *Daubert v. Merrell Dow Pharmaceuticals, Inc.*, 509 U.S. 579 (1993), the Supreme Court instructed that district courts are to perform a "gatekeeping" role concerning the admission of expert testimony. See id. at 589–93; see also *Kumho Tire Co. v. Carmichael*, 526 U.S. 137, 147–48 (1999). The admissibility of expert testimony is governed by Rule 702 of the Federal Rules of Evidence, which states: *2 A witness who is qualified as an expert by knowledge, skill, experience, training, or education may testify in the form of an opinion or otherwise if: (a) the expert's scientific, technical, or other specialized knowledge will help the trier of fact to under the evidence or to determine a fact in issue; (b) the testimony is based on sufficient facts or data; (c) the testimony is the product of reliable principles and methods; and (d) the expert has reliably applied the principles and methods to the facts of the case. Fed. R. Evid. 702. To determine that an expert's opinions are admissible, this court must undertake a two-part analysis: first, the court must determine that the witness is qualified by "knowledge, skill, experience, training, or education" to render the opinions; and second, the court must determine whether the proposed testimony is "reliable and relevant, in that it will assist the trier of fact." See *Conroy v. Vilsack*, 707 F.3d 1163, 1168 (10th Cir. 2013). To qualify as an expert, the witness must possess such "knowledge, skill, experience, training, or education" in the particular field as to make it appear that his or her opinion would rest on a substantial foundation and would tend to aid the trier of fact in its search for the truth. See *Life Wise Master Funding v. Telebank*, 374 F.3d 917, 928 (10th Cir. 2004). In determining whether the proffered testimony is reliable, the court assesses whether the reasoning or methodology underlying the testimony is scientifically valid and whether that reasoning or methodology can be properly applied to the facts in issue. See *Daubert*, 509 U.S. at 592–93. The *Daubert* Court identified testability, peer review and publication, the known or potential rate of error, and general acceptance among the factors relevant to assessing reliability. See id. at 592–94. In *Kumho Tire*, however, the Supreme Court emphasized that the *Daubert* factors *3 are not a "definitive checklist or test" and that a court's inquiry into reliability must be "tied to the facts of a particular case." See *Kumho Tire*, 526 U.S. at 150. In some cases, "the relevant reliability concerns may focus upon personal knowledge or experience," rather than the *Daubert* factors and scientific foundations. See id. The district court has "considerable leeway in deciding in a particular case how to go about determining whether particular expert testimony is reliable." See id. at 152. Plaintiff has retained Frank Burg, a certified safety engineer, to testify in this case about accident prevention, the pertinent standard of care and whether defendant's conduct fell below that standard of care. In its motion to exclude, defendant contends that Mr. Burg lacks sufficient facts, data and knowledge to provide reliable opinions and improperly assumes the truth of "facts" that are disputed; that Mr. Burg's opinions are the product of an "experimental methodology" that is not reliable and is unreliably applied; that Mr. Burg improperly relies on safety standards that are irrelevant and inapplicable to the facts of this case; and that Mr. Burg's testimony is not helpful to the jury. The court addresses each of these challenges in turn.1 1 While defendant's brief is dedicated primarily to challenging the reliability of Mr. Burg's opinions, defendant makes a threshold argument, in summary fashion, that Mr. Burg is not qualified to render opinions as to safety standards because he has not received an educational degree awarded by any department of engineering; he obtained his educational degree in psychology; and his home state of Illinois does not recognize professional "safety engineers" with any kind of certification. This argument is rejected and the court finds that Mr. Burg is qualified to render opinions in this case. As indicated in Mr. Burg's deposition, he is a registered professional safety engineer, licensed by the Commonwealth of Massachusetts, and his credentials are recognized by the American Society of Safety Engineers. In addition, Mr. Burg has more than 40 years of experience in safety and human factors engineering, including 18 years working for OSHA. He has served on the Standards Development Committee for the American Society of Safety Engineers and has served on committees for the American National Standards Institute (ANSI). *4

2. Does Mr. Burg Lack Sufficient Facts or Data to Render his Opinions?

In its motion to exclude, defendant contends that Mr. Burg's opinions are not reliable because Mr. Burg never visited the campus where the injury occurred; never spoke to any witnesses; never collected data or took field notes; and never performed any "measurements." Defendant further asserts that Mr. Burg's opinions are unreliable because those opinions are based on numerous erroneous factual assumptions—obtained only through plaintiff's deposition—that are disputed by defendant's evidence. Mr. Burg, for example, asserts that defendant plowed snow behind the dumpsters in the loading dock area; that plaintiff was never advised to use the east exit door to reach the designated smoking area; that the east exit door was equipped with "panic hardware" rendering it unusable; and that plaintiff was required to walk through the loading dock area to reach the smoking area. The fact that Mr. Burg has never visited the accident site or that his investigation is arguably deficient in other respects does not render his opinions unreliable in the absence of any suggestion that first-hand knowledge of the scene or other data is required. In other words, defendant has not explained how Mr. Burg's failure to speak to witnesses or failure to take "field notes" has had any effect on his opinions. See *Smith v. Ingersoll-Rand Co.*, 214 F.3d 1235, 1244 (10th Cir. 2000) (affirming district court's admission of testimony by human factors engineer and safety expert regarding procedures in developing milling machine, although expert did not have first-hand knowledge of the machine at issue; witness' testimony was limited to matters within the witness' field of expertise). Of course, defendant may cross-examine Mr. *5 Burg on these issues in an effort to demonstrate to the jury the asserted inadequacies of

Mr. Burg's investigation and opinions. With respect to the factual assumptions utilized by Mr. Burg (that snow was plowed against the building; that plaintiff was not told to use the east exit door to reach the smoking area; that panic hardware was on the east exit door; and that plaintiff was required to traverse the loading dock to reach the smoking area), he is entitled to rely on plaintiff's deposition to obtain those factual assumptions even though those facts are vigorously disputed by defendant. See *Williams v. Illinois*, 132 S. Ct. 2221, 2228 (2012) ("Under settled evidence law, an expert may express an opinion that is based on facts that the expert assumes, but does not know, to be true. It is then up to the party who calls the expert to introduce other evidence establishing the facts assumed by the expert."). However, Mr. Burg must explain in his testimony to the jury that he is assuming the truth of plaintiff's version of events and that he has not independently determined those facts. In other words, Mr. Burg cannot testify that defendant plowed snow against the building or that defendant failed to tell plaintiff about the east exit door. Rather, Mr. Burg may testify only that "if in fact defendant plowed snow against the building, then" such conduct was unreasonable (or whatever the opinion might be) or "if in fact plaintiff was not told about the east exit, then" defendant failed to utilize appropriate safety practices (or whatever the Defendant also contends that Mr. Burg's opinions are unreliable because Mr. Burg does not even know where plaintiff's fall occurred. According to defendant, Mr. Burg's report repeatedly refers to a "parking lot" when, in fact, the fall occurred in the loading dock area. As Mr. Burg clarified in his deposition, his use of the phrase "parking lot" in his report was intended to refer to the loading dock area where plaintiff fell. The court, then, will not exclude opinion might be). See id. (expert can explain facts on which his opinion is based without testifying to the truth of those facts).

3. Is Mr. Burg's Methodology Unreliable and Unreliably Applied?

Defendant asserts that Mr. Burg's opinions should be excluded because his methodology is unreliable and has been unreliably applied to the facts of the case. In support of this argument, defendant states that Mr. Burg's conclusions are based solely on his subjective beliefs as to whether the parties acted reasonably and that his methodology fails under the four factors identified by the *Daubert* Court. The crux of Mr. Burg's testimony is that specific standards exist regarding workplace safety and that, assuming plaintiff's version of the facts, defendant's conduct fell short of satisfying those safety standards in various respects. This aspect of Mr. Burg's testimony is not based on his subjective beliefs and is admissible. The fact that the nature of his testimony is "not scientific," as argued by defendant, or that it is not susceptible to a rigid *Daubert* reliability analysis is of no consequence. *United States v. Medina-Copete*, 757 F.3d 1092, 1101, 1103 (10th Cir. 2014) (*Daubert* factors are not to be applied woodenly in all circumstances; *Kumho Tire* expands *Daubert* to expert testimony that is not purely scientific; the key is whether the expert uses in the courtroom the same level of intellectual rigor that characterizes the practice of an expert in the relevant field).

4. Does Mr. Burg Rely on Irrelevant and Inapplicable Safety Standards?

In his opinion, Mr. Burg asserts that defendant, in failing to provide its employees with a safe, suitable and unobstructed path to the designated smoking area, violated various safety *7 standards such that defendant breached its duty to provide a safe workplace. Specifically, Mr. Burg opines that defendant's conduct violated the safety standards set forth by the American National Standards Institute (ANSI); fire code standards developed by the National Fire Protection Association (NFPA); OSHA's general duty clause and a specific OSHA regulation; and specific provisions of the International Building Code. Defendant moves to exclude Mr. Burg's reliance on these standards on the grounds that these standards are irrelevant and inapplicable.

a.   Alleged Violation of Fire Code

In his deposition, Mr. Burg testified that defendant's conduct violates the National Fire Protection Association's Life Safety Code which, according to Mr. Burg, requires defendant to In the heading of this particular section of its memorandum in support of the motion to exclude, defendant also states that Mr. Burg's opinions regarding these safety standards are "legal conclusions" and "confusing." Defendant does not address or pursue these additional assertions in any respect in its memorandum and the court, then, declines to address them here. Similarly, defendant references the ANSI standards in the heading but does not address ANSI in its brief. The court, then, need not consider at this juncture whether Mr. Burg's reliance on the safety standards set forth by ANSI is appropriate but notes that several courts have deemed such testimony admissible. See *Paul v. ASTEC Indus., Inc.*, 25 Fed. Appx. 623 (9th Cir. 2002) (district court properly allowed expert to testify as to product defect where expert's methodology included application of ANSI provisions to his observations); *Cospelich v. Hurst Boiler & Welding Co.*, 2009 WL 8547607, at *3 (S.D. Miss. 2009) (expert testimony as to ANSI standards was admissible to show reasonableness of defendant's actions or whether those actions were consistent with industry standards); *Cunningham v. District of Columbia Sports and Entertainment Comm'n*, 2005 WL 4898867, at *9 (D.D.C. 2005) (expert could properly compare defendant's conduct with ANSI standards; defendant free to argue that ANSI standards do not define applicable standard of care or that deviation was not cause of injury); *Miller v. Chicago & N.W. Transp. Co.*, 925 F. Supp. 583, 590–91 (N.D. Ill. 1996) (expert retained by plaintiff asserting negligence claims under FELA could testify about whether railroad complied with safety recommendations published by ANSI). *8 provide "free and unobstructed egress from all parts of the building" at all times. Defendant moves to exclude any reference to the Fire Code on the grounds that, among other things, plaintiff's claim in this case does not concern an attempt to exit a burning building such that the Fire Code simply does not apply to the facts presented here. While plaintiff responds to other arguments made by defendant regarding the applicability of the Fire Code, he does not address this particular argument in any way. There has been no showing, then, that the Fire Code speaks to whether it is dangerous to plow snow against a building with limited drainage or whether defendant should have treated or removed snow and ice in the parking lot where plaintiff slipped and fell. Similarly, there has been no showing that the Fire Code was intended to prevent the type of injury sustained by plaintiff or, stated another way, that plaintiff's injury result from a condition that the fire code was designed to prevent. In the absence of such a showing, the court will not permit Mr. Burg to reference the Fire Code in his testimony. See *Kilgore v. Carson Pirie Holdings, Inc.*, 205 Fed. Appx. 367, 370–71 (6th Cir. 2006) (expert retained by plaintiff who fell while descending stationary escalator could not rely on NFPA's fire code to support opinion that stationary escalator is dangerous; NFPA was irrelevant to facts of case); *Whelan v. Royal Caribbean Cruises Ltd.*, 2013 WL 5595938, at *4 (S.D. Fla. 2013) (plaintiff's expert could not rely on NFPA's fire code standards in slip-and-fall case; fire code evidence was irrelevant and possibly prejudicial where plaintiff did not allege that injury results from failure to conform with fire safety standards); see also *Elliott v. Food Lion, LLC*, 2014 WL 1404562, at *2 (E.D. Va. 2014) (reliance on purported violations of fire code for negligence per se claim was futile in slip-and-fall case in absence of suggestion that *9 plaintiff belonged to class of persons meant to be protected by fire code or that injury was the type the fire code was intended to protect against).

a. Alleged Violation of OSHA

Mr. Burg opines that defendant's conduct violated OSHA's general duty clause, see 29 U.S.C. § 654(a), requiring each employer to furnish a place of employment free from recognized hazards, as well as an OSHA regulation requiring employers to ensure that all "places of employment, passageways, storerooms and service rooms" be kept clean and orderly and that all floors be maintained in a clean and dry condition. See 29 C.F.R. § 1910.22. In its motion to exclude, defendant does not address Mr. Burg's reliance on the OSHA regulation, but contends only that OSHA's general duty clause does not apply to this case because it concerns "recognized" hazards and it is undisputed in this case that the ice upon which plaintiff slipped was "invisible." According to defendant, an invisible hazard "cannot be recognized" and, thus, is not covered by OSHA's general duty clause. Defendant's argument, however, ignores plaintiff's theory of the case—that is, that defendant's actions caused or contributed to plaintiff's injury (e.g., failure to designate a walkway) and/or caused or contributed to the presence of the invisible ice (e.g., negligent snow removal, failure to inspect and treat the loading dock area). In any event, defendant's argument assumes that OSHA's general duty clause is implicated only when an employer has actual knowledge of a personally recognized hazard—an interpretation that conflicts with the requirement that an employer exercise due diligence to proactively discover legitimate safety concerns and which this court rejects. See *Transportation Ins. Co. v. Citizens Ins. Co.*, 2013 WL 856641, at *5 & n.2 (E.D. Mich. 2013) (expert testified that a *10 "recognized hazard" under OSHA refers to a safety problem that OSHA has officially recognized or a hazard that is generally accepted throughout the industry as a standard or practice that can cause an injury to workers). In the absence of any other argument for exclusion, the motion is denied on this issue. See *Taylor v. TECO Barge Line, Inc.*, 642 F. Supp. 2d 689, 692–93 (W.D. Ky. 2009) (expert could testify as to whether employer violated OSHA's general duty clause as that issue was relevant to whether defendant provided a reasonably safe place to work).

   b. Alleged Violation of International Building Code

   Mr. Burg opines that defendant's conduct violates certain provisions contained in the 2000 edition of the International Building Code. In its motion, defendant contends that this opinion should be excluded because Mr. Burg admitted in his deposition that he was not aware of more recent editions of the International Building Code; he could not explain any differences between the 2000 edition and the more recent editions; and he did not know whether the 2000 International Building Code applied to an incident that occurred in 2011. According to defendant, then, Mr. Burg's reliance on the International Building Code is unreliable and must be excluded. While defendant may properly raise these issues on cross-examination, they do not serve as a basis to exclude Mr. Burg's reliance on the International Building Code in the absence of any evidence or suggestion from defendant that subsequent editions of the International Building Code materially changed the provisions upon which Mr. Burg relies. *Stephan v. Continental Cas. Ins. Co.*, 2003 WL 21032042, at *3 (D.N.J. 2003) (rejecting *Kumho* challenge *11* that expert had relied on improper edition of book in the absence of any evidence that the edition utilized by expert was inaccurate in any way).

5. Is Mr. Burg's Testimony Helpful to the Jury?

   Finally, defendant, in its reply brief, contends that Mr. Burg does "nothing more than suggest that an ice patch should have been treated" and that a jury does not require expert testimony to draw that conclusion. A review of Mr. Burg's testimony, however, reveals that it will assist the jury in determining significant issues in this case. Specifically, Mr. Burg has knowledge of reasonable and customary safety practices in the industry in light of specific, nationally recognized safety standards. To this extent, then, his testimony will assist the jury in determining the applicable standard of care and in determining whether defendant's conduct complied with that standard of care. This aspect of defendant's motion, then, is denied.

IT IS THEREFORE ORDERED BY THE COURT THAT defendant BNSF Railway Company's motion to exclude the opinions of Frank Burg (doc. 63) is denied in part and granted in part. IT IS SO ORDERED.

Dated this 14th day of November, 2014, at Kansas City, Kansas.

John W. Lungstrum

United States District Judge

Reference Page – Law Cases

Bachelder v. America West Airlines, Inc., 259 F.3d 1112 (2001)
Barsch v. Nueces County, WL 4785169 (2016)
Connelly v. Lane Construction Corp., 809 F.3d 780 (2016)
ConocoPhillips Co. v. United Steelworkers, local, WL 199268 (2010)
Grutzmacher v. Howard County, 851 F.3d 332 (2016)
Lawson v. Sessions, 271 F.Supp.3d 119 (2017)
Marshall v. Barlow's, Inc., 98 S.Ct. 1816 (1978)
Murphy Oil USA, Inc. v. National Labor Relations Board, 808 F.3d 1013 (2015)
Norceide v. Cambridge Health Alliance, 814 F.Supp.2d 17 (2011)
Perez v. Clearwater Paper Corp., 184 F.Supp.3d 831 (2016)
Secretary of Labor v. Sanderson Farms, Inc., OSHRC Docket No. 15-1928 (2015)
Toyota Motor Manufacturing, Kentucky, Inc. v. Williams, 122 S.Ct. 681 (2001)
United States v. Mar-Jac Poultry, Inc., WL 4896339 (2018)

Index

Note: Page numbers with 'n' denotes notes

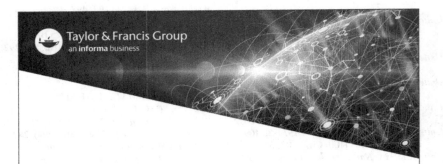

Printed in the United States
by Baker & Taylor Publisher Services